Kim Latham   Index 113/184

Spin glasses:
an experimental
introduction

Schräg in dem eisernen Gitter
der drehende Spin
am Drehsinn
errätst du die Seele

Frei nach Paul Celans 'Sprachgitter'

Spin glasses: an experimental introduction

J. A. Mydosh

Taylor & Francis
London • Washington, DC
1993

UK	Taylor & Francis Ltd, 4 John St., London WC1N 2ET
USA	Taylor & Francis Inc., 1900 Frost Road, Suite 101, Bristol PA 19007

© J. A. Mydosh 1993

All rights reserved. No part of this publication may be reproduced, stored in a retrieval system, or transmitted, in any form or by any means, electronic, electrostatic, magnetic tape, mechanical, photocopying, recording or otherwise, without the prior permission of the copyright owner and publishers.

British Library Cataloguing in Publication Data
are available from the British Library
ISBN 0-7484-0038-9

Library of Congress Cataloging-in-Publication Data
are available

Cover designed by Amanda Barragry

Printed in Great Britain by Burgess Science Press, Basingstoke on paper which has a specified pH value on final paper manufacture of not less than 7.5 and is therefore 'acid free'.

Contents

PREFACE		viii
1	**INTRODUCTION TO RANDOMNESS AND FREEZING**	**1**
1.1	What is a spin glass?	1
1.2	How to create randomness?	4
1.3	Magnetic interactions (are necessary)	6
	1.3.1 Direct exchange	6
	1.3.2 RKKY	7
	1.3.3 Superexchange	9
	1.3.4 Dipolar	10
	1.3.5 Mixed superexchange (random bonds)	10
1.4	Magnetic anisotropy (plays a role)	11
1.5	Frustration (a far-reaching concept)	13
1.6	The freezing process (a first look)	15
1.7	Some historical background	16
	Further reading	18
2	**DESCRIPTION OF RELATED CONCEPTS**	**20**
2.1	Virtual-bound-state model	20
2.2	Kondo effect and weak moments	24
2.3	Giant moments	28
2.4	Spin glasses (a second visit)	30
2.5	Mictomagnetism (cluster glass)	32
2.6	Percolation and long-range magnetic order	34
2.7	Concentration regimes	35
2.8	Metallic glasses: amorphous magnets	37
2.9	Re-entrant spin glasses	39
2.10	Superparamagnetism	40
2.11	Magnetic insulators and semiconductors	42
	Further reading and references	44

3	**BASIC SPIN-GLASS PHENOMENON**	45
	3.1 High-temperature ($T \gg T_f$) experiments	45
	3.1.1 Susceptibility	46
	3.1.2 Specific heat	50
	3.1.3 Resistivity	53
	3.1.4 Neutron scattering	57
	3.2 Experiments spanning T_f	63
	3.2.1 Ac-susceptibility	64
	3.2.2 Specific heat	76
	3.2.3 Resistivity	78
	3.2.4 Neutron scattering	80
	3.2.5 μSR (muon spin relaxation)	83
	3.2.6 Mössbauer effect	86
	3.3 Low-temperature ($T \ll T_f$) experiments	88
	3.3.1 Magnetization	89
	3.3.2 Specific heat	98
	3.3.3 Resistivity	100
	3.3.4 Neutron scattering	101
	3.3.5 Torque and ESR	105
	3.4 Spin glasses in a field	111
	3.5 Experimentalist's (intuitive) picture	113
	References	116
4	**SYSTEMS OF SPIN GLASSES**	118
	4.1 Transition-metal solutes	119
	4.1.1 Noble metals with transition-metal impurities	119
	4.1.2 Transition-metal/transition-metal combinations	120
	4.1.3 Collection of characteristic temperatures for AuFe and CuMn	122
	4.2 Some rare-earth combinations	125
	4.3 Amorphous (metallic) spin glasses	127
	4.4 Semiconducting spin glasses	129
	4.5 Insulating spin glasses	131
	4.6 What is a good spin glass?	133
	References	134
5	**MODELS AND THEORIES**	135
	5.1 Some historical perspectives	136
	5.2 The Edwards–Anderson model	139
	5.3 Sherrington–Kirkpatrick model	144
	5.4 Instability of the SK solution	147

Contents

5.5	The TAP approach	148
5.6	How to break the replica symmetry	150
5.7	Physical meaning of RSB	153
5.8	Dynamics of the mean-field model	161
5.9	Droplet model	164
5.10	Fractal-cluster model	167
5.11	Non-Ising spin glasses	171
5.12	Computer simulations	174
	Further reading and references	179

6 IDEAL SPIN GLASSES: COMPARISONS WITH THEORY 181

    6.1 What is an 'ideal' spin glass? 182
    6.2 $Rb_2Cu_{1-x}Co_xF_4$, an ideal 2D spin glass 184
    6.3 Ideal 3D spin glasses 194
    6.4 Future possibilities 200
        References 203

7 SPIN-GLASS ANALOGUES 204

    7.1 Electric dipolar and quadrupolar glasses 205
    7.2 Superconducting-glass phases 212
    7.3 Random-field Ising model 216
    7.4 Other 'spin-offs' 222
        Further reading and references 231

8 THE END (FOR NOW) 233

    8.1 Symmetry breaking and phase space 233
    8.2 Mesoscopic spin glasses 237
    8.3 Recent theoretical progress 241
    8.4 A final word 246
        References 248

INDEX 251

Preface

Phenomena that exist in nature should not only be observed and classified but describable by models or theories which then lead to a better understanding and predictability of the phenomenon. Often, as is the case with the spin glasses, experiment has serendipitously uncovered some unusual or unexpected effect which at first glance baffles the anticipations of our physical intuition. As a result of the novel experimental properties, the phenomenon requires further study and a preliminary picture or concept to represent it. Here a name should be given. With the accumulation of data mathematical notions are needed to quantify the results and an incipient theory is born. The phenomenon, its various experimental manifestations, the many (hopefully) specific examples which can be found, and the embracing theory, all developing over the course of many years, create a problem area, in our case the 'spin-glass problem'. Fresh input comes from the drawing of various analogies with other areas, new anomalies are discovered by more sophisticated experiments, and theoretical progress leads to a deeper understanding and predictive power. Hence the problem is nicely on its way to being solved.

Frequently there is a long pause after an important discovery and the recognition of its significance. Time and ideas are necessary to set the stage for further advancement and the delay can be great. With some luck the given phenomenon will become a major effect, i.e., an important topic of contemporary research. Hectic efforts, measured by the large number of workers presenting their results at countless conferences and workshops, are demanded. And all this could continue for many years before a (near) solution is found or the ideas and curiosity run out. Thus, for a particular period the area becomes highly fashionable or 'in-physics', and then after its fame and topicality diminish, the problem, if still unsolved, may remain as an interesting, yet *passé* research topic. Nevertheless during these intense years, a great deal was learned and this knowledge will remain as an integral part of physics and even appear in tomorrow's text books.

So it was and is with the spin glasses. Around the late 1960s there were some unusual effects observed on a series of magnetic alloys (rather simple and traditional ones such as iron in gold and manganese in copper).

However, then the time was not right but at least a name was coined. Only in the mid 1970s did theory using some radically new ideas come to grips with the experiments and the possibility of a sharp phase transition. Around 1975 'all hell broke loose' and this began the fame period of the spin glasses with great activity and much progress. Here analogies and spin-offs increased the pace. After 1985 the tempo slowed down and a third phase of tranquility appeared with more sophisticated theories, subtle new experiments and 'ideal' spin-glass materials. The phenomenon had run its course to establishment so that now is perhaps a good time to try and tell its story, naturally in a phenomenological way.

The purpose of this book is indicated in the title. It is meant to introduce the phenomenon and physics of spin glass (Chapters 1 and 2) via an experimental approach (Chapter 3). The method is to use simple physical pictures and concepts based on the empirical observations, rather than employing theoretical models with the associated mathematics. Hence, figures and tables will be more common than equations and derivations. Materials, i.e., the real spin glasses, will also play a major role (Chapter 4). In spite of this particular emphasis the theory of spin glasses is an essential part of the story and its ideas, techniques and advances will be portrayed in an elementary fashion (Chapter 5). How close does experiment come to theory? Chapter 6 considers the recent advances in finding and measuring spin glasses which closely mimic the theoretical models.

Why write such a book on spin glasses? At least four good reasons can be given.

1   The spin glass has become a fundamental and general form of magnetism; its general occurrence as 'magnetic ordering' occupies the third place after ferromagnetism and antiferromagnetism.
2   Randomness, frustration, glassiness and amorphousness represent very important phenomena in contemporary physics.
3   Radically new concepts and ideas are needed to describe the basic experimental behaviours.
4   There is a richness of analogies with many other areas extending from astrophysics via molecular evolution to zoology. A few of these analogies are examined in Chapter 7.

The spin glasses were surely one of the first precursors to the physics of complexity – a new area which has recently been established.

For the above reasons a simple, less theoretical treatment of the problem should be made available for the non-theorist or even non-physicists, and also the younger (graduate) students in order to serve as an introduction to the subject and its basic physics. Thus, the level of the book is low; only a general background in solid-state physics and elementary magnetism is required beyond the undergraduate courses, for the remaining necessary concepts are explained, many in Chapter 2 and others as needed. Emphasis is placed on experimental facts and physical notions with always a real spin-glass system in mind. The powerful developments in the theory – the new statistical mechanics – will be sketched and compared to experiment but not discussed in mathematic details or via long derivations. Concerning the theory of spin glasses there are a number of texts and review articles already available here and it would be unwise and unnecessary to compete with these excellent theoretical exposés. What is missing is the elementary level, rudimentary explanation of the spin-glass problem for just about everyone. More so the novice or newcomer to the field and the expert from another area are the welcome readers. Perhaps the real spin-glass connoisseurs would value the book as a reference to the salient experiment features and an indication of the theoretical lines and directions of progress.

The author has been too long occupied with the phenomenon of spin glasses. First there were the quiet years of experimentation, then the hectic decade when it seemed like everyone was working on the problem, and finally, now that the dust has settled a bit, a chance to look back at the past 25 or so years and see what knowledge has been gained and what we really understand. Although my research emphasis has changed to other topics and other problems, nevertheless, my 'first love' remains the spin glasses. Before forgetting everything it's time to pause for a moment and in retrospection write this book as quickly as possible. The season is certainly ripe, the need is great and a new generation of students is moving into this research area. But such an undertaking is not meant as the final word; the field is not finished. The spin-glass problem has not been fully solved, there is still movement, albeit slow, towards a more complete theoretical description with yet newer ideas. Also experimentation is continuing with novel forms of spin glasses and even more subtle effects. Hopefully then a more finalized version of this spin-glass text will appear in the next ten years. Thus my apologies now for presenting the yet incomplete and modifiable 'truth'.

Although the main ideas and threads of thought for this book were conceived at the Kamerlingh Onnes Laboratory in Leiden, most of writing took place at the University of California in Santa Barbara during a sabbatical visit of six months. I must thank V. Jaccarino for his kind hospitality at UCSB. Also I would like to acknowledge the support and encouragement of M. H. Vente during the rather long period of gestation, writing and correcting. M. Nieuwenhuijzen-Verschoor did a superb job in typing the manuscript. There were simply too many discussions and interactions some of them long-forgotten during the past 25 years which moulded my views on the spin glasses. I should mention the experimental years at Fordham University in New York and Institut für Festkörperforschung in Jülich, and the changing times at Leiden. All these places and their staffs, my colleagues, contributed to a rather particular and sometimes personal interpretation of the spin glasses which is related in this book. I hope it reads a bit like an adventure novel as so many parts have seemed to me, none the least of which was the actual writing. Read well.

1

Introduction to randomness and freezing

All beginnings are difficult, but we must somehow start, so let us do it as simply as possible. Pictures are worth many words, thus, some oversimplified sketches are used to illustrate what a spin glass is and how we create the all-important randomness. In addition to the disorder we must have magnetic interactions, and these are also treated in a pictorial way. The other important ingredients of a spin glass, anisotropy and frustration, are introduced via easy examples. After formulating a working definition of a spin glass we take a first look at the freezing process. The spin glass does indeed constitute a new state of magnetism, distinctly different from the long-range ordered ferro- and antiferromagnetic phases. Yet similar to these magnets, the spin glass also has a co-operative or collective nature in the frozen state. And this is what we must try to understand and put into a clear physical presentation with the aid of 'solvable' theoretical models. In order to satisfy the biographic aspects we give a very brief account of the historical development of the spin-glass problem and how the name was coined in 1970. With such an outline of the goals of Chapter 1, let us plunge into the spin glasses.

1.1 WHAT IS A SPIN GLASS?

We shall begin by giving a simple physical picture of a frozen spin glass. This can be accomplished by playing the 'coupling game' between randomly distributed magnetic moments, i.e., **spins** on a simple, two-dimensional (2D), square lattice. Once we know what the frozen configuration (say at temperature $T = 0$ K) looks like, we can attempt to track down the basic ingredients which have created this state. Afterwards we can increase the temperature to study the formation and precursors of our frozen spin glass. At least then, we shall know the type of physical phenomenon we are dealing with.

Let us imagine the square lattice whose sites are represented by dots as depicted in Fig. 1.1. A certain fraction x of the sites are occupied by a magnetic moment or spin. These are randomly distributed throughout the lattice and are designated by arrows. We keep x small ($\frac{1}{4}$) to avoid too many nearest-neighbour occupancies. Now let us 'turn on' a magnetic interaction among the moments. The coupling algorithm is ferromagnetic (parallel) between nearest neighbours and antiferromagnetic (antiparallel) between next-nearest neighbours. Such a situation is illustrated in Fig. 1.1 where we have removed any thermal disorder by reducing the temperature to a negligible value. To begin our coupling game we arbitrarily choose the alignment of the top right moment to point upwards. As the reader can verify by proceeding downwards, the given configuration then results. Note there are some spins or arrows which are completely decoupled from the larger clusters (3 fully alone and 2 pairs). Most spins enjoy a happily connected environment that has propagated from top to bottom. Further a nice balance (zero net moment) exists between the connected up and down spins (9 up versus 9 down). However, a small number of spins are unhappy, for they cannot satisfy their second-neighbour bonds. In particular, for the marked spin (large dot) at the lower right of Fig. 1.1 there is one first-neighbour coupling with strength $J_1 > 0$ (ferromagnetic) and two second-neighbour couplings with strength $J_2 < 0$ (antiferromagnetic). If $|J_1| = 2|J_2|$, this spin will not know which way to point. Thus, it is 'frustrated' and will communicate its discontent to the two spins below itself.

A striking example of a completely frustrated spin-lattice is a triangular

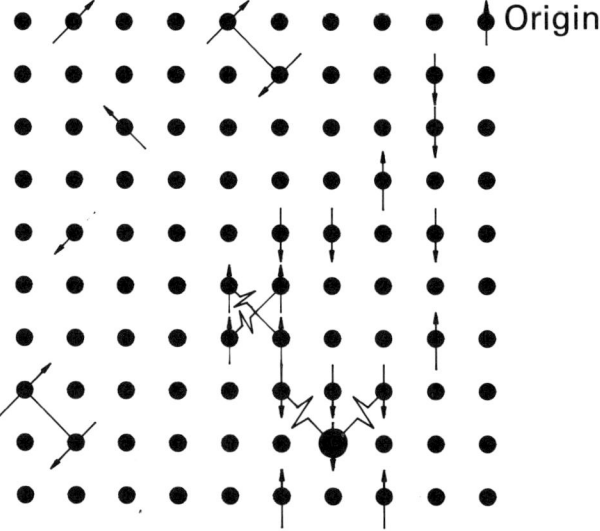

Fig. 1.1 Frozen spin glass with $J_1 > 0$ (ferromagnetic) and $J_2 < 0$ (antiferromagnetic). The zig-zag links indicate a broken bond; the spin in the heavy circle is frustrated.

1.1 What is a spin glass?

one with all antiferromagnetic interactions as shown in Fig. 1.2. Here there is total confusion among the surrounding spins once the direction of the centre spin is chosen. In terms of energy consideration, it will not make any difference whether the frustrated spin points up or down. You can imagine the complicated 'frozen-frustrated' situation in a real 3D lattice.

With such simple pictures we can visualize the ground ($T = 0$ K) or frozen state of a spin glass. Its basic ingredients are randomness (here in site occupancy), mixed interactions (we shall return to this point later), and frustration: many possible ground state configurations. In order to gain a deeper insight into this new state of magnetism, we should consider how it is formed out of the randomly churning spins at high temperature. This 'freezing process' and its associated dynamics will be introduced in a later section of this chapter.

Let us now give a working definition of a spin glass based upon the above schematic picture. We may define a spin glass as a random, mixed-interacting, magnetic system characterized by a random, yet co-operative, freezing of spins at a well-defined temperature T_f below which a highly irreversible, metastable frozen state occurs without the usual long-range spatial magnetic order. A key word in our definition is 'co-operative' which usually implies 'ordering' or even a phase transition. The question that the spin glasses pose is a paradoxical one: order out of randomness? How? As we shall see later there are a number of caveats to our definition, but in any event we now have a foundation. From now on, the term *spin glass* is used to generically represent the class of materials exhibiting the frozen-state transition. We shall become more specific as we progress.

The two most important prerequisites of our spin glass are firstly the randomness in either position of the spins, or the signs of the neighbouring couplings: ferro (↑ ↑) or antiferromagnetic (↑ ↓). There must be disorder, site or bond in order to create a spin glass, otherwise the magnetic transition will be of the standard ferromagnetic or antiferromagnetic type of long-

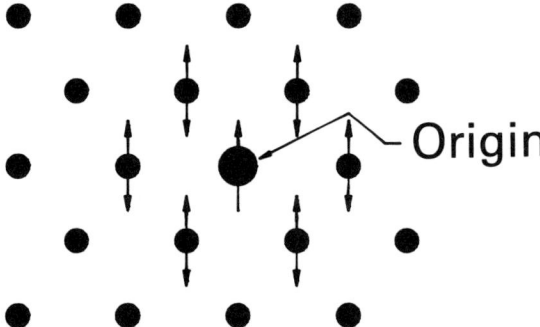

Fig. 1.2 Completely frustrated triangular (hexagonal) lattice with all antiferromagnetic interactions ($J_1 < 0$).

range order. The combination of the randomness with the competing or mixed interactions causes frustration, as illustrated in Fig. 1.1. This second prerequisite of frustration is a unique attribute of the spin-glass ground state, which is disordered, but still exhibits many interesting properties related to its co-operative nature. For example, a macroscopic anisotropy is created and peculiar metastabilities (time and aging dependences) occur especially in reaction to an applied magnetic field. These matters will be treated further by examining experiments in Chapter 3.

1.2 HOW TO CREATE RANDOMNESS?

As so often happens and long before the name spin glass was coined, nature already provided us with the magnetic alloys. They are composed of magnetic impurities bearing a moment or localized spin and occupying random sites in a non-magnetic host metal. The spin glasses were accidentally discovered with such binary alloys. Here we want to control the concentration x of these impurities so that they can interact with each other. Now if we take Fig. 1.1, make it a 3D metal, with a real lattice structure, e.g. face-centred cubic, and then randomly distribute the little arrows on a proper fraction of sites, we will most probably find a spin glass at low temperatures. The host can be just about any non-magnetic metal that dissolves the 'good moment' elements such as Mn, Fe, Gd, Eu, etc. The archetypal specimens of the metallic (site-random) spin glasses are $Cu_{1-x}Mn_x$ and $Au_{1-x}Fe_x$. These noble-metal alloys are also called canonical spin glasses. Site substitution of magnetic solute for non-magnetic solvent should occur completely randomly, i.e., without even short-range order of a chemical or atomic nature. For then we can treat the system as statistical and use gaussian probabilities. With a proper choice of elements this can be accomplished to a reasonably fair approximation.

Chemical compounds both insulating and conducting can also be made random. This is attained by diluting into one of the sub-lattices a magnetic species in place of a non-magnetic one. Some prime examples here are $Eu_xSr_{1-x}S$ (a semiconductor) and $La_{1-x}Gd_xAl_2$ (a metal). Many, many hundreds of additional combinations can be formed in the laboratory. And this plethora of systems has made the spin glasses a general and common type of magnetic ordering.

Moreover, the above **random-site** occupancy of the alloys, mixed insulating compounds and 'pseudo-binary' intermetallic compounds can be mimicked in a more modern and violent way. One takes a standard intermetallic compound, e.g. $GdAl_2$ and YFe_2, and destroys the crystalline lattice by making it amorphous. Techniques such as splat-cooling, quench-condensation and sputtering are employed to de- or un-crystallize the compound. Figure 1.3 illustrates the loss of a periodic lattice with an integer ratio (or fraction) magnetic to non-magnetic sites. A 1-to-1 ratio was chosen for simplicity in

1.2 How to create randomness?

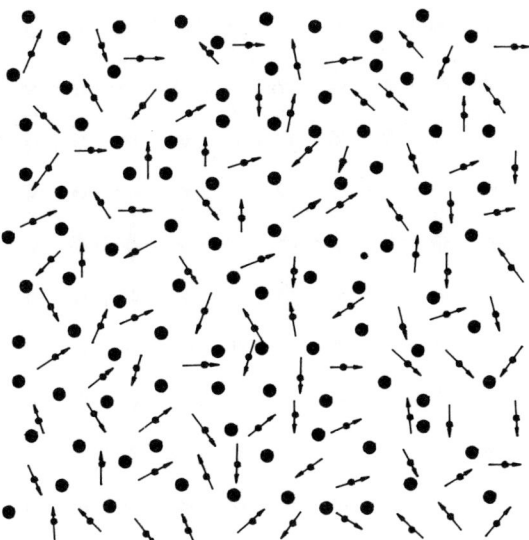

Fig. 1.3 Amorphous spin glass for which the disordered lattice sites are 50% occupied with magnetic moments.

Fig. 1.3. Really what matters now is the random apportionment of distances between the magnetic impurities. The same gaussian distribution function $P(\mathbf{r})$ occurs here as in our original random-site alloys and the mixed compounds. The variance or distribution width may be narrower for the amorphous materials sketched in Fig. 1.3, nevertheless, it is the positional randomness that creates an impurity-spin distance distribution which is the first and basic ingredient of our would-be spin glass.

So much said for site disorder. Let us now consider another possibility to produce a spin glass. Suppose we view the perfect lattice of Fig. 1.4, every site dot is occupied with a spin. Assume that the magnetic interaction is only between nearest neighbours (short-range) and imagine that this single coupling J alternates randomly in sign with $\pm J$ values as we move through the lattice. Thus, there are only parallel or antiparallel bonds as illustrated by the 'dash' or 'zig-zag' in Fig. 1.4. This is the **random-bond** type of system and it was only recently that their existence in real materials, e.g. the compounds $Rb_2Cu_{1-x}Co_xF_4$ and $Fe_{1-x}Mn_xTiO_3$, were unearthed to give reasonable approximations of $\pm J$ couplings. But more about these random-bond spin glasses when we arrive at the theoretical models for a spin glass. Suffice it to say, there must be disorder in the constitution of a spin glass: either **site randomness** with a distribution of distances between the magnetic spins, or **bond randomness** where the nearest-neighbour interaction varies between parallel coupling $+J$ and antiparallel coupling $-J$. If the system is truly random, the average value of all the exchange bonds is zero. So we now know how to create the all-important randomness.

Fig. 1.4 Random-bond spin glass where -- ⇒ ferromagnetic coupling and ⌇ ⇒ antiferromagnetic coupling. The number of ferromagnetic bonds is equal to the number of antiferromagnetic ones.

And once established, the randomness remains forever fixed or constant, consequently we give it the name **quenched-disorder**.

1.3 MAGNETIC INTERACTIONS (ARE NECESSARY)

1.3.1 Direct exchange

The spin-glass problem is one of interactions between the spins and it is through these exchanges and the above randomness that a unique type of ground state is formed. Magnetic-exchange interactions go back to the early days of quantum mechanics and are usually written in the form of a spin hamiltonian as $\mathcal{H}_{ij} = -J_{ij}\mathbf{S}_i \cdot \mathbf{S}_j$ parameter which couples the spins at lattice sites i and j. Direct exchange involves an overlap of electronic wave functions from the two sites and the Coulomb electrostatic repulsion. The Pauli exclusion principle keeps the electrons with parallel spin away from each other, thereby reducing the Coulomb repulsion. The energy difference between parallel and antiparallel configurations is the exchange energy. Usually an antiparallel configuration is favoured as in the simplest case of the hydrogen molecule and as often found in practice. However, since the wave functions of the magnetic d or f electrons decrease exponentially with distance from the nucleus, the J_{ij} obtained from the overlap integral is far too small to provide the necessary coupling. There was an unsuccessful

1.3 Magnetic interactions (are necessary)

attempt to create a direct exchange spin glass by doping phosphorus into silicon and trying to observe the freezing of the P-moments at very low (mK) temperatures. So another mechanism must be involved for our spin glasses.

1.3.2 RKKY

For the magnetic alloys the conduction electrons lead to stronger and longer-range **indirect-exchange interaction**. This is the now-famous Ruderman, Kittel, Kasuya, Yosida (RKKY) interaction whose hamiltonian is $\mathcal{H} = J(r)\mathbf{S}_i \cdot \mathbf{S}_j$. Embedding a magnetic impurity, i.e., a local moment (spin $\mathbf{S}_i$), in a sea of conduction electrons with itinerant spin $\mathbf{s}(r)$ causes a damped oscillation in the susceptibility of the conduction electrons, and thereby a coupling between spins ($\mathbf{S}_i$ and $\mathbf{S}_j$) according to

$$J(r) = 6\pi Z J^2 N(E_F) \left[\frac{\sin(2k_F r)}{(2k_F r)^4} - \frac{\cos(2k_F r)}{(2k_F r)^3} \right] \quad (1.1)$$

where Z is the number of conduction electrons per atom, J the s-d exchange constant, $N(E_F)$ the density of states at the Fermi level, k_F the Fermi momentum and r the distance between two impurities. This reduces to

$$J(r) = \frac{J_o \cos(2k_F r + \phi)}{(2k_F r)^3} \quad (1.2)$$

at large distance. A phase factor ϕ has been included to account for the charge difference between impurity and host and the former's angular momentum. Such oscillatory behaviour of $J(r)$ or really the Pauli susceptibility, which in the free-electron model has spherical symmetry, is illustrated by the two coupling schemes in Fig. 1.5. Notice that the $(1/r)^3$ fall-off is sufficiently long-ranged so that it can effectively reach a number of nearest-neighbour sites. Now suppose we put a second magnetic impurity with spin $\mathbf{S}_j$ at one of the neighbouring sites. This spin will produce its own RKKY polarization and the two conduction-electron-mediated polarizations will overlap in such a way as to establish a parallel or an antiparallel alignment of the two spins. These situations are sketched in Fig. 1.5 by plotting the Pauli susceptibility of the conduction electrons. Note that the sign (+ = ↑↑ and − = ↑↓) of the impurity coupling varies with distance. If we combine this property with the site disorder (various separations between the spins), discussed in the previous section, we have generated a random distribution of coupling strengths and directions. A computer simulation for some half-a-million RKKY bonds, J_{ij}, coupling a random-site magnetic alloy generates the probability function $P(J_{ij})$ shown in Fig. 1.6. Important is the near symmetry in the number of + and − bonds. Here the required factor of 'competition' among ferro and antiferromagnetic exchanges is obtained in

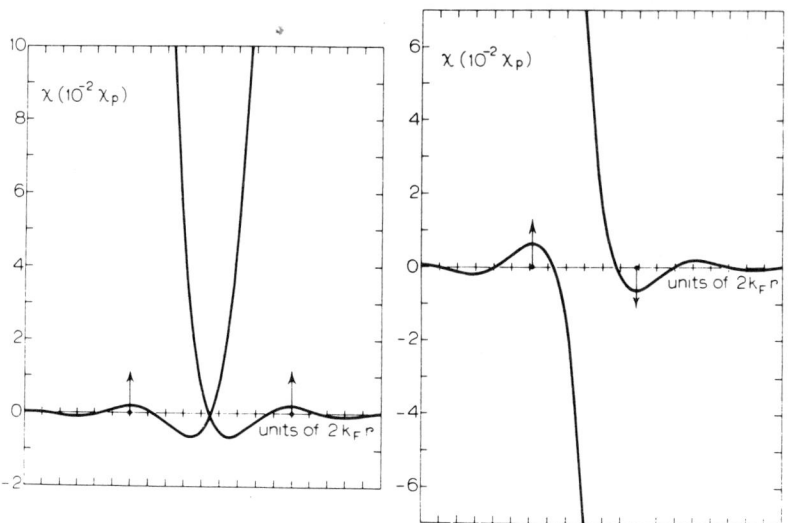

Fig. 1.5 Ruderman–Kittel–Kasuya–Yosida (RKKY) interaction between two impurities in terms of the Pauli susceptibility χ_P.

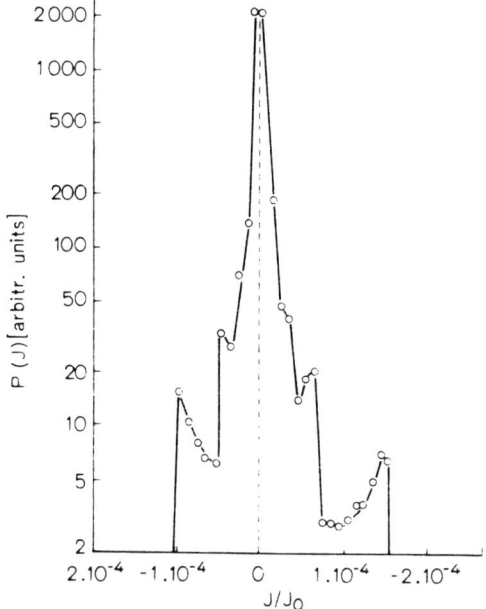

Fig. 1.6 Probability distribution of coupling strengths for about 5×10^5 bonds resulting from a RKKY interaction in a dilute magnetic alloy. From Binder and Schröder (1976).

1.3 Magnetic interactions (are necessary)

a natural and common way, namely via the oscillating RKKY. This is why the whole spin-glass problem started with the magnetic alloys. We must once again emphasize the combination of site disorder with the $+$ and $-$ RKKY interaction causing a mixture of competitive bonds that will eventually lead to frustration in some of these bonds.

1.3.3 Superexchange

How about insulating or semiconducting materials? Since there are no conduction electrons, in what way can we couple the spins if the direct exchange is so weak? The prime mechanism in insulators is that of superexchange. In this case an intervening ligand or anion transfers an electron (usually in a p state) to the neighbouring magnetic atom. A sort of covalent mixing of the p and d (or f) wave functions occurs with spins pointing in the same direction. Because the two anion p-spins must be opposite in direction (Pauli exclusion principle), they will cause antiparallel pairing with the d-electrons on the magnetic atoms to the left and to the right. This situation is shown in Fig. 1.7 and leads to an antiferromagnetic coupling via the ligand situated between the two magnetic atoms. The net result is a stronger and more long-ranged [usually at some large power n of $(1/r)^n$] interaction than the exponential fall-off of the direct coupling. Superexchange can also create a ferromagnetic coupling if we take into account the exchange polarization of the anion orbitals. In the magnetic

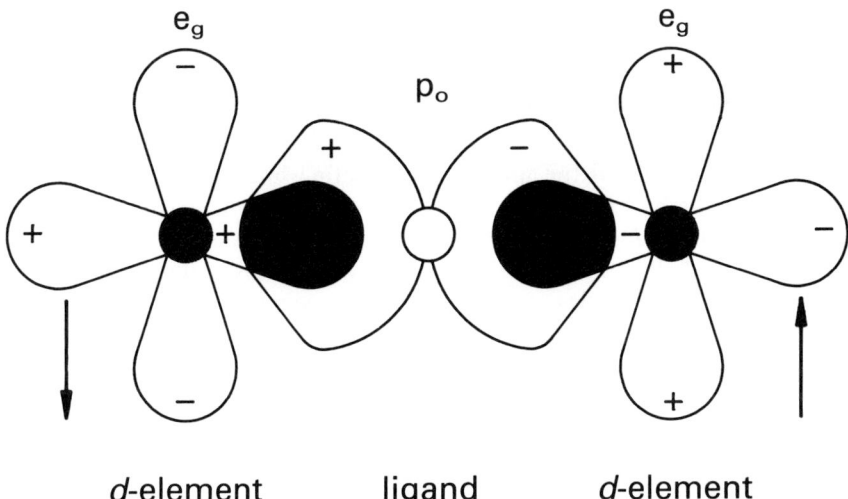

Fig. 1.7 Superexchange between two magnetic d-ions and a p-state ligand (σ-transfer). The shaded regions represent the covalent mixing of the different wave functions.

semiconductors the proximity of the valence band can lead to non-localized states which can propagate the exchange interaction even further. So, once again we can have a parallel or antiparallel orientation between the magnetic spins which, if randomly spaced, results in the competing interactions necessary to form the spin-glass state.

1.3.4 Dipolar

Another mechanism that causes spins to interact is the dipolar energy. Although weak it is always present and takes the form:

$$\mathcal{H}_{ij}^{\text{dip}} = \frac{1}{r_{ij}^3} [\boldsymbol{\mu}_i \cdot \boldsymbol{\mu}_j - 3(\boldsymbol{\mu}_i \cdot \hat{r}_{ij})(\boldsymbol{\mu}_j \cdot \hat{r}_{ij})] \tag{1.3}$$

Note that in addition to the $(1/r_{ij})^3$ distance dependence there is also a built-in *anisotropy* to the dipolar interaction which can orient the spins ferro- or antiferromagnetic. For example, if the spins are oriented along $\mathbf{r}_{ij}$, they will couple parallel; however, if the spins are oriented perpendicular to $\mathbf{r}_{ij}$, then they will couple antiparallel. Try it by seeing which configuration gives the lowest (negative) energy in the above equation. The anisotropy of the interaction plays an important role in the resulting ordering. Up until now our RKKY and superexchange couplings were considered as isotropic interactions, i.e., only a function of distance r with no angular dependences. The dipolar interaction introduces an angular dependence into the coupling scheme and we will delve more into magnetic anisotropy in the next section.

1.3.5 Mixed superexchange (random bonds)

Thus far we have listed the mixed or competing interactions which are to be combined with the random site occupancy to form a spin glass. But what about the $\pm J$ random-bond mechanism mentioned at the end of section 1.2 and illustrated in Fig. 1.4? Now we must rely on using two different magnetic species, in our previous examples Cu (which in an insulator, but *never* in a metal, can be magnetic $S = \frac{1}{2}$) and Co, and Fe and Mn. Let us examine the insulating, superexchange $Rb_2Cu_{1-x}Co_xF_4$ system. Here the tetragonal crystal electric field experienced by the Co ($S = \frac{3}{2}$) splits the various spin states such that only the lowest $S = \frac{1}{2}$ doublet is populated at low temperatures. Hence both the Cu and Co ions have effectively $S = \frac{1}{2}$. The crystal-field splitting also causes a single-ion anisotropy (see next section) which forces the spins to point only up or down along the z (c-axis) direction. This simple, up or down, collinear orientation of the moments we call an **Ising character**. At this point we bring the nearest

1.4 Magnetic anisotropy (plays a role)

neighbour superexchange into action. Because of the different ground state orbitals of the Cu, their wave-functions alternate lobes along the x and y axes. This generates a ferromagnetic coupling. By substituting the Co on the Cu sites, two possibilities exist for the Co–Cu superexchange. Firstly a Cu-lobe pointing along the distance to the Co ion results in an antiferromagnetic coupling; and secondly a Cu-lobe perpendicular to the Co site forms a ferromagnetic pair. Finally, if two Co atoms happen to be nearest neighbours the usual antiferromagnetic superexchange is the outcome. Therefore, we have four distinct interactions among the mixed Cu–Co system, two are parallel and two antiparallel. Figure 1.8 shows the bond-probability function $P(J)$ for a particular choice of $x = 0.29$. While not exactly the $\pm J$ distribution used in the creation of Fig. 1.4, we do have a favourable approximation to a random-bond spin glass with a realizable material. More will be said about this system when we deal with the 'ideal' spin glasses in Chapter 6.

1.4 MAGNETIC ANISOTROPY (PLAYS A ROLE)

We have already been forced into using the notion of magnetic anisotropy which offers a preferred direction for aligning the spins. The simplest form of this anisotropy previously used in Fig. 1.2 is the Ising model in which all the spins are constrained to point either up or down. This is caused by a very strong uniaxial anisotropy in one direction, usually taken as the z-axis. Alternatively, the spins can also be forced to lie in a 2D plane, then we have x-y anisotropy. Remember that the Heisenberg model is, of course, fully isotropic, and the spins can align in any direction, even non-collinear, of the 3D space. These various types of magnetic anisotropies will become

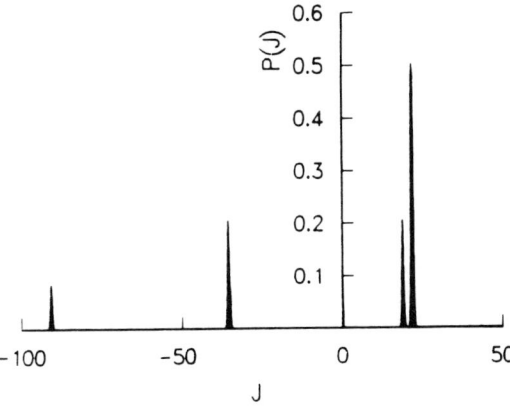

Fig. 1.8 Bond probability function $P(J)$ for $Rb_2Cu_{1-x}Co_xF_4$ with $x = 0.29$. From Dekker (1988).

very important when we treat the possibilities of a phase transition using the predictions of the theoretical calculations. In addition the anisotropy plays a major role in governing the frozen-state properties of a spin glass.

Let us briefly review the various mechanisms for generating the magnetic anisotropy. As previously stated the dipolar interaction gives an intrinsic preferred coupling as a function of angle, i.e., orientation of the moments at sites i and j relative to $\mathbf{r}_{ij}$. This will certainly have a bearing on the particular type of magnetic ordering at low temperatures. Since the dipolar energy is usually so weak compared to the various exchange energies, it is a secondary effect and must be used in conjunction with an (isotropic) exchange interaction to invoke its influence, for instance, on the particular direction of the moments as the spin glass freezes.

Single-ion anisotropy, produced from the local crystalline electric fields, is another important source of anisotropy. We already used this mechanism in explaining the random-bond Rb_2CuCoF_4 system. On a microscopic basis, it is directly related to the crystal symmetry of the given material and the spin and orbit angular moment of the magnetic species. Certain preferred directions are imposed on the (non-S state) magnetic ions by the crystalline electric field, thus creating the magnetocrystalline or single ion anisotropy. Its energy is usually written in spin-only hamiltonian form as

$$\mathcal{H}_{SI} = -D\sum_i (S_i^z)^2 \qquad (1.4)$$

for an uniaxial crystal,

$$\mathcal{H}_{SI} = -D\sum_i [(S_i^x)^4 + (S_i^y)^4 + (S_i^z)^4] \qquad (1.5)$$

for a cubic crystal; and

$$\mathcal{H}_{SI} = -\sum_i D_i (S_i^z)^2 \qquad (1.6)$$

for an amorphous material. The prefactor D is the local anisotropy constant, except for the amorphous material, where D_i has a distribution of magnitudes and directions (axes) which are different at every site. Consequently, in a non-crystalline solid there can exist a random anisotropy defined through a locally varying 'easy axis'.

Yet another form of anisotropy arises in the metallic spin glasses and is due to the Dzyaloshinskii–Moriya (DM) interaction. Its energy (hamiltonian) is given by $\mathcal{H}_{ij}^{DM} = -\mathbf{D}_{ij} \cdot (\mathbf{S}_i \times \mathbf{S}_j)$ and describes the following process: a conduction electron is initially scattered by spin $\mathbf{S}_i$, then it interacts with a non-magnetic scatterer having a large spin-orbit coupling, and finally, the conduction electron scatters off another spin $\mathbf{S}_j$. It is this intermediate non-magnetic, spin-orbit scattering that is the essence of the DM anisotropy. Here there appears in $\mathbf{D}_{ij}$ a cross product $(\mathbf{R}_i \times \mathbf{R}_j)$ of the positions of the two spins with respect to the non-magnetic scatterer placed at the origin.

1.5 Frustration (a far-reaching concept)

The net result of this anisotropy is to establish a unidirectional character which aligns the spin in a *single* direction, quite different from the uniaxial (parallel/antiparallel) forms previously discussed. If we perform the summations for these two distinct types (DM plus uniaxial) of anisotropy, we arrive at a macroscopic anisotropy energy

$$E_{AN} = -K_1 \cos\Theta - \tfrac{1}{2}K_2 \cos^2\Theta \qquad (1.7)$$

where Θ is the angle which rotates the spins from their frozen direction. K_1 is the **unidirectional anisotropy constant**, which changes sign with a rotation of π, and K_2 is the **uniaxial anisotropy constant**, which remains the same under a rotation of π. For the CuMn spin glass K_1 can be made predominant and a unidirectional character appears.

To summarize this section we have tried to collect and outline the relevant forms of anisotropy or freezing-direction preference. Unique to the spin glasses are the possibilities of random anisotropy and the DM interaction creating a unidirectional anisotropy. Also noteworthy is that they can manifest themselves as a macroscopic energy in certain experiments (Chapter 3). We shall use the notion of anisotropy frequently as we dig further into the mysteries of the spin glasses.

1.5 FRUSTRATION (A FAR-REACHING CONCEPT)

The word frustration and its illustration have already appeared at the very beginning of this chapter. Figure 1.2 sketched the basic triangle and antiferromagnetically bonded Ising spins. The result was six degenerate ground states (all with the same energy) for which there is always one unsatisfied bond. Try this exercise yourself. The concept of frustration, implicit in many physical and biological phenomena, was first clearly used and quantified in the description of the spin glasses. It is the essential ingredient to establish the special ground state. So let us look a bit closer at frustration.

Suppose we have a square or plaquette of four Ising spins with four $\pm J$ bonds as shown in Fig. 1.9. For the left-hand configuration (Fig. 1.9(a)) of

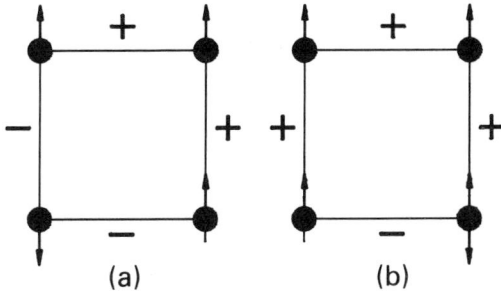

Fig. 1.9 Square lattice with mixed interaction $+$:F, $-$:AF. (*a*) unfrustrated plaquette $\Phi = 1$ and (*b*) frustrated plaquette $\Phi = -1$.

two + bonds and two − bonds, all the bond energies are satisfied and there will only be a two-fold-degenerate ordered state. This arbitrariness is caused by the initial choice of the first spin which could be set either up or down. Here we say the plaquette is **unfrustrated**. Moving over to the right-hand configuration in Fig. 1.9(b) (three + and one − bonds), all the bond energies cannot simultaneously be satisifed. One spin remains frustrated or one bond is broken no matter what we do. Notice how the degeneracy increases to 8-fold (count them for yourself). For this trivial example eight energetically equivalent ground states are possible. In a mathematical way we may write the frustration function as $\Phi = \Pi$ sign $(J_{ij}) = \pm 1$, where the minus sign denotes a frustrated system. For the case shown in Fig. 1.9, $\Phi = J_1 J_2 J_3 J_4 = +1$ for the unfrustrated left-hand side and -1 for the frustrated right-hand side.

If there is a great deal of frustration in our system, many of the plaquettes will be frustrated. Connecting such plaquettes with a set of minimal distance lines gives a measure of the excess energy caused by the frustration. The task, and it is not an easy one, is to find the minimum length between the various frustrated plaquettes and thereby determine the lowest ground-state energy. There is great similarity between this 'optimization' problem and others such as the intriguing 'travelling salesman' problem (see Chapter 7).

Note that the frustration creates a multidegenerate, metastable, frozen ground state for the spin glass. And as we shall soon see such a ground state leads to many interesting and unique experimental properties. But now we should ask the question − 'is frustration alone enough to generate a spin glass according to our working definition?' The answer is '*No*, it must be coupled with disorder or randomness, site or bond, our first basic ingredient'. An antiferromagnetically coupled, regular triangular lattice (which is full frustrated) will not exhibit the **co-operative freezing transition** of a spin glass. More likely, a slow 'blocking' will occur to a ground state which is dominated by large magnetic fluctuations. Furthermore, there seems also to be the need for mixed (+ and −) interactions supplementary to the randomness in order to attain our spin glass. Thus, and this aspect is still under contention, a system with only antiferromagnetic interactions plus disorder will not yet be a good, co-operative spin glass. The 'extra necessary' ingredient would be to include some ferromagnetic bonds. Support for our standpoint comes from very recent experiments on Kagomé (triangular-like) lattices with only antiferromagnetically coupled ions which do not show generic spin-glass behaviour.

Let us recapitulate the essence of the previous sections to secure our point of view. Randomness or disorder is indispensable and can be created by varying the distances between the magnetic species or by randomizing the bonds among a periodic set of magnetic ions. The mixed (ferro–antiferromagnetic) interactions are essential to install the competition and ensure co-operativeness of the freezing process. Anisotropy seems to be required to establish preferred directions along which the local spins can freeze. And frustration is a direct consequence of the disorder and mixed interaction, while a necessary condition for a spin glass, it is not a sufficient one.

1.6 THE FREEZING PROCESS (A FIRST LOOK)

At this point it would be instructive to try to present our first naive picture of the evolution of a spin glass as we reduce the temperature from far above the freezing temperature T_f to far below. Our description is based on experimental observation which will be explicated and discussed in Chapter 3. We now shall hope to gain some rudimentary insights into the underlying physics of the freezing process.

Of course, as $T \to \infty$, there will simply be a collection of paramagnetic spins, i.e., independent and rapidly rotating arrows in the 'chaos' caused by the high temperature. When the temperature is lower from $T \gg T_f$ many of these randomly positioned and freely rotating spins build themselves into locally correlated units or clusters, even domains, which can then rotate as a whole. These 'clusters' may simply be ferromagnetically coupled resulting in 'giant' or superparamagnetic moments. Or they may form from a strongly localized overlapping of RKKY interactions, in which case, because of the (+) parallel, (−) antiparallel oscillations, the net moment will be proportional to the square root of the number of spins taking part. The remaining spins, those not belonging to any cluster, are independent of each other, but help to transmit interactions between the clusters allowing for changes in the cluster sizes and their relaxation or response times. This formation of clusters can be tracked with various experimental techniques and is a direct consequence of the randomness and mixed interactions.

Now as $T \to T_f$ the various spin components begin to interact with each other over a longer range, because the temperature disorder is being removed. The system seeks its ground-state ($T = 0$) configuration for the particular distribution of spins and exchange interactions. This means a favourable set of random alignment axes generated by the local anisotropy into which the spins or clusters can lock. However, as we learned in the previous section frustration plays its role and a multi-degenerate array of ground states presents itself for the system to choose from. Since there is a spectrum of energy differences between the frozen states, the system may become trapped in a metastable configuration of higher energy.

What type of transition or transformation occurs at the freezing of a spin glass? Is it a phase transition with T_f a critical temperature (T_c), treatable within the theories of critical phenomena? Or is there a gradual blocking of the cluster rotations so that the system slowly settles into the frozen state? A third possibility exists somewhere in between these two extremes, where a new type of 'random' transition occurs requiring a very different theoretical description. How are the spin glasses and their freezing related to the real 'window' or 'wine' glasses and their transition? Is it possible to combine the glassy analogies into one unified treatment? Yet, at the moment, the real glass transition does not have a generally acceptable microscopic model and new efforts (e.g., mode coupling theories) are constantly going into creating one. Consequently, we cannot answer these questions completely, because the exact nature of the freezing process both in real glasses and spin glasses is presently unresolved. There are, however, certain directions, which became clear during the intervening twenty years of research, leading to the conclusion that the spin glasses are unique and necessitate an unconventional description both phenomenologically and mathematically. And the whole purpose of this monograph is to introduce you to these strange and intriguing properties and their underlying physics.

Before concluding this chapter with some historical background, we should mention the low-temperature ($T \ll T_f$) frozen state. Here unusual magnetic behaviour appears related to the 'glassy' condition. Most of these effects are caused by the application and cycling of an external field which also probes the anisotropy. Additional metastabilities and relaxations have been attributed to the rate dT/dt at which the sample is cooled through T_f, and the 'waiting time' t_w below T_f, with T constant, before the field is changed.

Another salient feature of the frozen state is the appearance of a unidirectional (see section 1.4) anisotropy which has a microscopic basis and creates randomly distributed *local* orientations throughout the sample. With various field-cooling or cycling procedures this quantity can be made to manifest itself as a macroscopic variable and is of particular importance in governing the low-temperature collective properties, such as magnetization, torque, ESR, etc. The excitation spectra of the frozen state can also be probed via various scattering techniques and more localized than propagating modes are found.

Finally it should be emphasized again that no usual ferro- or antiferromagnetic state with long-range magnetic order is formed. Hence, one does *not* observe magnetic **Bragg peaks** with scattering experiments due to a lack of translational invariance. The spin-glass condition for the randomly frozen state is $\Sigma_i S_i = 0$ where the sum is over all spin sites. Another convenient way of describing a frozen spin glass is by its time correlation function $\langle S_i(t)S_i(0)\rangle$ which remains finite for $T < T_f$ as $t \to \infty$.

1.7 SOME HISTORICAL BACKGROUND

Magnetic alloys have a long history dating back a hundred or so years. Especially interesting are the noble-metal/transition-metal combinations in which the first spin-glass effects were discovered. Already in the 1930s experimental investigations on AuFe and CuMn indicated some strange

1.7 Some historical background

magnetic properties. The first had to do with the magnetism or lack of it caused by the introduction of a magnetic impurity into a non-magnetic matrix. Would the moment remain, or would it lose its magnetism when dissolved in the sea of conduction electrons? Moment formation was the issue and the virtual-bound-state model (c. 1960) resolved the matter. On the other hand, the transport properties exhibited anomalous effects; resistivity minimum and giant thermoelectric power were discovered and given the name Kondo problem. The main research emphasis during the 1960s was on the moment formation and the Kondo effect. There were indeed studies of more concentrated alloys, but the inherent interactions were often incorrectly neglected to get at the **single-impurity** Kondo behaviour.

A number of investigations in the 1950s did consider the interaction of well-formed moments; some experimental behaviour was unusual but not striking. For example, supplementary linear term in the specific heat at low temperatures was observed in various magnetic alloys. In addition, the susceptibility measured in rather large fields showed a broad maximum at a temperature where roughly the magnetic remanence disappeared. The latter and its hysteresis were different from the conventional ferromagnetism; more like a collection of mutually interacting ferromagnetic and antiferromagnetic domains. Mössbauer experiments with its advent during the same period indicated there was some type of random magnetic order developing at low temperature. These were the empirical fruits of the 1960s. But then the main concern especially theoretically was with the single (Kondo) impurities.

The situation changed around 1970, at that time the sharp cusps in the low-field susceptibility of spin glasses were discovered and correlated with a magnetic-ordering temperature as determined from other measurements. Moreover, the RKKY interaction was brought to bear, and from its $1/r^3$ decrease, the invariance of the product xr^3 ($=$ const.), where x is the magnetic-impurity concentration, was discerned and enabled the magnetization M/x and the specific heat C_p/x to be represented as universal functions of the reduced variables H/x and T/x. A 'scaling' (or reduction of much data to a single curve) of many experiments was thereby possible. These advances marked the beginning of the spin-glass problem as we know it today.

But how did the name 'spin glass' arise and adhere? The term was first suggested by B. R. Coles in 1970 to be applied to the strange magnetic behaviour of the weak moment AuCo system. Simultaneously, this expression appeared again at Coles' instigation in a 1970 paper by P. W. Anderson linking localization in disordered systems with the magnetic-alloy problem. How it took hold is more difficult to answer. Other terms were competing, **mictomagnetism** was suggested around 1971 by P. A. Beck. Later the designation 'cluster glass' was used. But anyway what's in a name? It's more important to learn what physical phenomenon the nomenclature represents and how we can understand its observable facts. Then we can give a 'catch-all' appellation to this body of knowledge.

Returning to our history and the 1970s, finally the theorist became interested around 1975 and 'all hell broke loose' with a number of exciting new models and their attempted solutions, viz., the Edwards–Anderson model and the Sherrington–Kirkpatrick solution. In Chapter 5 we shall delve into these theories. The five year interval between 1975 and 1980 was a time of great activity – theory led the way with experiment offering a series of new surprises. Naturally many papers (thousands) were published. Around the end of this period, the order parameter describing the mean-field phase transition of spin glasses was shown to be inadequate. The frozen state according to theory was unstable. Henceforth, there began a renewed quest for the proper ground state or **replica-symmetry-breaking** scheme – a term with which we shall later become acquainted. A suitable scheme was discovered a few years later, but its physical meaning was initially obscure. Nevertheless, a new form of statistical mechanism with 'ultrametricity' was created requiring a novel, unconventional order parameter. And with great popularity the spin-glass models and mathematics were applied to other diverse areas. One especially interesting new field is that of neural networks. Dynamics (time and frequency dependences) was also treated within the mean-field theory and recognized as a key element. But as more and more materials were called spin glasses, the contact between theory and experiment diminished. It was difficult to compare the empirical behaviour of the many real spin glasses to the new idealized theories. Circa 1985 the post-replica age began, for, another theory based upon a 'droplet' (2-state) model was proposed for short-range spin glasses. Here the insulating materials could be treated and compared with the model's predictions, if *ideal* glasses could be fabricated in the laboratory.

After 1985, the pace of spin-glass research slowed down and while the basic phenomena have been established and catalogued, a thorough understanding or full description via the existing theories is still lacking. Into the 1990s there remain many unresolved or grey areas of the spin-glass problem. As we conclude our treatise in the final chapter we could perhaps look into some of these. If history is our guide, usually in times of peace and quiet, complete solutions are obtained for difficult questions. So let us wait and see what happens during the writing/reading of this book.

FURTHER READING AND REFERENCES

General solid state physics

Kittel, C. (1986) *Introduction to solid state physics*, (6th edn), Wiley.
Ashcroft, N. W. and Mermin, N. D. (1976) *Solid state physics*, Holt, Rinehart and Winston.
Kittel, C. (1964) *Quantum theory of solids*, Wiley.

General magnetism

Morrish, A. H. (1965) *Physical principles of magnetism*, Wiley.
Crangle, J. (1977) *The magnetic properties of solids*, Arnold.
White, R. M. (1983) *Quantum theory of magnetism* (2nd edn), Springer.
Mattis, D. C. (1988) *The theory of magnetism* Vol. I and II, Springer.

Introductory articles to spin glasses

Stein, D. L. (1989) Spin glasses, *Scientific America*, July, 36.
Fisher, D. S., Grinstein, G. M. and Khurana, A. (1988) *Physics Today*, Dec.
Ford, P. J. (1982) *Contemp. Phys.*, **23**, 141.
Hurd, C. M. (1982) *Contemp. Phys.*, **23**, 469.
Mydosh, J. A. (1978) *J. Magn. Magn. Mater.*, **7**, 237.

Historical background

Anderson, P. W. *Physics Today*. Series of six articles in Reference Frame, Jan. 1988, March 1988, June 1988, Sept. 1988, July 1989, and Sept. 1989.

References

Binder, K. and Schröder, K. (1976) *Phys. Rev.*, **B14**, 2142.
Dekker, C. (1988) PhD Thesis, University of Utrecht.

2

Description of related concepts

In order to better present the spin glasses and establish their place in magnetic phenomenon, we should first consider how they evolved and where they belong with respect to other allied magnetic effects. Some perspective at this point is important and a number of cognate terms are in need of introduction. This chapter will briefly survey the related or surrounding phenomena from which the spin glasses developed. The main portion of Chapter 2 is devoted to the dilute magnetic alloys. Moment formation via the virtual-bound-state model and the Kondo effect along with the giant moments will be introduced. Afterwards we will discuss the various types of behaviour which appear in the different concentration regimes extending from very dilute to fairly concentrated. This is to show exactly where the spin glasses lie with respect to their disorder. Here a valuable new concept for us is percolation which distinguishes the spin glasses from the more-concentrated, long-range-ordered, random systems. Then we shall view the magnetic glasses, since amorphous magnetism is certainly a part of the spin-glass problem and many of these materials exhibit the novel 're-entry' phenomenon. Finally, the notion of superparamagnetism is explored and its applicability to the spin glasses perused, and to conclude, the insulating and semiconducting systems will be mentioned and identified.

2.1 VIRTUAL-BOUND-STATE MODEL

The simplest case of a magnetic ion in a solid is that of a rare-earth magnetic insulator. Most rare-earth ions have similar chemical properties with valency +3 and the $4f$ (magnetic) electrons are situated well to the interior of the atom and surrounded by $5s$, $5p$ and $5d$ electrons. Therefore, the ions in many insulating compounds are non-interacting and simply paramagnetic. This means the rare earths nicely obey the Curie law

2.1 Virtual-bound-state model

$$\chi = \frac{NJ(J+1)g^2\mu_B^2}{3k_BT} \qquad (2.1)$$

$$p_o(\text{THEO}) = g[J(J+1)]^{\frac{1}{2}} \quad \text{and} \quad p_o(\text{EXP}) = \frac{(3k_BT_\chi)^{\frac{1}{2}}}{\sqrt{N}\,\mu_B}. \qquad (2.2)$$

Here N is the number of paramagnetic ions, g is the Landé g-factor, μ_B the Bohr magneton and k_B the Boltzmann constant. For rare-earth ions J is a good (total) angular-momentum quantum number and can be constructed out of the atomic ground-state L and S quantum numbers invoking Hund's rules. Once we know J it is an easy task to define a theoretical effective Bohr magneton number $p_o(\text{THEO})$ and compare it with experiment via the measured $p_o(\text{EXP})$. The agreement is usually good and the few exceptions Pm^{+3}, Sm^{+3} and Eu^{+3} can be corrected for by using the concept of induced-moment (Van Vleck) magnetism. Crystalline electric fields will influence the ionic behaviour at low temperatures. However, such effects can be taken into account knowing the crystal symmetry and the energy splitting.

The situation is somewhat different regarding the other source of magnetic ions – the 3d transition metals. In this case there is a much larger radius for the 3d-orbital wave functions and the 4s electrons are usually lost to the ionic bonding of the compounds. Thus, no outer electrons exist to shield the 3d electrons and the crystalline electric fields cause significant energy splittings, much larger than that induced by the spin-orbit coupling. So J is not a good quantum number and the ionic model no longer works. There is, however, one simplifying hope, namely that the lowest orbital level, when split by a cubic crystal field, is a singlet $L = 0$. In many cases this is the only level populated and there can be no further splitting due to the spin-orbit coupling. Accordingly, we say the orbital moment has been quenched and we have 'spin-only' magnetism. In these instances we can compare theory to experiment merely using

$$p_o(\text{THEO}) = g[S(S+1)]^{\frac{1}{2}} \qquad (2.3)$$

and observe sound agreement. Nevertheless, there are many exceptions to the above simplified singlet case. Sometimes doublet or triplet orbital levels are lowest, and these are further split by the spin-orbit coupling. Also incomplete quenching can arise and the spin-orbit interaction could be large. All these cases must be handled on an individual basis compound by compound, and while complicated combinations of crystalline electric field and spin-orbit coupling arises, notwithstanding, they are all treatable within an individual *atomic* model which gives the local atomic moment.

The metals are completely different, for now we have an interaction between the magnetic ions and the sea of conduction electrons. Whether a magnetic moment survives is a fundamental question. In order to treat this problem we must use the virtual-bound state (vbs) or Anderson model.

Let us consider a single 3d magnetic ion immersed in the sea of conduction electrons produced by a host metal. To keep things simple we regard the host as a free-electron-like metal. There are four terms which arise in the hamiltonian describing such a system:

$$\mathcal{H} = \text{KE}(s\text{-elec}) + \text{KE}(d\text{-elec}) + V_{sd}(\text{mixing}) + U_{dd}(\text{Coulomb}). \quad (2.4)$$

The first two terms represent the individual kinetic energies of s and d electrons, respectively. The third term is a mixing or hybridization interaction between the s and d electrons, and the final term gives the intra-atomic Coulomb repulsion of the lowest d orbitals of opposite spin. Finding the solution of this vbs-hamiltonian is a long and arduous task. We shall not go into the mathematical details, but simply view some sketches of the physical situations.

Figure 2.1 (left-hand side) shows a two-band (spin up/spin down) density of states $N(E)$ plot for the host's s-band – filled up to Fermi level E_F and a localized (atomic-like) d-impurity level with energy less than E_F. Now we turn on the mixing or hybridization and the intra-atomic Coulomb repulsion. The former causes the 'localized' level to be firstly shifted downwards in energy with respect to E_F: $E_d \approx 2N(E_F)|V_{sd}|^2/D$ (D is the s-bandwidth), and then to be broadened with width $2\Gamma \approx 2\pi N(E_F)|V_{sd}|^2$. The latter (Coulomb repulsion) causes an energy shift U between the spin-up level and the spin-down level. The right-hand side of Fig. 2.1 delineates the case where $|V_{sd}| \ll |U|$. Note how the two virtual-bound states (vbs) are formed on each of the sub-bands, one filled, one empty. Here the spin-down vbs-level is simply pushed above the Fermi energy and is therefore unoccupied. Thus, we have created a strongly magnetic case via the filling of the spin-up vbs, i.e., a localized magnetic moment is formed.

But what happens if we make $|V_{sd}|$ comparable to U? The spin-up and spin-down vbs levels become broader and could partially overlap with the

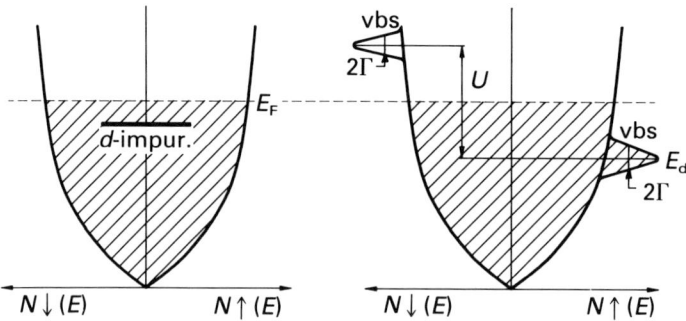

Fig. 2.1 Schematic diagram (left-hand side) of a 3d impurity immersed in a sea of conduction electrons separated into spin-up and spin-down electrons. Right-hand side illustrates the formation of the vbs. An occupied spin-up vbs at E_d with width 2Γ is attached to the conduction electron density of states. A similar, but unoccupied, vbs split by energy U above E_F occurs with down spin.

2.1 Virtual-bound-state model

Fermi energy. This means they will be occupied with different amounts of spin-up *d*-electrons and spin-down *d*-electrons. A reduced Coulomb repulsion results in an upwards shift of the up-vbs relative to the down-vbs. This further increases the overlap and is shown in the left-hand side of Fig. 2.2. The net vbs occupancy of up-spin minus down-spin will determine the 'local' *d*-moment which could easily be non-half integral. Thus, any value of spin or moment could appear in a dilute magnetic alloy. Or, if we even push our parameters further $|V_{sd}| \gg U_{NET} \approx 0$, we arrive at the symmetric case of up-spin occupancy equal to down-spin occupancy, and no net spin, i.e., the impurity loses the magnetism; it is effectively *dissolved* in the sea of conduction electrons. Figure 2.2 (right-hand side) depicts this non-magnetic possibility.

The main consequence of the virtual-bound-state model is that an atomic magnetic impurity may not even form a moment when introduced into a metal. Everything depends on the given set of parameters. If it does, then *any* (not just half integer) value of spin or moment may be generated. The model had its heyday in the early 1960s, when the problem of moments formation was a paramount issue. What the model does not answer is how the moment forms as a function of temperature and what effect this temperature dependence will have on the experimental properties.

We shall look further into these questions in the next section. But before concluding this brief sketch of the vbs model, let us mention the Schrieffer–Wolff (S–W) canonical transformation which allows us to combine the various model parameters into an effective exchange interaction J. In 1966 S–W showed that in the strongly magnetic limit [the case displayed in Fig. 2.1 (right-hand side)]

$$J = -2|V_{sd}|^2 \frac{U}{E_d(E_d + U)} < 0 \qquad (2.5)$$

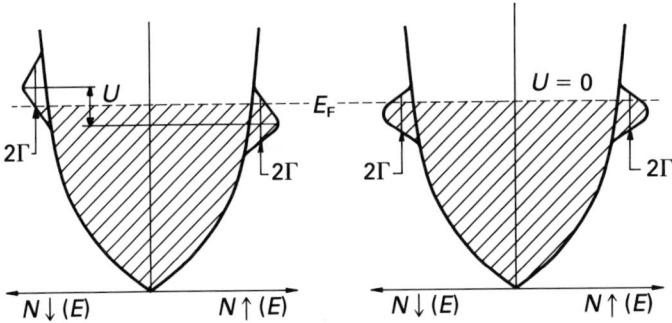

Fig. 2.2 Schematic diagram (left-hand side) of partially magnetic vbs. Note how the magnetic moment can form with any spin value. Right-hand side represents the symmetric case of equal spin-up and spin-down occupancies ($U = 0$) and thus no net magnetic moment.

The underlying physics of this antiferromagnetic coupling between local-moment and conduction electrons may be gained from the intuitive picture given in Fig. 2.3. However, to make the draught realistic we should place the vbs levels close to E_F so we can invoke the process of 'covalent mixing'. Now what happens when, due to the covalent mixing, there is an electron transfer from the occupied vbs to the up-conduction electron band and from the down-conduction electron band to the unoccupied vbs? The answer is that a difference in the Fermi levels or fillings will appear between up and down spin bands. In order to correct for this non-equilibrium situation, up-spin conduction electrons flow into down-spin conduction-electron states (see Fig. 2.3). The net result of this mixing and siphoning process of up-spin electrons is to produce slightly more down-spin conduction electrons than up-spin conduction electrons at the expense of the occupied local moment. If we realize that the covalent mixing between unlike spins is 'attractive', then a small conduction-electron polarization of opposite spin will be created surrounding the local moment (up-spin vbs). The corresponding admixture of like spins is forbidden by the Pauli exclusion principle and a repulsion or delocalization results. Although the processes described in Fig. 2.3 are rather schematic, they do lead to an antiferromagnetic coupling between the local moment and the conduction electrons. And this indeed is the origin of the Kondo effect.

2.2 KONDO EFFECT AND WEAK MOMENTS

The subject of dilute magnetic alloys may be divided into two classes of interest. Firstly, non-interacting or very-dilute magnetic impurities dissolved into a non-magnetic host can be classified under the general heading 'Kondo effect' – a localized antiferromagnetic interaction of an isolated or single-

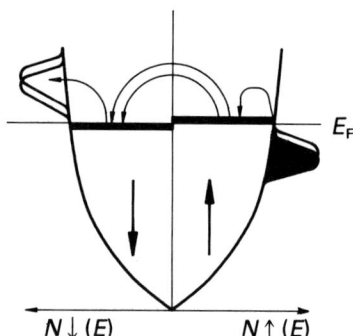

Fig. 2.3 The effect of covalent mixing on the vbs which siphons off the moment and causes antiferromagnetic coupling ($J < 0$) between the local moment and conduction electrons.

2.2 Kondo effect and weak moments

impurity spin with the surrounding conduction electrons. Secondly and of prime importance in this book are the impurity spin-spin interactions which lead to spin-glass freezing and long-range magnetic order.

Returning to the first (Kondo) class, we must treat the magnetic spins as isolated from each other. In practice, such a condition is rather difficult to realize, since the previously mentioned RKKY interaction is present and falls off rather slowly as $(1/r)^3$. Hence, there is always a weak coupling between even far-distant spins. A reasonable empirical approximation is to work at impurity concentrations below 100 ppm and not too low a temperature. Assume now that the RKKY is negligibly small. The only remaining interaction (since the dipolar one can be completely disregarded) is between the local-moment spin **S** and the conduction-electron spins **s**, and as just described it may be treated by the so-called s-d or s-f exchange hamiltonian $\mathcal{H} = -J\mathbf{S} \cdot \mathbf{s}$. Here $J(< 0)$ is the antiferromagnetic exchange coupling introduced in the previous section. At high temperatures the impurities behave like free (paramagnetic) moments. However, below a characteristic temperature, specific for each alloy system (see Table 2.1) the isolated impurity becomes non-magnetic due to its interaction with the conduction electrons. This temperature is known as the Kondo temperature T_K and it theoretically signifies the breakdown of higher-order perturbation theory which is used to calculate the physical properties from the s-d or s-f hamiltonian. In the calculations

$$T_K = (D/k_B) \exp\left[-1/N(E_F)|J|\right] \qquad (2.6)$$

where D is the bandwidth (usually taken to be of order of the Fermi energy E_F or Fermi temperature T_F) and $N(E_F)$ is the conduction-electron density

Table 2.1 Isolated impurity Kondo (spin fluctuation) temperatures for important dilute alloy systems

Host	Impurity:	V	Noble metal-transition metal				
			Cr	Mn	Fe	Co	Ni
Cu		1000 K	2 K	10 mK	30 K	500 K	>1000 K
Ag		–	10 mK	<10⁻⁶ K	5 K	–	–
Au		300 K	1 mK	<10⁻⁶ K	0.2 K	500 K	>1000 K

	Transition metal-transition metal				
Mo	–	10 K	1 K	25 K ⎫	simple
Rh	–	50 K	50 K	1000 K ⎭	
Pd	100 K	10 mK	20 mK	0.1 K ⎫	exchange
Pt	200 K	0.1 K	0.3 K	1 K ⎭	enhanced

of states at the Fermi surface. For $T < T_K$ there is a gradual loss of local moment as the conduction electrons begin to form a surrounding cloud of oppositely (antiferromagnetically) polarized spin. This situation, i.e., that of a quasi-bound state, is sketched in Fig. 2.4. Its formation is not a phase transition, but a slow transformation on a logarithmic temperature scale. In the calculations, this broad-temperature transition to the quasi-bound state is heralded by the appearance of troublesome logarithmic divergences which reduce the response of the moment to external fields and enhance the electron scattering cross-section. Note that the entity comprising the quasi-bound state does not remain fixed in its orientation. Instead, it churns about as a whole, a rotating, yet compensated, moment.

Experimental manifestations of the Kondo effect $T < T_K$ are:

(i) a loss of magnetic moment, the magnetization falls below its free-moment value and the susceptibility is less than its Curie value

$$\chi_K = \frac{N\mu^2}{3k_B(T + T_K)} < \chi_C = \frac{N\mu^2}{3k_B T} \qquad (2.7)$$

where $\mu = g\mu_B[S(S+1)]^{\frac{1}{2}}$.

(ii) The enhanced scattering rate creates a logarithmic upturn in the resistivity at low temperature.

Accordingly, the total resistivity becomes for $T > T_K$

$$\rho_{TOT}(T) = \rho_o + AT^5 + x\rho_1 + x\rho_m[1 + 4J\,N(E_F)\ln(k_B T/D)] \qquad (2.8)$$

when the phonon T^5 dependence is included (ρ_o is the residual resistivity of the host metal and ρ_1 the residual resistivity of the impurity). For $J < 0$ a minimum appears in $\rho_{TOT}(T)$ since with $T \ll D$ the last term in the above equation diverges logarithmically as the temperature is lowered. This now-famous Kondo minimum is illustrated in Fig. 2.5 and signifies the competition between decreasing phonon and increasing Kondo contributions for $T \to 0$. Notice that $T_{MIN} \propto x^{\frac{1}{5}}$ where x is the magnetic-impurity concentration.

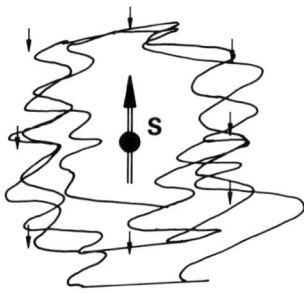

Fig. 2.4 Static picture of the Kondo bound state with local moment surrounded by oppositely polarized conduction electrons.

2.2 Kondo effect and weak moments

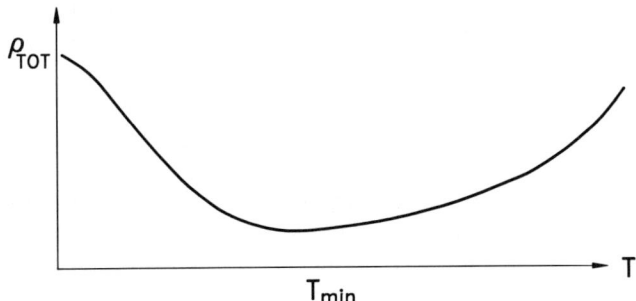

Fig. 2.5 The resistivity minimum in the temperature dependence of a dilute alloy. Notice that the Kondo contribution becomes saturated ('unitary scattering limit') as $T \to 0$.

An alternative approach to the interaction between a local moment and the conduction electrons is the localized spin-fluctuation model. Here the local spin fluctuates in *amplitude* (recall that spin also changes its orientation due to temperature disorder – Curie's law) at a rate $(\tau_{sf})^{-1}$ where τ_{sf} is the spin-fluctuation lifetime. When this rate is greater than the orientational changes produced by thermal fluctuations, the impurity spin appears non-magnetic and there is additional conduction-electron scattering. As before, a broad temperature interval separates the magnetic from the non-magnetic regime, and an analogous temperature may be defined as $T_{sf} = h/(k_B \tau_{sf})$. Table 2.1 collects both Kondo and spin-fluctuation temperatures for some dilute alloy systems. Note how these temperatures span many orders of magnitude.

These two models, quasi-bound state and localized spin fluctuation, give quite similar experimental predictions, and for the present treatise we can assume the two characteristic temperatures to be equal. Thus, for $T \gg T_K$ good magnetic moments exist, i.e., $\mu = g\mu_B/[S(S+1)]^{\frac{1}{2}} S(S+1)$, while for $T \ll T_K$, the moments become 'weak' and lose their magnetic character. It is not just temperature which can remove or install the magnetic moment, an external magnetic field H has a similar effect. Applying such large fields to a Kondo system below T_K will gradually restore the magnetization to its high temperature value. The physics is as if the field breaks up and aligns the antiparallel quasi-bound state or slows down the spins fluctuations making $T_{sf} \to 0$. Only quite recently (early 1980s), after most interest in the Kondo effect had subsided, an exact solution was found for the Kondo problem. Perturbation theory cannot fully treat a single magnetic impurity weakly coupled to a Fermi sea of electrons at low temperature. A rather complicated and mathematical Bethe ansatz treatment was employed to diagonalize exactly the Kondo hamiltonian and derive its thermodynamic quantities. The contemporary physical picture is that of a many-body *singlet* ground state and corresponding resonance in the density of states situated at $k_B T_K$ above the Fermi energy with a height $\propto 1/T_K$. It is this impurity (Kondo) resonance which creates the many interesting and anomalous properties.

What we neglected to mention before is that the specific heat of a Kondo system is large and has a broad maximum at approximately T_K. This along with the magnetization and susceptibility has now been calculated precisely from the above exact theory. The trouble was then to find a suitable, real, ideal material whose thermodynamic properties could be directly compared to theory. A similar difficulty exists, as we shall later see, with the spin glasses. Finally, $(La_{1-x}Ce_x)Al_2$, where $T_K = 0.2$ K and $S = \frac{1}{2}$ for the crystal-field split ground state, was deemed appropriate. And a very nice, quantitative fit between experiment and theory resulted, at least for the measurable thermodynamic quantities, magnetization, susceptibility and specific heat. Yet what remains to be calculated are the transport properties: the resistivity increase and the giant thermoelectric power associated with the Kondo effect. Up to now there have been no exact entire-temperature-range solutions to compare with the large amounts of experimental data available.

The occurrence of the Kondo effect is a hindrance to spin-spin interactions and magnetic ordering. In Table 2.1 a collection of Kondo temperatures is given for some transition-metal alloy systems which do indeed exhibit interactions and ordering effects at sufficiently high concentrations. At very low concentrations and below T_K the moments are 'weak' and cannot simply interact with each other. However, the trick here is to bring the impurities sufficiently close to each other and turn on the RKKY interaction. For, we can greatly eliminate the role of the Kondo effect and weak moments by increasing the concentration and creating good moments via a strongly magnetic **local environment**, since when interactions are taken into account $T_K(\text{single impurity}) \gg T_K(\text{pair of impurities}) \gg T_K(\text{triple})$, etc.

2.3 GIANT MOMENTS

The term giant moment describes the unique behaviour of dissolving a single localized magnetic moment in a non-magnetic host and thereby producing a net moment much greater than that due to the bare magnetic impurity alone. This overall moment is the sum of the localized moment plus an induced moment in the surrounding host metal. Notice now the two important differences with our preceding Kondo effect. First of all the interaction with the conduction electrons is dominantly ferromagnetic, i.e., a parallel coupling, and secondly, the host must be able to support this positive coupling over quite a long range (*not* the $+/-$ oscillations of the RKKY).

For the description of a large number of experimental properties, these entities (impurity moment plus polarization cloud) can be considered as

2.3 Giant moments

one magnetic moment which is then called 'giant'. A typical example of such a system is *Pd*Fe in which the Fe moment is approximately $4\mu_B$, yet experiments on *Pd*Fe reveal moments $\simeq 12\mu_B$ per Fe atom. Thus, via a strong, parallel, itinerant-electron polarization extending for about 10 Å, the Fe moment induces a small average ($\approx 0.05\mu_B$) moment on the 200 or so surrounding Pd atoms. Figure 2.6 sketches a static giant moment. Since this interaction is ferromagnetic, the Kondo effect is unimportant and long-range ferromagnetism occurs at very low (≈ 0.1 at. % Fe) concentrations due to the extended, indirect, ferromagnetic coupling. This behaviour represents the simplest type of ferromagnetic ordering to appear in a dilute, random alloy.

Sometimes in the literature the term giant moment is also used for the large magnetic entities which are caused by chemical clustering of magnetic atoms or groups or nearest neighbours needed to produce a moment via the local environment. Here we shall reserve the term giant moment only for those moments associated with isolated 'good moment' impurities supported by large host polarization.

A very interesting feature of the giant moments is that at very low concentrations ($\approx$ few ppm Fe) and large separations (20–30 Å) between the Fe impurities, the ubiquitous RKKY oscillations of almost insignificant amplitude will be felt, if the temperature is reduced sufficiently. Hence, a spin glass is possible with the presence of mixed interactions. *Tour de force*

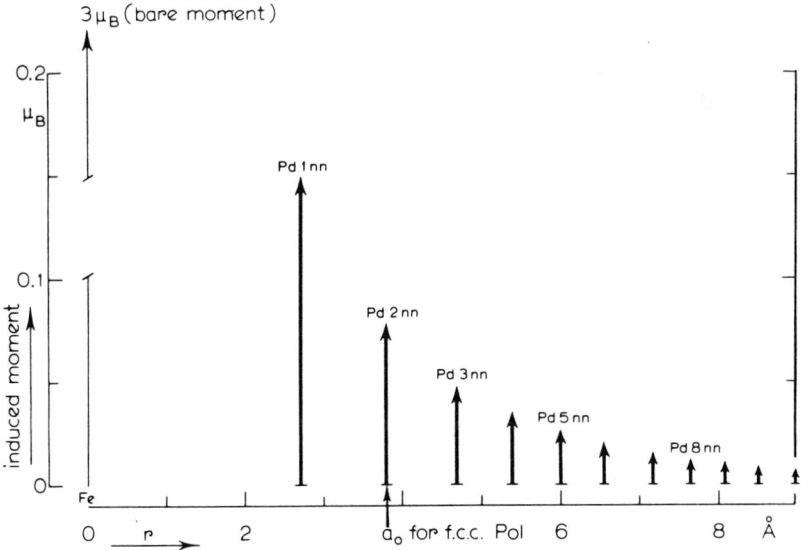

Fig. 2.6 Schematic picture of the giant-moment polarization cloud for an Fe impurity in Pd.

experimentation has verified this with a freezing temperature in the micro-Kelvin range.

2.4 SPIN GLASSES (A SECOND VISIT)

Since we had our first glimpse at what the spin glasses were in Chapter 1, let us now consider what they are not. The moments must be good, not weak – no significant Kondo effect. The concentration must be large enough so that there is enough competition from the mixed interactions – not the ferromagnetic predominance of the giant moments and not long-range, spatial ordering of the usual ferro- or antiferromagnetic transitions.

This last condition is quite interesting. Mathematically, it may be expressed as $\Sigma_i S_i = 0$ (no ferromagnetism), but $\langle S_i \rangle_t \neq 0$ where $\langle \ \rangle_t$ represents a time average over a long observational time t, and in addition

$$\frac{1}{N} \sum_i \langle S_i \rangle_t \exp(i \mathbf{k} \cdot \mathbf{R}_i) = 0 \quad \text{as } N \to \infty \tag{2.9}$$

This equation states that there is *no* long-range antiferromagnetic ordering, i.e., the directions of the spins are *not* related to a particular wave vector $\mathbf{k}$, as is true for even the most complicated of antiferromagnets with lattice vectors $\mathbf{R}_i$.

Another way of viewing the spin ordering is via their Ising ± 1 spin–spin correlation function

$$\langle S_i S_j \rangle_t = 0 \quad \text{but} \quad \langle S_i S_j \rangle_t^2 \approx \exp\left(\frac{-|\mathbf{R}_i - \mathbf{R}_j|}{\xi_{SG}}\right) \tag{2.10}$$

where ξ_{SG} is the spin-glass correlation length. The crucial questions here are: does the spatial correlation length $\xi_{SG} \to \infty$ as $T \to T_f$, and does the decay of the 'order parameter' governed by a characteristic time become infinitely long signifying a phase transition? The first would imply the second as in a conventional second-order phase transition. However, the spin glasses are different and we do not really know the final answers to the above questions. Long is not infinite. So let us now hypothesize a phase transition and study its consequences. It follows that we can replace the time averages by the ensemble averages of statistical mechanics and define an order parameter for the frozen spin-glass phase:

$$q = \langle \langle S_i \rangle_T^2 \rangle_C \neq 0 \quad \text{for } T \leq T_f \tag{2.11}$$

where the inner brackets represent ensemble (or *t*hermal) averages while the outer brackets are for a configurational average over the set of random interactions. q is called the Edwards–Anderson spin-glass order parameter and will appear often as we tread further into the models and mysteries of spin-glass theory.

2.4 Spin glasses (a second visit)

Since the magnetic impurities have random positions in the lattice, the effective molecular field, **H**, may be calculated as a superposition of contributions from all the impurities in the alloy, due to the infinite range of the RKKY interaction. Furthermore, **H** has a distribution of magnitudes, $P(H)$, and a random distribution of directions because of the oscillatory nature of the RKKY interaction. A computer simulation has been made for $P(H)$ using an fcc lattice with about 100 000 sites, 324 being occupied with spins (x ≈ 0.3%), and an RKKY interaction performs the coupling among these vector spins S_i. The results for $P(H)$ are shown in Fig. 2.7. The reader should compare this figure to Fig. 1.6 which gives the probability of RKKY interaction strengths. Notice that $P(H) \to 0$ as $H \to 0$ meaning that at each lattice site occupied by a spin there will be a finite internal field having both a magnitude and direction. Although $P(H)$ generates the same function for a large enough spin system, it is not required that all the spins point in exactly the same direction for each simulation or 'freezing'.

The freezing temperature T_f of a spin glass may be viewed from a very simple model. By considering only the *envelope* of the RKKY oscillations, $J'(r) = J_o/r^3$, we have the strength of the interaction as a function of distance. For a low-concentration alloy the average distance of impurity separation is $\bar{r} \propto (1/x)^{\frac{1}{3}}$. Now we use the thermal disorder energy, $k_B T$, to destroy the correlations between the spins, on average, $\bar{r}$ apart. In other words, the exchange energy caused by the RKKY interaction from a spin S_1, acting at a distance $\bar{r}$ on another spin S_2 must be greater than thermal disordering for freezing to occur. At the freezing temperature,

$$J'(r) = k_B T_f \quad \text{or} \quad T_f = x \frac{J'_o}{k_B} \qquad (2.12)$$

This means that no matter what the concentration is, there will always exist a temperature for which the RKKY interaction will prevail and produce this random (due to the oscillations) freezing. Here only good moments are assumed, i.e., no Kondo effect or $T_K = 0$.

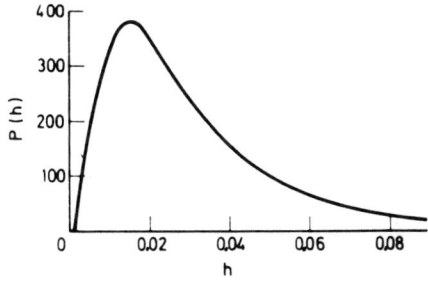

Fig. 2.7 Probability distribution of ground-state, internal-fields magnitudes for an RKKY interacting spin glass. From Walker and Walstedt (1977).

A further remark about these dilute alloys and the RKKY interaction has to do with the relation $(x\bar{r}^3)$ = constant. If a measure of length is used which is related to the system of impurities, then we can 'scale' the properties for one concentration x_1 to those of another x_2. Our length scale is now the average distance between impurities $\bar{r}$, so the volume $\bar{r}^3$ contains a constant number of magnetic impurities equal to $x\bar{r}^3$. This has particular significance for the thermodynamic properties. Magnetization, susceptibility and specific heat can be expressed in the form

$$\frac{M}{x} = f\left(\frac{H}{x}, \frac{T}{x}\right) \quad , \chi = f\left(\frac{H}{x}, \frac{T}{x}\right) \quad \text{and} \quad \frac{C_p}{x} = f\left(\frac{H}{x}, \frac{T}{x}\right) \quad \text{with } T_f \propto x .$$
(2.13)

Experiments in the dilute limit [just above the concentration, x_o, where the Kondo effect plays a major role, viz. $T_K \simeq T_f(x_o)$] verify these 'scaling laws' and thereby prove the efficiency of the RKKY interaction in creating the spin-glass state. We shall return to the roots of this scaling at the beginning of Chapter 5.

2.5 MICTOMAGNETISM (CLUSTER GLASS)

As the concentration of magnetic impurities is increased, there is a greater statistical chance of the impurity being first or second nearest neighbour to another impurity. Since the wave functions for the 3d electrons (transition metal) have a finite extent, they can also carry a type of RKKY-polarization. A similar situation occurs for the more localized 4f electrons (rare earths), but now through an intermediate polarization of 5d electrons. Consequently, there is a short-range RKKY-like interaction which can strongly couple neighbouring impurities ferro- or antiferromagnetically depending upon the particular magnetic element involved and the neighbour position. Accordingly, magnetic clusters can form as a result of concentration fluctuations in a random alloy. In addition, at these larger concentrations short-range chemical or atomic order, and even chemical clustering of longer-range (due to solubility problems of the given alloy) may occur. Such deviations from randomness can greatly influence the local magnetism.

It has been experimentally found for a number of typical spin-glass alloys (e.g. *Cu*Mn and *Au*Fe) that both of the above mechanisms (statistical fluctuations and chemical ordering) produce a predominance of ferromagnetic couplings and very large effective moments of order 20–20 000 μ_B. When the magnetic behaviour is dominated by the presence of such large magnetic clusters, the terms 'mictomagnet' or 'cluster glass' have been used in the literature. Figure 2.8 shows some of these clusters within the dotted lines embedded in the spin-glass matrix. At sufficiently low temperature, these clusters will freeze with random orientation in a manner analogous to the

2.5 Mictomagnetism (cluster glass)

Fig. 2.8 Spin glass with mictomagnetic cluster for a 2D square lattice.

spin-glass freezing. The presence of such large ferromagnetic entities simplifies detection of the freezing process, for, the magnetization and susceptibility are very large. Also enhanced are the remanences and hysteretic behaviour of the frozen state. However, the mictomagnetic phase is especially sensitive to its metallurgical state. Heat treatments, with various times, temperatures and cooling rates, along with plastic deformation can greatly affect the cluster formation and magnetic behaviour. Of special interest here are the hard magnetic properties such as the peculiar displaced hysteresis loops and the remanences and irreversibilities in the magnetization.

The origin of mictomagnetism goes back to a series of early investigations in the 1960s where an 'ensemble of domains' model was used. In the 1970s the name mictomagnetism was coined for a continuation of such experiments where careful consideration was given to the metallurgical state and the effects of random freezing below a certain temperature. Other workers, a few years later, used the name cluster glass to describe similar very large moment effects in more concentrated alloys. Since mictomagnetic and cluster glass refer to alloys of higher concentration where long-range magnetic order is incipient, they represent a complicated mélange which is difficult to study or to model. Therefore, this topic will not receive great attention in the following sections.

2.6 PERCOLATION AND LONG-RANGE MAGNETIC ORDER

For the sake of completeness, let us proceed and further increase the number of magnetic atoms. At some given concentration for the particular crystal structure, each magnetic site will have at least one magnetic nearest neighbour. When such a macroscopic connection or uninterrupted chain extends from one end of the crystal to the other, the **percolation limit** (that concentration for which there is a non-zero probability that a given occupied site belongs to an unbounded cluster) has been reached. The schematic 2D lattice shown in Fig. 2.8 is far below the *site* percolation concentration which, for a square lattice with only nearest neighbour coupling, requires 58% of the total sites to be occupied. Similar critical concentrations for percolation may be obtained for other 2D and 3D crystal structures with various near-neighbour interactions.

Another type of percolation is **bond percolation** where an interaction or bond which couples two particular sites is present or absent. Then the closed bonds must continue in a macroscopic path through the crystal. A simple physical picture for bond percolation is conduction through a network of on/off switches. How many switches between the sites of our square lattices (Fig. 2.8) need be closed before a current flowing in one end of the network will flow out the other? 50% is the answer.

Percolation applied to magnetic systems is more complicated. Let us begin with a 3D fcc lattice and simple nearest-neighbour interactions, and randomly fill the lattice sites with magnetic atoms; if they are nearest neighbours, they will align parallel; if not, they are uncoupled. When 17% of the sites are occupied an infinite ferromagnetic cluster will form. This represents long-range ferromagnetic ordering and a proper phase transition at a Curie temperature T_C. Certainly there will be modifications in the critical phenomenon due to the percolation behaviour and its fractal nature, and also because of the many 'free' spins which are not (or only weakly) coupled to the infinite cluster. With nearest-neighbour antiferromagnetic bonding in the same 3D fcc crystal, a much larger concentration is needed. Here the element of frustration or unsatisfied orientation enters, because the fcc triangular symmetry is unfavourable for long-range antiferromagnetic order and many more spins are needed to build this coherent two sub-lattice structure in fcc space. Now the percolation limit is $\approx 45\%$ site occupancy above which we have a long-range antiferromagnet with a distinct T_N.

We can take the following step in our percolation approach and allow mixed interactions, remembering this is the essential ingredient for our spin glasses. For example, return to Fig. 1.1 where the first neighbour was ferromagnetic and the second, antiferromagnetic. There does exist an 'infinite' coupled cluster, but now without long-range 'periodic' order. The spins are frozen randomly depending on the initial condition. Here we might say that the spin glass correlation length $\xi_{SG} \to \infty$, but recall the

2.7 Concentration regimes

example of Fig. 1.1 is at $T = 0$ K as usual for percolation problems. Nevertheless, we can still attempt to use the concept of percolation, even cutting off the envelope of the long-range RKKY interaction with the thermal disorder, as was done in section 2.3. Now we do not care how the spins are coupled, only if they are. The main difficulty with the percolation approach is that it explicitly neglects frustration – an essential element of the frozen state and something that is built into the spin-glass transition. So we conclude that percolation theory can only give an oversimplified and incomplete picture of a spin glass. However, it becomes quantitative and more useful in describing the various types of long-range magnetic order appearing at larger concentrations above the spin-glass phase.

2.7 CONCENTRATION REGIMES

After all this discussion we should try to summarize the various concentration regimes of different behaviour for a metallic magnetic alloy. It must now be clear that the concentration is most important in determining the magnetic state of the alloy. Figure 2.9 offers a schematic representation for such a concentration regime division. At the very dilute magnetic concentration (ppm) there are the isolated impurity-conduction electron couplings which result in the Kondo effect. This localized interaction (if $J < 0$) causes a weakening or fluctuation of the magnetic moment, and below the Kondo temperature, the moment disappears and the impurity appears non-magnetic. Thus, the Kondo effect prevents strong impurity–impurity interactions which are a basic necessity of the spin glasses. Nevertheless, the local-environment model tells us that there will be pairs (or triplets, etc.) which possess a much lower T_K than the singlets. These 'good' moment pairs may then interact with other *pairs* and give rise to spin-glass behaviour. Therefore, in principle, there is no lower concentration limit to spin-glass consequences, until we run out of measurement response or temperature. Usually, however, we take $T_K \simeq T_f(x_o)$ as the concentration, x_o, below which the Kondo effect plays a large role in modifying the spin-glass behaviour.

Up to a concentration of few thousand ppm (≈ 0.5 at. %), the scaling approach (see section 2.4), based on the $(1/r)^3$ decrease of the RKKY exchange is particularly appropriate for describing the problem of interacting impurities. This means that the measurable properties (magnetization, susceptibility, specific heat, and remanences) are universal functions of the concentration scaled parameters T/x and H_{ext}/x, and in addition $T_f \propto x$. While an effective way to scale the various data, such a treatment tells us nothing about the freezing process and the possibility of a phase transition. At low temperature $T \ll T_f$ and with the incorporation of anisotropic dipolar interactions, also $\propto (1/r)^3$ and thus scalable with x, the magnetic behaviour was initially explained within the existing model of Néel superparamagnetism. Superparamagnetism (see below section 2.10) is the temperature and field

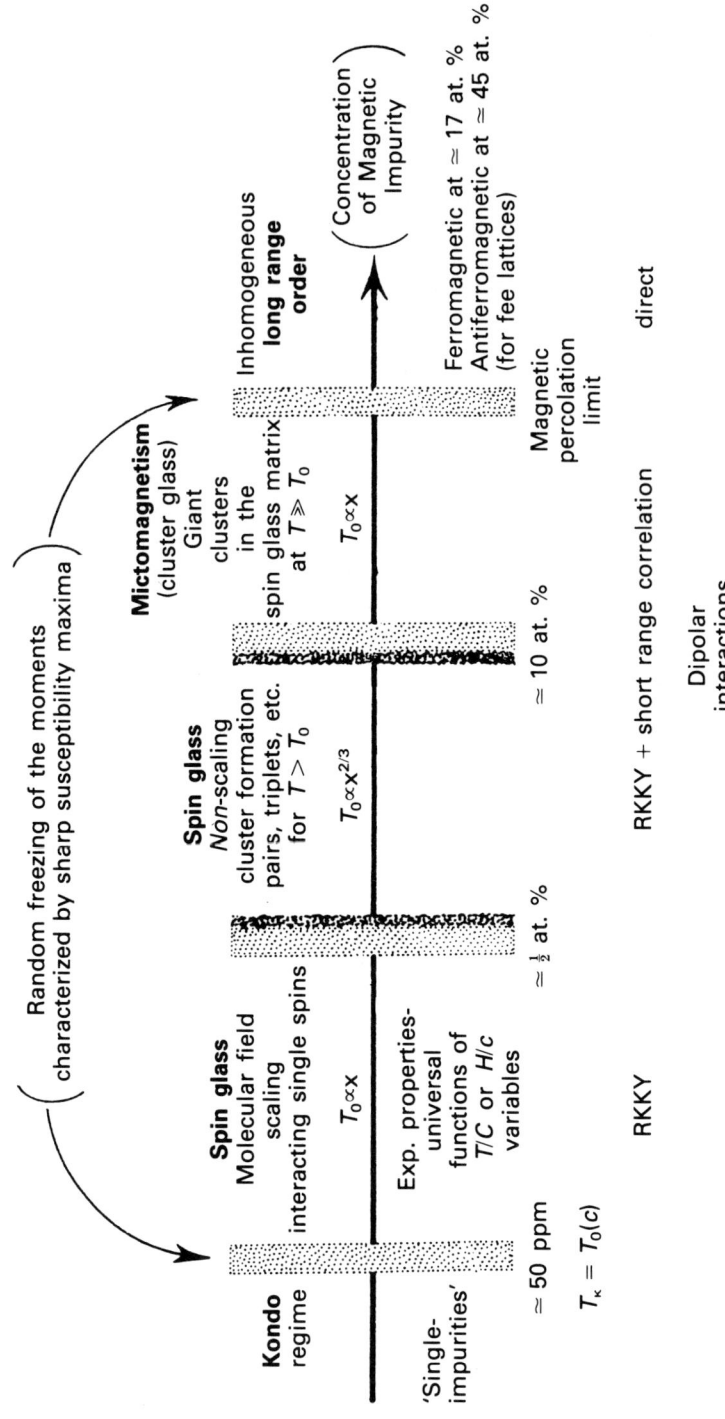

Fig. 2.9 Various concentration regimes for a canonical spin glass illustrating the different types of magnetic behaviour which occur.

activation of non-interacting 'fine' magnetic particles or large clusters with respect to their anisotropy energy barriers. This served as a first attempt to treat the frozen state and it will be considered in more detail at a later stage.

Although the scaling regime and its companion model represent a reasonable first-order approximation at low x, the random freezing persists over an extended concentration range which exhibits the more dramatic experimental effects. After the scaling laws break down around $\frac{1}{2}$ at. %, a non-scaling spin-glass regime occurs where $T_f \propto x^{\frac{2}{3}}$. Here clusters (pairs, triplets, etc.) begin to form and exert their influence. When the magnetic behaviour is dominated by such clusters, at around 10–15 at. %, the nomenclature mictomagnetism or cluster glass is used to emphasize the metallurgy and anomalies generated by these very large magnetic clusters.

Finally, the percolation limit is reached for long-range, but inhomogeneous, ferro- or antiferromagnetic order with a well-defined T_C or T_N. The particular characteristics of the infinite-cluster, long-range order, while of interest in itself, will not concern us any further. In the subsequent chapters we employ the generic term **spin glass** from the dilute limit almost up to the percolation limit for long-range order. Hereby we can avoid, for the sake of simplicity, the mostly unnecessary distinction of having three or more spin-glass regimes. Once again Fig. 2.9 illustrates the various concentration regions. Note that the different regions are not sharply separated; there is a smooth and gradual transformation from one to the other (except perhaps for a well-behaved alloy at its percolation threshold).

A temperature – second dimension – ordinate should be added to Fig. 2.9 in order to give it the character of a T-x phase diagram. In Fig. 2.10 such a general scheme is shown. T_K represents the 'average' Kondo temperature which decreases as a function of the concentration. For $T_K > T_f$ we have the weak moment concentration regime. Then as the local environment builds good moments ($x > x_o$), the spin-glass regime appears with first a linear, followed by a less than linear T_f dependence on x. Finally, when the percolation limit is surpassed $x > x_p$ there is a rapid rise of Curie or Néel temperatures as the long-ranged ordered state is formed and strengthened. So by properly adjusting the values of x_o and x_p and scaling the three temperatures of interest T_K, T_f and T_C (or T_N), a number of real alloy systems may be described with the generic phase diagram of Fig. 2.10.

2.8 METALLIC GLASSES: AMORPHOUS MAGNETS

This name has been given to a new class of non-crystalline solids which are conducting and contain magnetic elements. Most of the metallic glasses are naturally based on rather large concentrations of transition elements (TM), e.g. $Fe_{80}P_{20}$ and $Co_{80}B_{20}$, so long-range ferromagnetism is a common occurrence. It is only when the TM-concentration is reduced by non-

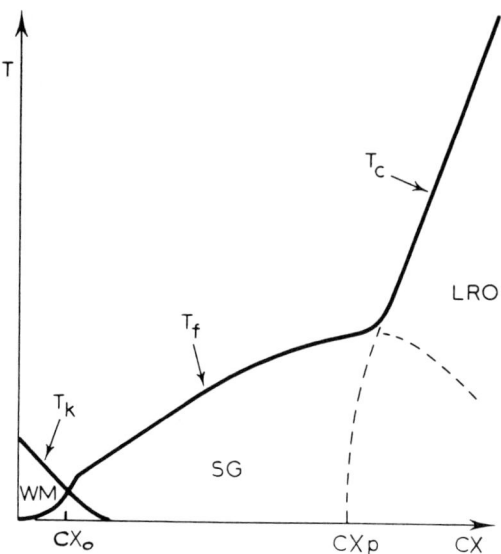

Fig. 2.10 General (schematic) T-x phase diagram for a dilute magnetic alloy.

magnetic substitutions, Pd or Ni, which now have lost their magnetic tendencies due to their hybridization with the metalloid (P or B) atoms, that a spin-glass phase appears. A similar situation may be created using rare-earth atoms as the magnetic impurity of the amorphous (a) compound. Some of the many examples here are a-$La_{80-x}Gd_xAu_{20}$ and a-$Dy_{1-x}Cu_x$.

The spin-glass ingredients in these amorphous magnetic systems are essentially the same as for our dilute magnetic alloys of the previous sections. Since they are also metals, the RKKY interaction is operative in its usual competing-exchange form. Very similar behaviour and regimes exist as a function of x – the magnetic concentration. The 'double disorder' of random site occupancy within the non-crystalline lattice certainly increases the overall randomness and eliminates the problem of chemical clustering. Look-alike T-x phase diagrams can be mapped out for the numerous amorphous magnetic alloys which further have the great advantage of being able to be prepared over a wide concentration range, in many different combinations, without change in (a) structure. We shall not discuss the various elaborate techniques employed to 'quench a liquid' and thereby form an 'amorphous' spin-glass.

Two novel features manifest themselves in the amorphous magnets. Firstly, due to the crystallographic disorder, the electron mean free path l is significantly reduced. This introduced a dampening factor, approximately $e^{-r/l}$, into the RKKY-interaction which effectively cuts off the $(1/r)^3$ polarization at a smaller length scale. Secondly, again due to the random environment of the non-magnetic species, the various rare-earth impurities (except Gd) will experience the effects of random anisotropy. Namely, the

2.9 Re-entrant spin glasses

direction of the uniaxial anisotropy becomes a random variable. This means we must write the anisotropy energy in our hamiltonian (see Section 1.4) as

$$\mathcal{H}_{RA} = -\sum_i D_i (\mathbf{n}_i \cdot \mathbf{S}_i)^2 \qquad (2.14)$$

where $\mathbf{n}_i$ represents different, randomly oriented, easy directions at the various sites i. The net result here is to enhance the spin-glass behaviour to the detriment of the ferromagnetic ordering. Thus, the amorphous magnets offer many 'modern combinations' (see Chapter 4) of spin glasses without really adding basically new effects or fundamentally differing insights.

2.9 RE-ENTRANT SPIN GLASSES

At the percolation boundary of phase diagram, recalling Fig. 2.10 and its dashed lines, a strange behaviour takes place, if the long-range order is ferromagnetic. The spin-glass phase does not simply stop but persists into the 'ferromagnetic' phase. This effect is shown for a real material $Au_{1-x}Fe_x$ in Fig. 2.11. The strangeness is related to the observation that the more disordered state appears from the more ordered one as the temperature is reduced. According to experiment it would seem that the $T = 0$ phase is a cluster glass created out of the 'percolated' *infinite* ferromagnetic cluster which has broken up into large but randomly frozen clusters. Hence, the apparent long-range order is destroyed although a gamut of large ferromagnetic clusters is still present – conforming to our previous definition of a cluster glass or mictomagnet. The word re-entrant, something of a misnomer, means here that the cluster-glass phase develops from a ferromagnetic state, and thus it re-enters the frozen (disordered) phase out of another, not paramagnetic, state.

There are many different systems especially among the amorphous metallic magnets which show this re-entrant behaviour. So the phenomenon is general, yet a good theoretical model is hard to find. Various suggestions have been proposed, one of which is the appearance of a temperature-dependent random anisotropy (dipolar?). At low temperatures this anisotropy grows sufficiently large that it severs many of the weak links holding together the rather tenuous $x \gtrsim x_p$ infinite cluster. Then the anisotropy points these divided units along the randomly preferred directions. Another model involves a freezing of transverse spin components. At the ferromagnetic transitions T_c the longitudinal components order, but because of the local mean field being inhomogeneous, a random freezing perpendicular to the applied field occurs at a lower temperature T_f.

Recent experimental work indicates that the 'infinite-cluster' ferromagnetic state is not truly long-ranged. Instead, it consists of many extremely large

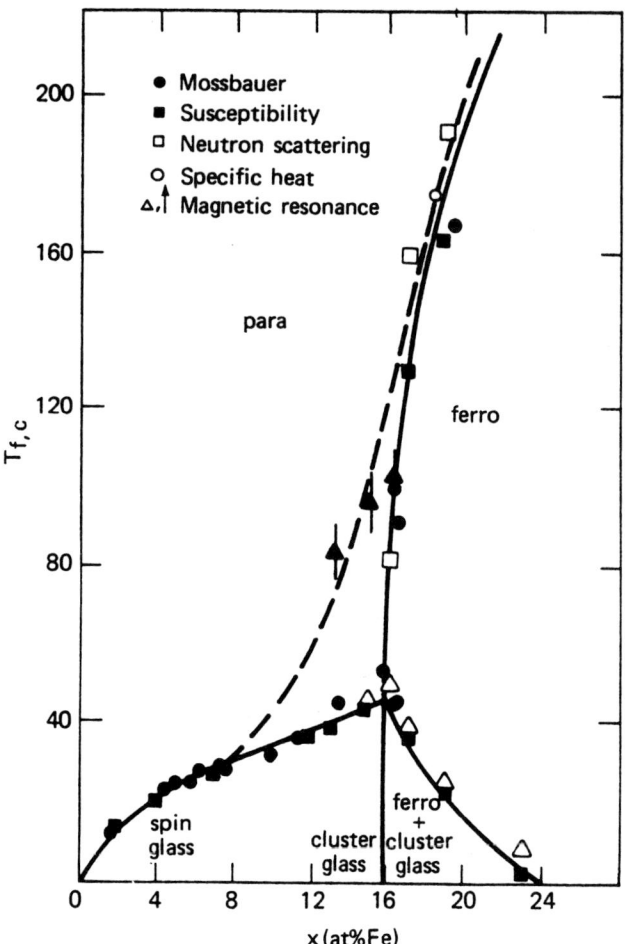

Fig. 2.11 Magnetic phase diagram of AuFe constructed from the anomalies observed in various experiments according to symbol; from Coles et al. (1978).

clusters which are hard to tell from infinite. These large, but disordered from each other, clusters then dissociate into the collection of smaller clusters at the re-entry transition. Such processes occur without the usual type of percolative phase transition.

2.10 SUPERPARAMAGNETISM

Suppose we ponder a family of magnetic clusters, also called domains or fine particles. They may be composed of any type of *internal* magnetic order, namely, ferromagnetism, antiferromagnetism, random, etc., and their

2.10 Superparamagnetism

ordering already has occurred at a temperature far above our considerations. The clusters, in contradistinction to the previous cluster glass, do not interact with each other. So, we ask, what happens to this collection of independent particles, constituting magnetic moments proportional to either n or Δn or $\sqrt{n}$ (n is the number of local moments within a cluster) as a function of applied field and temperature.

First of all at reasonably high temperature, the moments of the various particles fluctuate due to the thermal disorder in a paramagnetic way. One may, therefore, use the Brillouin function to describe the magnetization

$$M\left[\frac{g\mu_B SH}{k_B T}\right] = M_o(n)\left[(S + \tfrac{1}{2})\coth\left((S + \tfrac{1}{2})\frac{g\mu_B H}{k_B T}\right) - \tfrac{1}{2}\coth\left(\frac{g\mu_B H}{2k_B T}\right)\right] \quad (2.15)$$

Since each particle contains many spins, we require the limit of the Brillouin function as $S \to \infty$. Can you remember how this function depends upon S? For S becoming large the initial slope decreases and there is substantial rounding as the argument increases making saturation more difficult. Confirmingly, experimental data nicely fit the Brillouin function with $S = \infty$, at least if T is not reduced too far. Since the measured magnetization for our assemblage of particles is zero when no external field is applied, and follows the Brillouin function for infinite spin when a field is swept, the name superparamagnetism aptly fits.

Now we lower the temperature to look for additional effects. Here we must include the consequences of anisotropy in an energy-barrier model as depicted in Fig. 2.12. There are many possible sources of the anisotropy: magnetocrystalline, shape, dipolar, surface, etc. The anisotropy constant times the particle volume forms the energy barrier height, $E_a = KV$ which can be overcome with either field $\mathbf{M} \cdot \mathbf{H}$ or temperature $k_B T$ energy. Hence, two minima appear with oppositely directed spin, i.e., two easy orientations of magnetization. The relaxation time of the magnetization between these two states is given by thermal activation (Arrehenius law)

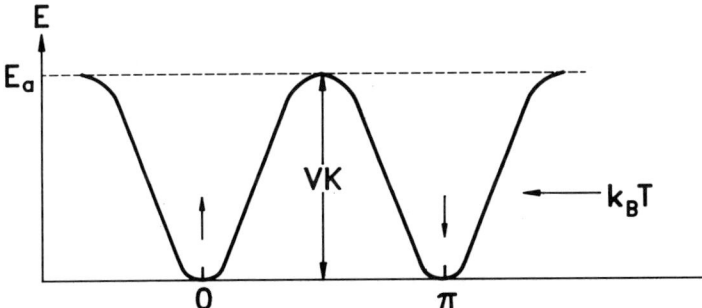

Fig. 2.12 Anisotropy barrier for an up/down superparamagnet. Note the thermal activation energy ($k_B T$) is too small to cause rapid barrier hopping.

$\tau = \tau_o \exp(E_a/k_B T)$, where E_a is the energy barrier separating the states and τ_o is a microscopic limiting relaxation time usually $\approx 10^{-9}$s. Below a blocking temperature T_B of order $E_a/25k_B$, the fluctuations between the two states are becoming long enough to be observable on a laboratory time scale. For a yet lower temperature the system will appear 'blocked' in *one* of the states. With the energy of this state lowered by the external field a stable remanence would emerge when averaged over the many particles. Commensurate with the blocking process is a drop of the magnetization; this manifests itself as a broad temperature-dependent maximum in $M(T)$, since the various particles are blocked out at different temperatures.

For spin glasses there are strong interactions between the moments. So the concept of independent superparamagnetic particles is incorrect. Certainly the freezing process of a spin glass is co-operated and phase-transition-like, in stark contrast to the gradual blocking of the superparamagnetic particles. Nevertheless, as a naïve and crude approximation we can still use some of the above ideas for the frozen state. Historically this was the first comparison. There are always some correlated spin clusters which still have a random distribution of barrier heights and anisotropy directions. The frozen spin glass locks into this array and its ground state exhibits the magnetic remanence while relaxation is associated with the thermal activation. At low temperatures $T \ll T_f$ the relaxation is slow, proportional to logarithm of the time, and a metastable random state arises. As we shall soon see it sometimes becomes arduous to distinguish experimentally between an insulating spin glass and a superparamagnet.

2.11 MAGNETIC INSULATORS AND SEMICONDUCTORS

While we can discriminate reasonably well among the metallic materials as to what is a spin glass and what is not, there are more difficulties with doing the same for non-conducting materials. The main problems with the latter systems are the short range and weakness of the interactions which are present, and the predominance of only antiferromagnetic exchange coupling in many of the disordered insulators. Of course, for the very few materials, which have disorder, mixed (ferro and antiferromagnetic) interactions and frustration, we have no qualms about grouping them alongside the canonical RKKY spin glasses (see listing of spin glasses in Chapter 4 and the *ideal* spin glasses in Chapter 6).

In a manner similar to the dilute magnetic alloys or the amorphous magnets we can create site disorder or a distribution of impurity distances in the form of insulating glasses. So there is no problem with randomness. Yet for short-range exchange $J(r)$ is proportional to either $\exp(-r)$ or $(r)^{-p}$ where the power p is 8 to 10, thus a strong cut-off in the range of coupling occurs after the first or second nearest neighbours. Here above the percolation limit x_p for the given crystal structure, the long-range antiferromagnetism

2.11 Magnetic insulators and semiconductors

manifests itself, albeit with reduced T_N and modified critical behaviour compared to the pure (nondiluted) system. Many of these cases are without any exchange-frustration effects. The study of such alterations in the phase-transition properties is called the random-exchange problem. One particularly suitable case for experiment and comparison with theoretical predictions is when the system has a simple Ising critical phenomenon; accordingly the name random-exchange Ising model is used. Or, if a similar system is placed in a large magnetic field, it becomes an experimental demonstration of the so-called random-field Ising model (see Chapter 7).

What takes place below x_p (the magnetically diluted limit)? Well, most measurements indicate that the system behaves analogous to the superparamagnets. The infinite *antiferromagnetic* cluster is broken up in many various size domains of antiferromagnetic order which are independent, i.e., do not interact with each other, and are therefore not correlated. Consequently, even though the many clusters are composed of antiparallel spins, there are still small moments to be associated with their irregularities and surfaces, and gradually these are blocked out according to the model introduced in the previous section. Note, as already mentioned, there are a number of similarities with the frozen state of a spin glass and the blocked superparamagnet. However, some major dissimilarities are that the time scales become more smeared out and the blocking temperature T_B cannot be compared to the sharp and well-defined spin-glass freezing temperature T_f. To see these differences clearly, requires a lot of fine experimentation – it's a tough job here to sort out the true spin glasses.

What happens if we do have a lattice favouring frustration and sufficient antiferromagnetic exchange to form an infinite cluster? Will then the system be a spin glass? These questions are genuine, for, real systems do exist especially with the semimagnetic semiconductors such as $Hg_{1-x}Mn_xTe$ and $Cd_{1-x}Mn_xTe$. Even when doped their conductivity is not large so the RKKY mechanism is inoperative and the antiferromagnetic-only exchange is mediated by the valence electrons – a type of superexchange (see section 1.3.3). Once again these systems do have many spin glass properties particularly in the frozen state. Yet they do not possess the sharp freezing behaviour of the canonical spin glasses. It would seem that they are spin-glass-like at low temperatures, but somehow enter this state via another and more gradual type of freezing process. To know what we are talking about we must distinguish between superparamagnets, canonical (RKKY) or good spin glasses, spin glass-like and completely frustrated non-random materials. These distinctions will only become clear when we consider in detail the experimental properties of the mixed interacting spin glasses. And such is our task for the next chapter.

FURTHER READING AND REFERENCES

Related concepts

Kondo, J. (1969) in *Solid state physics*, Vol. 23 (eds F. Seitz and D. Turnbull) Academic Press, New York, 183.
Heeger, A. J. (1969) *Ibid*, 293.
Rado, G. T. and Suhl, H. (eds) (1973) *Magnetism V*, Academic Press, New York.
Nieuwenhuys, G. J. (1975) *Adv. Phys.*, **24**, 515.
Mydosh, J. A. and Nieuwenhuys, G. J. (1980) in *Ferromagnetic materials*, Vol. 1 (ed. E. P. Wohlfarth) North Holland, Amsterdam, 71.
Moorjani, K. and Coey, J. M. D. (1984) *Magnetic glasses*, Elsevier, Amsterdam.
Stauffer, D. and Aharony, A. (1992) *Introduction to percolation theory*, Taylor and Francis, London.

Spin glasses

Mezard, M., Parisi, G. and Virasoro, M. A. (1987) *Spin glass theory and beyond*, World Scientific, Singapore.
Fischer, K. H. and Hertz, J. A. (1991) *Spin glasses*, Cambridge.
Chowdhury, D. (1986) *Spin glasses and other frustrated systems*, World Scientific, Singapore.
Binder, K. and Young, A. P. (1986) *Rev. Mod. Phys.*, **58**, 801.
Maletta, H. and Zinn, W. (1986) in *Handbook on the physics and chemistry of rare earths* (eds K. A. Gscheidner, Jr and L. Eyring) North Holland, Amsterdam.
Rammal, R. and Souletie, J. (1982) in *Magnetism of metals and alloys* (ed. M. Cyrot) North Holland, Amsterdam.
Mydosh, J. A. (1981) in *Magnetism in solids: some current topics* (eds A. P. Cracknell and R. A. Vaughan) SUSSP, Edinburgh.
van Hemmen, J. L. and Morgenstern, I. (eds) (1983) *Heidelberg colloquium on spin glasses*, Vol. 192 Lecture Notes in Physics, Springer, Berlin.
van Hemmen, J.L. and Morgenstern, I. (eds) (1987) *Glassy Dynamics* Vol. 275, Lecture Notes in Physics, Springer, Berlin.

References

Coles, B. R., Sarkissian, B. V. and Taylor, R. H. (1978) *Phil. Mag. B*, **37**, 489.
Walker, L. R. and Walstedt, R. E. (1977) *Phys. Rev. Lett.*, **38**, 514.
Walker, L. R. and Walstedt, R. E. (1980) *Phys. Rev. B.*, **22**, 3816.

# 3
Basic spin-glass phenomenon

We have been considering for the past two chapters the spin glasses and their entourage of cognate phenomena. After having spent so much time and energy introducing the encompassing physics and basic principles of the spin glasses in a rudimentary and qualitative way, we now should turn to experiment in order to find out exactly how it supports and confirms these notions, i.e., our physical picture, of a spin glass. Accordingly, measurements on real systems and their proper interpretation are the windows through which we gain our knowledge. Let experiment be our guide – after all it did come first.

We shall attempt in this chapter to survey salient experimental features of a spin glass beginning with the high-temperature measurements and then reducing the temperature to $T \approx T_f$. Here we examine the novel critical behaviour and the various manifestations of a phase transition. Finally, the important experiments characterizing the low-temperature frozen state ($T \ll T_f$) are viewed with respect to its unique properties. Our three temperature regime approach is arbitrary, yet effective, since three very different types of behaviour are encountered. Our choice of a few from the many thousands of experiments published on spin glasses is geared to bring our physical picture into focus and keep it generic – not a collection of exceptions. Thus, we want to embed our notion of a spin glass firmly in reality before proceeding with models and theories which, of course, should give the ultimate explanation of the phenomenon and lead to a final understanding.

3.1 HIGH-TEMPERATURE ($T \gg T_f$) EXPERIMENTS

Perhaps the dullest part of the spin glasses is their behaviour far above the freezing temperature. For, we would expect simple paramagnetic effects in this limit, or a collection of freely rotating, independent magnetic moments. Using a paramagnetic model we can compare this basis to the generic spin-glass measurements of susceptibility, specific heat, resistivity and neutron scattering. Systematic data have been tediously accumulated over the past years for these experimental quantities. We know what to anticipate so let's see what will be found as T_f is slowly approached from far above.

3.1.1 Susceptibility

For a paramagnetic, the susceptibility χ is just $\chi = M/H$ since we usually measure it in the low-field limit of the magnetization's Brillouin function. There exist many techniques with sufficient sensitivity to detect either χ directly or M in a given small H. Fortunately, for the majority of the canonical spin glasses, the host susceptibility is small, e.g. the noble metals have a tiny temperature-independent diamagnetic contribution which is known and can be subtracted.

Now we carry out a systematic study on many different concentrations of a spin glass alloy, say CuMn. Figure 3.1 shows the results plotted as

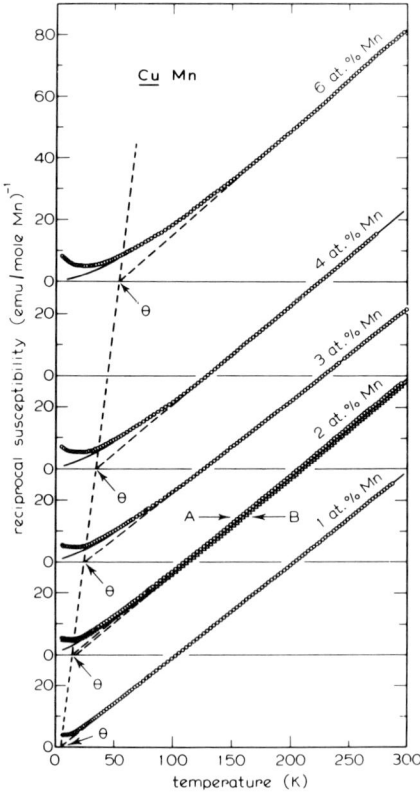

Fig. 3.1 The reciprocal susceptibility as a function of temperature: the dashed lines are a linear extrapolation to determine Θ and the solid line represents the fit to the model calculations; from Morgownik and Mydosh (1981).

3.1 High-temperature ($T \gg T_f$) experiments

inverse susceptibility versus temperature, recall equation (2.1). Such a plot permits a Curie–Weiss type of analysis

$$\chi = \frac{N\mu_B^2 g^2 S(S+1)}{3k_B(T-\Theta)} \quad (3.1)$$

Notice that the temperature dependence is asymptotically good as T increases. There is the expected straight line behaviour whose extrapolation gives the Curie paramagnetic temperature Θ. The freezing temperature is not seen clearly in these measurements (it is smeared out), since they were performed in a rather large field (0.6 T) in order to increase the sensitivity at high temperatures.

Further analysis is done by calculating the effective moment; recall equation (2.2)

$$p(T) \equiv \frac{N\mu_B^2}{3k_B}\left[\frac{d(\chi^{-1})}{dT}\right]^{-\frac{1}{2}} \quad (3.2)$$

In the case of a Curie–Weiss behaviour $p(T)$ is simply the usual effective Bohr magneton number as given in equation (2.3)

$$p(T\to\infty) \equiv p_o = g[S(S+1)]^{\frac{1}{2}} \quad (3.3)$$

Figure 3.2 presents the results of this analysis for the same CuMn concentrations. Note how all samples have very similar p_o values ≈ 5 in units of μ_B; see Fig. 3.3. Also, returning to Fig. 3.2, pay attention to the upturns in p beginning at a 'deviation temperature' T_d. At $T < T_d (\gg T_f)$ the effective moment begins to increase. The meaning is clear; ferromagnetic clusters are beginning to develop out of the paramagnetic background. The greater the concentration of Mn, the higher T_d and the larger $p(T)$ becomes at low temperatures. This growth of local ferromagnetism is further reflected in $\Theta(x)$ and its rapid increase with concentration. In standard theory the Curie paramagnetic Θ represents a sum of all the exchange interactions which are present

$$\Theta(x) = x\frac{S(S+1)}{3k_B}\sum_r J(\mathbf{r}) \quad (3.4)$$

for a completely random alloy. Thus, the ferromagnetic exchanges are becoming more probable as x increases. And Fig. 3.3 shows this behaviour with its roughly linear x dependence of Θ (except at low concentrations).

One can repeat the same series of measurements on many different systems with various concentrations. Interestingly, most of the canonical (RKKY) spin glasses show a lot of local ferromagnetism developing far above T_f. With so much data available one could even attempt a simple ad-hoc cluster model and calculate with the aid of a computer the effects of first and second nearest-neighbour pairs and triplets on the susceptibility $\chi(T, x)$. Some of these clusters are illustrated in the right-hand panel of

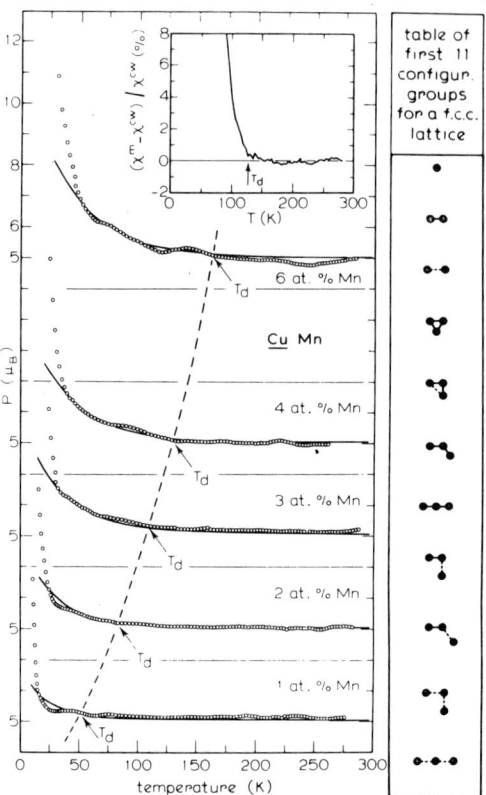

Fig. 3.2 P as defined in the text versus temperature. The deviation from C–W behaviours at T_d is illustrated in the inset; also shown are the first 11 configuration groups for an fcc lattice. From Morgownik and Mydosh (1981).

Fig. 3.2. The hope is to obtain a better than Curie–Weiss fit, at least a bit further down in temperature. But such a model cannot be accurate as T_f is approached due to the co-operative interaction effects – pairs and triplets will no longer suffice. A few years ago such a model was turned loose on a variety of dilute alloys ($\frac{1}{2} < x < 5$ at. %) and even included the deviations from randomness due to chemical clustering.

The model was a computer calculation of all possible energy levels for the various cluster configurations according to the hamiltonian

$$\mathcal{H} = -\sum_{i<j} s_i J_{ij} s_j - g\mu_B H \sum_i s_i \tag{3.5}$$

where $s_i = S, S-1, \ldots, -S$. Since the experimentally determined S-value ($p_o = 5.01\mu_B$) for CuMn is very close to 2, we set $S = 2$ with $g = 2$. The applied magnetic field H is equal to 0.6 T as in the experiment. The canonical ensemble is then calculated taking into account all states:

3.1 High-temperature ($T \gg T_f$) experiments

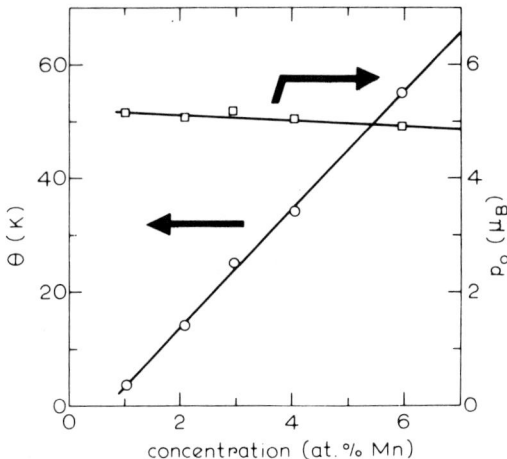

Fig. 3.3 Effective Bohr-magneton number p_o and the paramagnetic Curie temperature Θ as a function of concentration in CuMn: p_o and Θ are determined from least-squares fitting of the susceptibility for temperatures above T_d; from Morgownik and Mydosh (1981).

$5(=2S + 1)$ for the single impurity, 5×5 for each doublet group, $5 \times 5 \times 5$ for the various triplet groups, etc. The total ensemble Z is obtained as a product of the group ensembles weighted with occupational probabilities for the various cluster configurations at the given concentration and degree of short-range chemical ordering. Only for a limited x-range ($\approx \frac{1}{2}$ to 5 at. %) does the model work. Once the partition function Z is known, it is a straightforward task to calculate the thermodynamic properties from the usual equations.

$$\chi = \frac{M}{H} = \left(\frac{k_B T}{H}\right) Z^{-1} \frac{\partial Z}{\partial H} \tag{3.6}$$

$$C_m = -T\left(\frac{\partial^2 F}{\partial T^2}\right) \quad \text{and} \quad F = -k_B T \ln Z \tag{3.7}$$

The fitting parameters for this configurational cluster model were the various exchange interactions as a function of neighbour shell (or distance). The solid lines in Figs. 3.1 and 3.2 are the model calculations. They do indeed fit better than the Curie–Weiss law. However, they cannot be used too close to T_f. The net outcome was an estimate of the exchange parameters J_n, extended, with certain modifications, to fifth nearest neighbour. Such is illustrated in Fig. 3.4 along with the expected RKKY conduction-electron polarization for a Mn impurity in Cu. There is more ferromagnetism present in the model calculations than in the RKKY interaction. It would seem that for the close by impurities (first or second neighbours), d-like conduction electrons are mediating the indirect exchange interaction, thus causing the

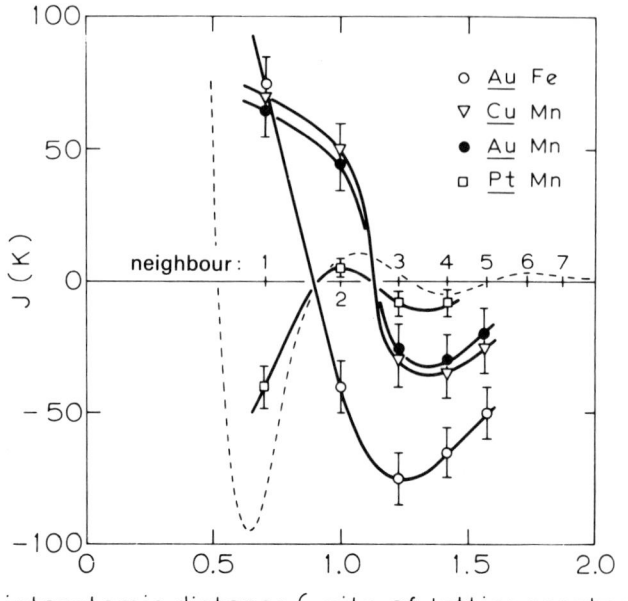

Fig. 3.4 Exchange parameters J_n as function of neighbour distance: the dashed line represents the RKKY-conduction electron polarization; from Morgownik and Mydosh (1983).

extra ferromagnetic couplings and discrepancy with standard free-electron (s-like only) RKKY theory, equation (11). Nevertheless, the oscillating ferro/antiferromagnetic interactions are clearly and happily at hand. So while experiment nicely verifies this competition of mixed couplings (except, recall section 2.1.1, for the *insulating* Mn and Co aluminosilicate glasses where J_1 and J_2 are both antiferromagnetic), it warns about the presence of ferromagnetic clusters – the building blocks out of which the spin-glass state is established. A similar conclusion was reached for the more concentrated CuMn alloys on the basis of a Brillouin function analysis. The small number of large ferromagnetic clusters determined from the fitting gave birth to the term mictomagnetism.

3.1.2 Specific heat

Another most-important thermodynamic property of the spin glasses is their specific heat, also at high temperatures. Unfortunately, the main problem with the specific heat is to separate out non-magnetic contributions, e.g. phonons and electrons (in a metal). Above a few kelvin these effects overwhelm the magnetic (Kondo) or spin-glass terms. Since the various contributions all superimpose, elaborate procedures have been developed

3.1 High-temperature ($T \gg T_f$) experiments

to subtract the unwanted components and obtain the magnetic specific heat. Both the measurement technique itself and the various subtraction procedures are laborious and imprecise, thereby making the high-temperature spin-glass specific heat a tough quantity to determine accurately. So we must work with low T_fs or dilute systems, and see how far we can increase the temperature before the error bars become too large.

Very careful measurements and extraction of the magnetic component do exist for a number of canonical spin glasses. Figure 3.5 shows a now classic set of data for CuMn 0.3 at. % where $T_f = 3.0$ K. We shall return to this figure a number of times in the coming sections. Suffice it for now to note the broad maximum in C_m above T_f ($T_{max} \approx 1.4\ T_f$) and the gradual fall off at increasing temperatures $T \to 4$ or $4.5\ T_f$. Too bad we cannot go yet higher in temperature. The entire decrease with increasing T becomes less pronounced, quite smeared, and the maximum disappears as an external magnetic field H is applied. Qualitatively the data exhibited in Fig. 3.5 are generic – always a broad maximum above T_f with a slow decrease at the highest temperatures and field smearing.

In order to compare these long-range RKKY spin-glass results with those of a short-range insulating spin glass, we display the measurements on

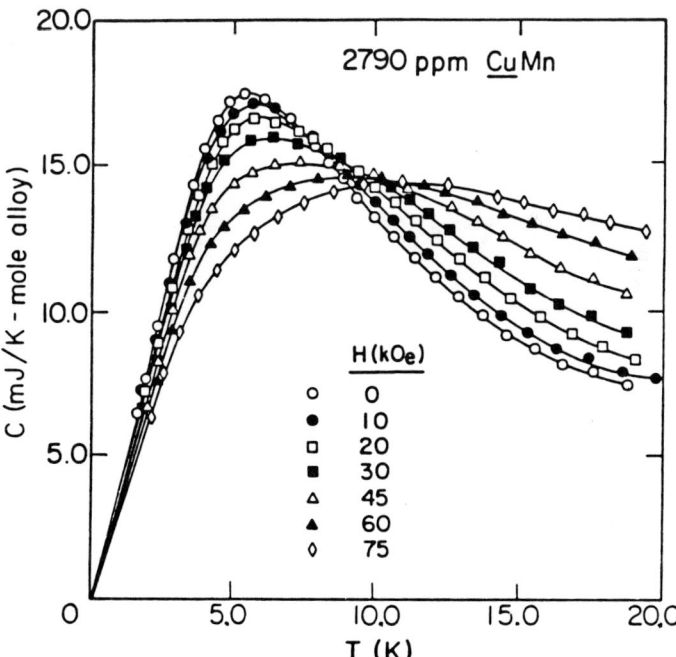

Fig. 3.5 Magnetic contribution of the specific heat of CuMn spin glasses with 0.3 at. % Mn plotted versus temperature in various magnetic fields ($T_f = 3.0$ K); from Brodale *et al.* (1983).

$Eu_{0.4}Sr_{0.6}S$ (T_f = 1.7 K) in Fig. 3.6. Be careful now for the data are plotted C_m/T versus T in the main part and log C_m versus log T in the inset. Once again the same general type of behaviour is found (see inset): a rounded peak at $\approx$ 1.8 T_f followed by a drop at higher T. Now this decrease seems stronger than the RKKY spin glass. The main portion of Fig. 3.6 exhibits the field smearing as before and also indicates the magnetic entropy S_m from the areas under the various field-dependent curves. Remember that

$$S_m = \int_0^T \frac{C_m}{T} dT = cR \ln (2S + 1) \quad (3.8)$$

Even a rough glance at the C/T versus T plots would tell that a great deal of the magnetic entropy is lost or frozen-out far above T_f. And the applied-field shifts even more entropy from the low-temperature portion to the high temperatures. We recall that there is a constant amount of entropy available in the sample via equation (3.8), due to the fixed number of magnetic degrees of freedom. The larger the field the more degrees of freedom lost at a given fixed $T(> T_f)$. Crude estimates suggest that even in zero field 70–80% of the degrees of freedom are already forfeited above T_f. This doesn't leave too much entropy, only 20–30%, for a big effect at T_f.

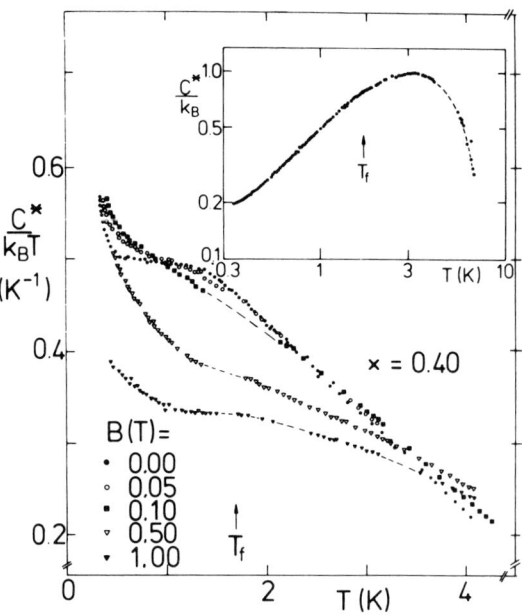

Fig. 3.6 Specific heat per Eu atom divided by $k_B T$ versus temperature T in various applied fields (units in tesla) for $Eu_xSr_{1-x}S$ with x = 0.40. Inset shows the zero-field behaviour in a log–log plot. The freezing temperature T_f is indicated by an arrow. The upturn of the curves for C/T, seen at very low temperatures, is interpreted as being caused by a Schottky term due to magnetically decoupled small clusters of Eu spins. From Meschede et al. (1980).

3.1 High-temperature ($T \gg T_f$) experiments

But returning to our higher-temperature specific heat, what can we say or how can we analyse its decrease? Perhaps the best, yet rough, fit to the CuMn data is a $1/T$ dependence. This was suggested, early on, by the scaling analysis and by a virial expansion of thermodynamic quantities for interacting dilute spins. However, the (EuSr)S spin glass seems to possess a more rapid high-T fall off – the $C_m(T)$ data unhappily do not lend themselves to a more quantitative resolution. Nevertheless, such a result is very reasonable. Since (EuSr)S has only two nearest-neighbour interactions and the second is already quite weak compared to the first, temperature should easily overcome these couplings. In other words, thermal disorder will quickly wipe out the short-range magnetic clustering. Thus, the full number of degrees of freedom are quickly reached and the magnetic specific heat returns to zero. On the other hand, the canonical CuMn spin glass has the stronger and longer-range RKKY interaction so one would expect the short-range order or clustering to persist to much higher temperatures, hence, the long tail.

In keeping with the above thinking we could at least try the configurational cluster model employed to treat the high-T susceptibility of the previous section. It should be easy, for, the specific heat is a thermodynamic quantity derivable from the same free energy which is always calculated from the partition function Z of the various configurations (see section 3.1.1). We simply take temperature derivatives instead of field derivatives of the free energy and determine C_m via equation (3.7). By using the same values of $J_1, \ldots, J_5$ found from the susceptibility we can fully calculate $C_m(T)$ for $H = 0$. The results are shown in Fig. 3.7 for CuMn (2.4 at. %). Unfortunately, here T_f is too large and the experimental data are too limited and scattered in the high-T-regime of interest. The 0.3 at. % CuMn of Fig. 3.5 is simply not concentrated enough for our model to work. At best, we can say the trend looks right, however, nothing quantitative can be gained.

Despite all these *experimental* (not model) difficulties with the high-temperature specific heat, we are drawn to the same conclusion as from the susceptibility. Namely, magnetic-cluster formation or short-range order is appearing far above T_f. And it is out of these clusters that the spin glass state is formed, i.e., our 'building blocks'. There is indeed a quantitative difference in high-T behaviour between the long-range and short-range interacting spin glasses. This can be understood by relating the interaction energy as a function of distance $J(\mathbf{r})$ to the thermal disorder energy $k_B T$ and noting that the RKKY materials require a much higher temperature to overcome their interaction energies.

3.1.3 Resistivity

We now move onto a transport property, viz., the electrical resistivity, which is easy to measure yet hard to extract the exact spin-glass contribution

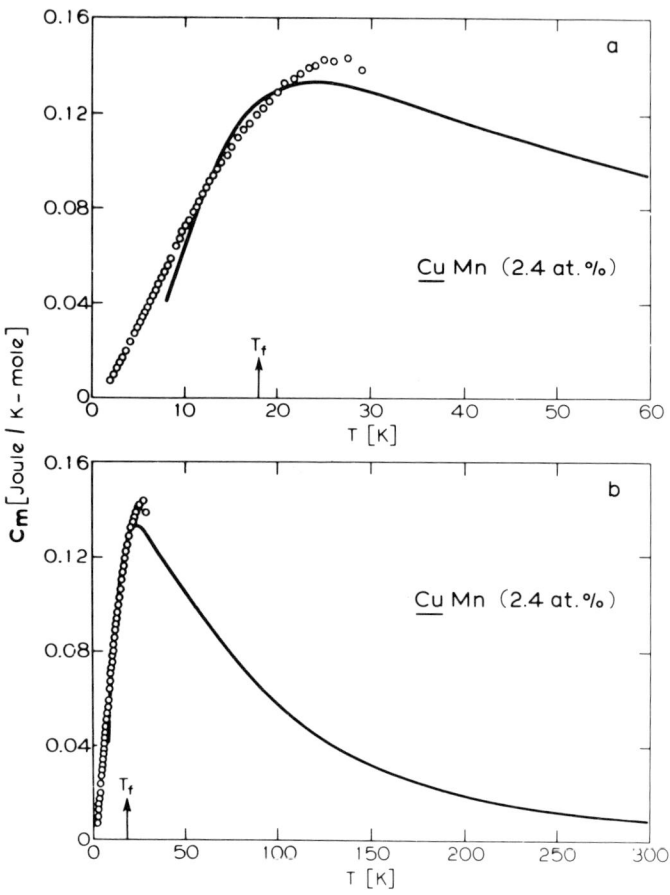

Fig. 3.7 Open circles: measured magnetic specific heat for CuMn (2.4 at. %). Experimental data from Wenger and Keesom (1976). Solid line: fit to the configuration cluster model with the J_ns determined from the susceptibility fit: (a) low temperature regime; and (b) model prediction up to room temperature.

and even more arduous to calculate at high temperatures. Of course, at very high temperature we have the constant spin-disorder scattering of the fully paramagnetic state. Accordingly, we choose for examination a highly conducting canonical spin glass.

Figure 3.8 shows the 'magnetic part', $\Delta\rho$, versus the temperature for a series of CuMn alloys. This so-called magnetic part is determined from a point-by-point subtraction of the pure-Cu resistivity. There are certain, but hopefully small, discrepancies with this simple superposition and subtraction of various contributions. For example, mixing or cross terms could occur and modify the temperature dependences especially when these are rather weak. However, note the large changes in $\Delta\rho$ over the entire temperature

3.1 High-temperature ($T \gg T_f$) experiments

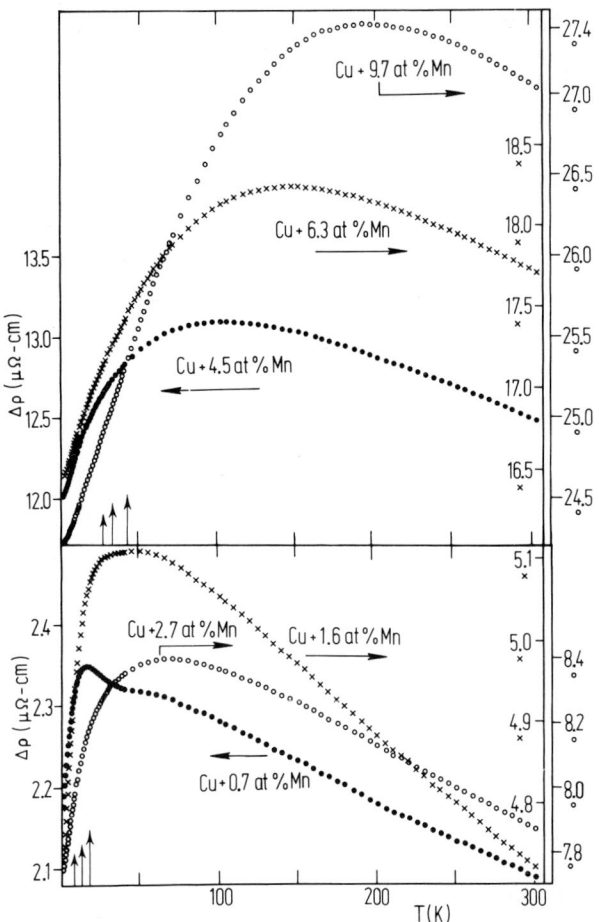

Fig. 3.8 Temperature dependence of the magnetic resistivity for spin glass CuMn: the arrows represent T_f determined from the low-field a.c. susceptibility; from Ford and Mydosh (1976).

range. There is always a temperature dependence, caused by the local-spin/conduction-electron scattering.

What we seem to be detecting, at very high temperatures, is the long tail (Fig. 3.8) of the Kondo effect which has a negative temperature coefficient ($d\rho/dT < 0$) and $\Delta\rho \propto \ln(k_B T/D)$, equation (2.8), extending over many decades in temperature. This means that there are still isolated magnetic impurities present and these can form the quasi-bound state. Yet we draw attention to the maximum which develops and the rapid drop as the temperature is lower towards T_f. Here the change in sign of the slope (− to +) represents a competition between the single-impurity Kondo effect which causes a resonant scattering and the RKKY interaction which wants to couple the spins, and thereby reduce the spin-disorder scattering. All this occurs far above T_f and as a precursor to it.

In order to gain some insight into the physics of the resistivity maximum we must develop a theory of this competition, i.e., the Kondo effect in an interacting spin system. Recall our formula (2.6), for the single impurity Kondo temperature or energy

$$k_B T_K = D \exp\left(\frac{-1}{N(E_F)|J|}\right) \tag{3.9}$$

and let us introduce a mean RKKY-interaction energy

$$\Delta_{RKKY} \simeq \frac{xJ^2 S(S+1)}{E_F} \tag{3.10}$$

with the same exchange J parameter. A lengthy calculation was performed to relate T_m (the temperature of the maximum in $\Delta\rho$) to T_K and Δ_{RKKY}. The result for $\Delta_{RKKY} \gg T_K$ is

$$T_m \simeq \tfrac{1}{2}\Delta_{RKKY} \ln\left(\frac{\Delta_{RKKY}}{T_K}\right) \gg T_f \tag{3.11}$$

Thus, T_m is roughly proportional to the concentration x as seen in experiment. Note that the above theory is more sophisticated than our intuitive idea (section 2.7) that increasing the concentration will simply reduce T_K. However, to obtain a quantitative comparison with theory, a concentration variable, because of the above, makes everything too complicated. A better experimental variable is *pressure* which in a known manner can change J, then $J(P)$ can be inserted into the previous equation and T_m is directly calculable and measurable at various pressures.

What have we learned from the many resistivity studies of canonical spin glasses? First of all, there are the effects of magnetic correlation and clustering far above T_f which reduce the spin-disorder scattering. But we already knew this, for, the conclusion is perfectly consistent with the susceptibility and specific-heat experiments. However, to describe such cluster growth and short-range ordering is too complex for scattering theory. Transport properties except in well-defined limits or simplified models are usually very hard to calculate. Secondly, we have the Kondo effect present. This is still offering a theoretical challenge, and while the complete temperature dependence of $\Delta\rho$ could not be obtained, at least the temperature of the maximum in $\Delta\rho$ and its pressure dependence was extractable and understandable in terms of this competition Kondo versus RKKY. Again, with the resistivity at high T, we must rely on qualitative pictures rather than even simple theories.

3.1 High-temperature ($T \gg T_f$) experiments

3.1.4 Neutron scattering

We proceed now to a more dynamic and microscopic technique which should open for our viewing the fine-scale ($\approx$ Å) **spatial** and short-time ($\approx 10^{-12}$ s) **temporal** correlations in a spin glass. Neutron scattering is completely different from the previous measurements of thermodynamic or transport properties because it can directly probe these spin-spin correlations on an atomic basis, and thus it should be easier to construct and verify interpreting models.

A bit of theoretical background is needed before we discuss the spin-glass experiments. The cross-section for neutron scattering due to a momentum transfer between incoming and outgoing neutron $\mathbf{q} = \mathbf{k}' - \mathbf{k}_o$ and an energy transfer $\hbar\omega = (\hbar^2/2m_N)(k'^2 - k_o^2)$ is

$$\frac{d^2\sigma}{d\Omega d\omega} = N\left(\frac{\gamma e^2}{m_N c^2}\right)^2 \frac{k'}{k_o} S(\mathbf{q}, \omega) \tag{3.12}$$

where N is the number of spins causing the scattering, $(\gamma e^2/m_N c^2)$ is the neutron coupling constant, and $S(\mathbf{q}, \omega)$ is the neutron-scattering function, which we can write, for an isotropic spin system, as

$$S(\mathbf{q}, \omega) = \frac{F^2(q)}{N\pi} \int_{-\infty}^{+\infty} \left[\sum_{i,j} \langle S_i(0) S_j(t) \rangle \exp(i\mathbf{q} \cdot \mathbf{r}_{ij})\right] \exp[-i\omega t] \, dt \tag{3.13}$$

The pair correlation function $\langle S_i(0)S_j(t) \rangle$ is for the local-moment spins separated by a distance $\mathbf{r}_{ij}$, and $F(q)$ is the atomic-spin form factor which is usually known. It is helpful to split up this function by adding and subtracting $\langle S_i \rangle \langle S_j \rangle$. Accordingly, we can distinguish a dynamic and a static component.

$$S(\mathbf{q}, \omega) = S_D(\mathbf{q}, \omega) + S_S(\mathbf{q})\delta(\omega) \tag{3.14}$$

where

$$S_S(\mathbf{q}) = \frac{2F^2(q)}{N} \sum_{i,j} \langle S_i \rangle \langle S_j \rangle \exp[i\mathbf{q} \cdot \mathbf{r}_{ij}] \tag{3.15}$$

This static part gives rise to an **elastic scattering** ($\omega = 0$) which may be further divided into **coherent** (Bragg) at particular q-values or **diffuse** (smeared over q-space) scattering contributions. If there is long-range magnetic order, we find sharp Bragg peaks at the wave vector for the ordering $\mathbf{q}$. For a ferromagnetic $\mathbf{q} = 0$ so we simply obtain Bragg peaks commensurate with the given crystal structure. This is how the percolation concentration for AuFe was established by detecting the appearance of these Bragg peaks. Note they are never seen in a spin glass. If there is short-range magnetic or chemical ordering, diffuse scattering will emerge at different q-value or regions in the Brillouin zone which represents q-space for the given solid. (For CuMn various scattering contours related to both chemical clusters and magnetic short-range order were observed throughout the Brillouin zone even at high temperatures. We discuss this situation further below.) If there is a randomly frozen spin glass, the Edwards–Anderson order parameter should be reflected in $S_S(q) = 2F^2(q)\overline{\langle S_i \rangle^2}$. Unfortunately, a direct determination of the spin-glass order parameter will be masked by several other q dependences especially those related to the short-range order.

We now return to the dynamic or inelastic part which is

$$S_D(\mathbf{q}, \omega) = \frac{F^2(q)}{N\pi} \int_{-\infty}^{+\infty} \sum_{i,j} [\langle S_i(0)S_j(t)\rangle - \langle S_i\rangle\langle S_j\rangle]$$

$$\exp[i(\mathbf{q}\cdot\mathbf{r}_{ij} - \omega t)]\, dt \qquad (3.16)$$

The correlation function within the square brackets can be written in terms of a Bose–Einstein thermal factor and a generalized susceptibility $\chi(\mathbf{q}, \omega)$

$$S_D(\mathbf{q}, \omega) = \frac{2F^2(q)}{\pi g^2 \mu_B^2} \left[\frac{1}{1 - \exp(-\hbar\omega/k_B T)}\right] \mathrm{Im}\,\chi(\mathbf{q}, \omega) \qquad (3.17)$$

This in turn may be simplified for an isolated magnetic atom in a sea of conduction electrons becoming

$$S_D(\mathbf{q}, \omega) = \frac{2F^2(q)}{g^2 \mu_B^2} \left[\frac{1}{1 - \exp(-\hbar\omega/k_B T)}\right] \omega \chi(q) f(\omega) \qquad (3.18)$$

where $\chi(q)$ is the static q-dependent susceptibility and $f(\omega)$ is a q-independent lorentzian,

$$f(\omega) = \frac{1}{\pi}\left(\frac{\Gamma}{\Gamma^2 + \omega^2}\right) \quad \text{with } \Gamma = \frac{\hbar}{\tau} \qquad (3.19)$$

and τ is the Korringa relaxation time.

We can test these formulas through a particularly simple experiment. For a series of CuMn (1 to 7 at. %) the total scattering cross-section has been measured for a constant q-value at 300 K. The data are exhibited in Fig. 3.9 where the central elastic peak related to $S_S(\mathbf{q})\delta(\omega)$ has been omitted. At low concentrations a lorentzian fit gives good agreement using the single relaxation time, derived from Γ (the half-width parameter), of order $\tau \approx 10^{-13}$ s, a reasonable Korringa value. However, as the concentration of Mn is increased there are noticeable deviations from the single-τ lorentzian fit even at this high temperature. A distribution of relaxation times $N(\tau)$ is needed (see below) to recover the accord. So now we have necessitated an important new parameter, the spin relaxation time τ and its distribution function $N(\tau)$, which is directly obtainable from neutron-scattering measurements. Such are critical for the characterization of our

3.1 High-temperature ($T \gg T_f$) experiments

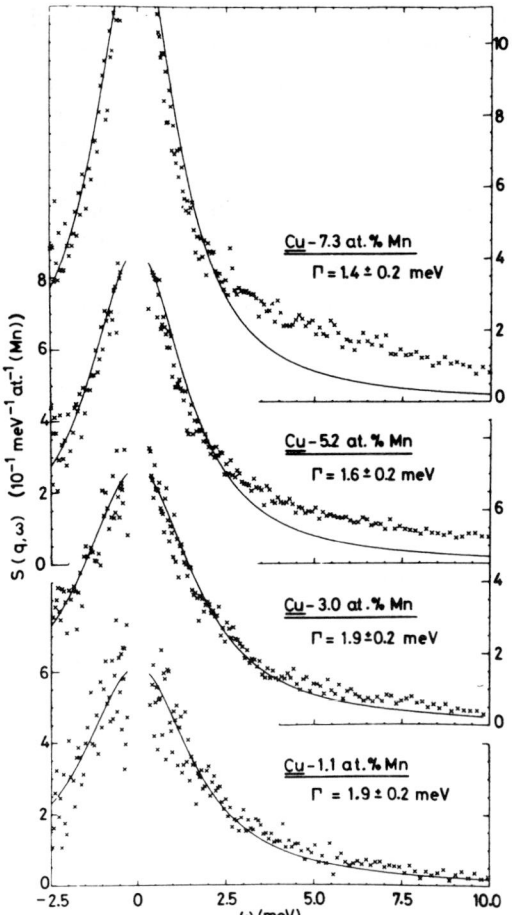

Fig. 3.9 $S(q, \omega)$ versus ω for $q = 0.08$ Å^{-1} at 300 K for several CuMn alloys. The central region of the elastic peak is omitted in the diagram. The continuous curves represent best fits to a lorentzian spectral function in the energy range -2.5 meV $< \omega < 2.5$ meV. From Murani (1978a).

spin-glass freezing. Let us incorporate into our dynamical structure factor this distribution of relaxation times $g(q, \Gamma)$. The $S_D(\mathbf{q}, \omega)$ formula (3.18), therefore, becomes

$$S_D(\mathbf{q}, \omega) = \frac{2F^2(q)}{g^2 \mu_B^2} \left[\frac{\hbar \omega}{1 - \exp(-\hbar\omega/k_B T)} \right] \frac{1}{\pi} \int_0^\infty g(q, \Gamma) \frac{\Gamma}{\Gamma^2 + \omega^2} \, d\Gamma \tag{3.20}$$

There is an upper theoretical limit (time of passage of a neutron through the sample) to the longest relaxation time which can be measured with neutrons. Without energy analysis and for a typical sample size, a relaxation process longer than $\approx 10^{-6}$ s will be seen as a static event. The situation is even worse with energy analysis, for, now the specific instrumental energy-resolution, ΔE, determines the static onset. Using the uncertainty relation, any relaxation process with characteristic time $\tau \gtrsim h/\Delta E$ will be indistinguishable from a completely static or elastic one. For today's highest resolution spectrometers (smallest ΔE), τ of 10^{-7}–10^{-8} s is achievable. Consequently, the dynamic or inelastic neutron-scattering technique is a 'fast observer' of the spin glasses. In order to take into account the above distinction, we separate our dynamic structure factor into a truly dynamic part and an 'instrument-caused' static or elastic part. Accordingly,

$$S_D^T(\mathbf{q}, \omega) = \frac{2F^2(q)}{g^2\mu_B^2}\left[\frac{\hbar\omega}{1-\exp(-\hbar\omega/k_BT)}\right]\frac{1}{\pi}\int_{h/\tau}^{\infty} g(q, \Gamma)\frac{\Gamma}{\Gamma^2+\omega^2}\,d\Gamma \tag{3.21}$$

and

$$S_S^T(\mathbf{q}, \omega) = \frac{2F^2(q)}{g^2\mu_B^2}\left[\frac{\hbar\omega}{1-\exp(-\hbar\omega/k_BT)}\right]\frac{1}{\pi}\int_0^{h/\tau} g(q, \Gamma)\frac{\Gamma}{\Gamma^2+\omega^2}\,d\Gamma \tag{3.22}$$

Consider first of all the static structure factor $S_S^T(\mathbf{q}, \omega)$ of equation (3.22). For the $\Gamma\ (= h/\tau)$ values which most contribute to the integral, $\hbar\omega \ll k_BT$. Then

$$S_S^T(\mathbf{q}) = \int_{-\infty}^{+\infty} S_S^T(\mathbf{q}, \omega)\,d\omega = \frac{2F^2(q)}{g^2\mu_B^2}\frac{k_BT}{\pi}\int_{-\infty}^{+\infty}\int_0^{h/\tau} g(q, \Gamma)\frac{\Gamma}{\Gamma^2+\omega^2}\,d\Gamma\,d\omega \tag{3.23}$$

$$= \frac{2F^2(q)}{g^2\mu_B^2}k_BT\int_0^{h/\tau} g(q, \Gamma)\,d\Gamma \tag{3.24}$$

This part of the total structure factor we call the 'elastic' scattering, since due to the instrumental resolution ΔE, it appears as a static $\omega = 0$ process. Note that all relaxations, which are slower or longer than $\tau = h/\Delta E \approx 10^{-8}$ s, are lumped into this term. If there are no slow relaxations, then $S_S^T(\mathbf{q}) \to 0$ except if we are near a Bragg peak or a q-region of diffuse (short-range) scattering.

Now consider the dynamic structure factor $S_D^T(\mathbf{q}, \omega)$ in equation (3.21). Here we use the quasi-static approximation such that $\hbar\omega \ll k_BT$ and integrate over all ω. Thus

$$S_D^T(\mathbf{q}) = \int_{-\infty}^{+\infty} S_D^T(\mathbf{q}, \omega)\,d\omega = \frac{2F^2(q)}{g^2\mu_B^2}\frac{k_BT}{\pi}\int_{-\infty}^{+\infty}\int_{h/\tau}^{\infty} g(q, \Gamma)\frac{\Gamma}{\Gamma^2+\omega^2}\,d\Gamma\,d\omega \tag{3.25}$$

3.1 High-temperature ($T \gg T_f$) experiments

$$= \frac{2F^2(q)}{g^2\mu_B^2} k_B T \chi(q,\tau) \tag{3.26}$$

where we have defined a wave-vector, **q**, and relaxation-time, τ, dependent susceptibility as

$$\chi(q,\tau) = \int_{h/\tau}^{\infty} g(q,\Gamma)\,d\Gamma \tag{3.27}$$

The $S_D^\tau(\mathbf{q})$ we obtain after integrating over all energies is called quasi-elastic scattering. It represents all the fast fluctuations and rapid relaxation processes in the paramagnetic state. If this state is destroyed by spin–spin interactions, then the quasi-elastic contribution will decrease.

The key concept compelled by experiment in all this formalism is the distribution of relaxation times $N(\tau)$ being very fast and δ-function like at high temperature, and then evolving as $T \to T_f$ to longer time scales via a broad distribution. Well, what can we measure to get a better handle on $N(\tau)$? One usually picks a series of momentum transfers or q-values at certain arbitrary (see below) positions in the Brillouin zone. Very low q-values mean small-angle scattering and here one could also obtain information on the local spatial correlations and their distributions or ranges. With q constant we then determine the scattering cross-section by sweeping the energy transfer ω. Now the big problem is separating the elastic ($\omega = 0 \pm \Delta E$) central peak scattering from the surrounding quasi-elastic cross-section. By fitting the shoulders (outside the elastic peak) to a lorentzian function, we arrive at a distinction: quasi-elastic cross-section, namely, what is contained in the lorentzian; and elastic cross-section, namely, the remaining sharp central peak.

Figure 3.10 shows such a separation for the scattering cross-sections of a AuFe (10 at. %) sample. At high temperature (150 K), the elastic part is practically zero, yet there is a lot of quasi-elastic (or fast) scattering present. This increases with decreasing q as greater regions of real space, and thereby more spins become involved. Then as the temperature is reduced the elastic scattering grows at the expense of the quasi-elastic scattering because more slow relaxation times are fed into the time window of the instrumental resolution. The smooth elastic increase continues down across T_f, while the fast quasi-elastic scattering slowly decreases as the number of fast relaxation processes is eliminated.

We have been emphasizing in the present discussion spin relaxation times and their distribution which shifts to longer times and broadens. This is a natural consequence of our neutron scattering analysis in energy (ω) or time (t) space. However, we can also relate the physical process to cluster formation or the growth of correlated volumes of spins. The bigger the cluster size, the slower its relaxation time. Hence, as the fast single spins become bound into clusters, they lose their 'fastness'. In any case, scattering intensity in a spin glass is always transferred from quasi-elastic to elastic as

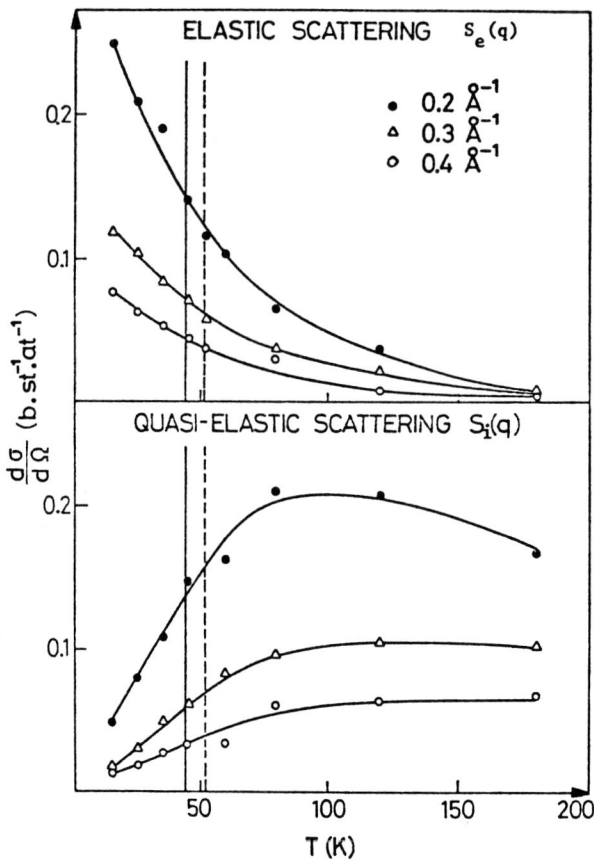

Fig. 3.10 The elastic scattering and the integrated quasi-elastic scattering $S_i(q)$ as a function of temperature for three q-values for the AuFe (10 at. %) alloy. The vertical solid line marks the temperature of the maximum in ac-susceptibility $\chi(0)$ and the dashed line the temperature of the maximum in $\chi(q)$ from the neutron results. From Murani (1978b).

these clusters are formed and slowed down. The neutron scattering sees them directly. And as we traverse T_f in the next section we can even arrive at some more quantitative results.

Similar neutron measurements have been performed for CuMn (5–25 at. %), but now the total (elastic plus quasi-elastic) scattering cross-section was determined at different temperatures as a function of momentum transfer q. Here the whole of q-space, i.e., the Brillouin zone, was scanned for both magnetic and nuclear (non-magnetic) scattering. This refinement was made possible by neutron-polarization analysis for which the magnetic scattering occurred with a spin-flip of the incoming polarized neutron and vice versa for the non-magnetic scattering. By examining the scattering contours one can distinguish regions of chemical (non-magnetic) clustering

3.2 Experiments spanning T_f

from regions of magnetic clustering. Figure 3.11 shows a 1D q-scan along the (10ζ)-direction for the *magnetic* cross-section of a rather concentrated (25 at. %) CuMn alloy. There is a lot of magnetic scattering smeared out over the Brillouin zone even at room temperature. A model calculation confirms this to be ferromagnetic short-range order which continues to exist at temperatures above 400 K. As the temperature is reduced two peaks at two distinct q-values emerge from the diffuse-scattering plateau. Something is certainly happening as T_f is crossed, but we must save this for a future section. At this point in our journey through the high-temperature regime of the spin glasses, we should content ourselves with the formation of these mainly ferromagnetic clusters. Their existence in either time or space has been established consistently from all four experiments considered in this section. Next we must study the behaviour of spin glasses and these clusters at the freezing temperature.

3.2 EXPERIMENTS SPANNING T_f

Here is where the action is! The possibility of a sharp freezing transition to a unique phase must be investigated through fine experimentation on the canonical spin glasses. While many different techniques have been

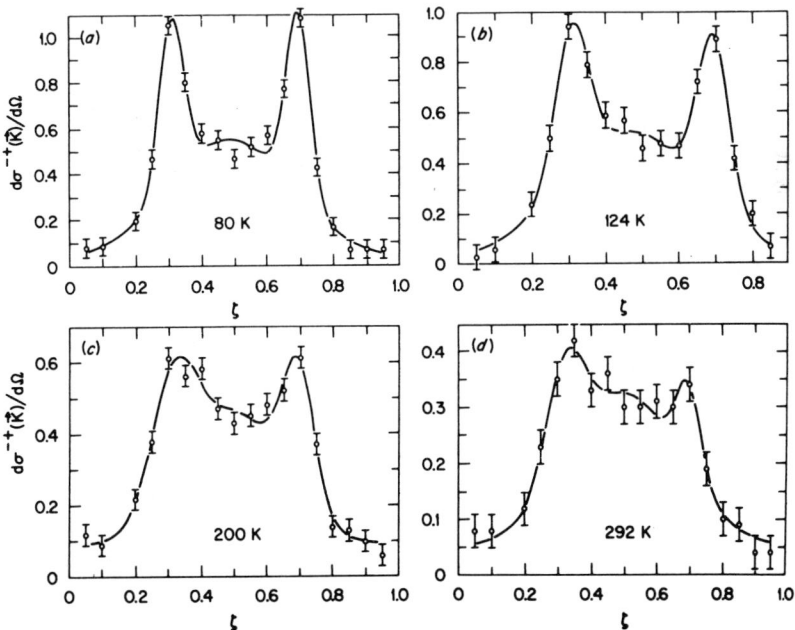

Fig. 3.11 Magnetic cross-section for CuMn (25 at. %) along (10ζ) at elevated temperatures; from Cable et al. (1984).

employed to study the 'phase transition', only a few come to grips with the dynamics (wide range of measurement times) and zero (or small) external-field requirement. For, as we shall soon see, dynamics plays a most essential role and a field of only a few hundred gauss wipe out the sharpness of the transition. We made a choice of four experimental methods in the last section, so let us now carry on with these. Our dc-susceptibility becomes mainly ac-susceptibility and its frequency dependence is the key variable. Specific heat and resistivity are extended to span T_f and certain types of critical analysis are performed. We view different relaxation-time extractions and interpretations of the neutron-scattering technique including the very novel spin-echo method. And to gain some experimental variation on a fine scale near T_f, we introduce, at the risk of increased complexity, two 'local-probes' experiments: muon spin relaxation (μSR) which tests the inter-atomic molecular fields both spatially and temporally, and the Mössbauer effect which is a local nuclear technique for measuring the hyperfine-field reflecting the magnetism of the electronic system.

3.2.1 Ac-susceptibility

There are many methods to determine the magnetic susceptibility χ. Usually a static magnetization M is measured and divided by the applied field H. However, when M is no longer directly proportional to χ, then this method fails. A different technique is to apply a small ac-field, called the driving field h, and measure χ by taking the derivative $\partial M/\partial h$ at some frequency ω and even with a dc-H applied. This ac susceptibility is especially important for spin glasses, since h can be made very small ≈ 0.1 gauss and ω can be varied over a rather large frequency range, thereby permitting a full determination of the real $\chi'(\omega)$ and imaginary $\chi''(\omega)$ parts.

It was in the early 1970s that sharp cusps were discovered in the ac-susceptibility of AuFe and CuMn. Both of these canonical spin glasses were studied for many years previous but only in large magnetic field. Since a few hundred gauss of applied field smears out the cusp to a broad maximum (see below), this is why the spin glasses, as an excitingly new and possibly unique phase transition, were not identified a decade or two sooner.

Figure 3.12 reproduces the first measurements on AuFe which combined the term spin-glass with the sharp cusps. The frequency was the low audio range (50–155 Hz) and the driving field 5 gauss. Notice how the peak increases in magnitude and shifts to higher temperatures as the concentration of Fe is increased. Such distinct temperature effects were totally unexpected from early speculation based upon a random molecular-field model and the previously measured high-field susceptibilities. The height of the peak is rather small, orders of magnitude below the typical 'step height' or demagnetization cut-off of a ferromagnetic transition. It is more or less comparable to that found in many long-range antiferromagnetic material,

3.2 Experiments spanning T_f

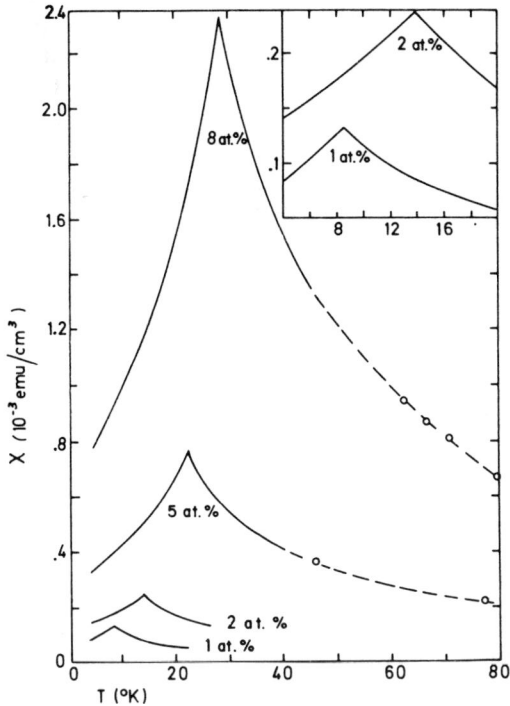

Fig. 3.12 Low-field susceptibility $\chi(T)$ of AuFe for $1 \leq x \leq 8$ at. %. The data were taken every $\frac{1}{4}$ K in the region of the peak, and every $\frac{1}{2}$ or 1 K elsewhere. The scatter is of the order of the thickness of the lines. The open circles indicate isolated points taken at higher temperatures. From Cannella and Mydosh (1972).

even through the concentration of magnetic moments for a spin glass is much less. The tail in Fig. 3.12, persisting to high temperatures, overlaps with the dc-susceptibility in larger fields which was discussed in the previous section. It may be fit with the same modified Curie–Weiss analysis as before. However, the ac-susceptibility rapidly loses its sensitivity at high temperature due to experimental reasons, and so, no longer has the desired accuracy. On the low temperature side

$$\chi(T) = \chi(0) + bT^n \quad \text{with } m \approx 2 \quad (3.28)$$

Thus χ extrapolates to a finite value for $T \to 0$ and a ratio $\chi(0)/\chi(T_f) \approx 0.5$–$0.6$ is roughly obtained for magnetic concentrations between 0.1 and a few atomic percent. These measurements have been extended to a great variety of spin glasses which show the same general characteristics.

Figure 3.13 shows how an applied dc-field rounds off the peak. Already in 200–300 gauss the peak is smeared away and only a broad maximum remains. The reader can imagine what will happen in the few thousand

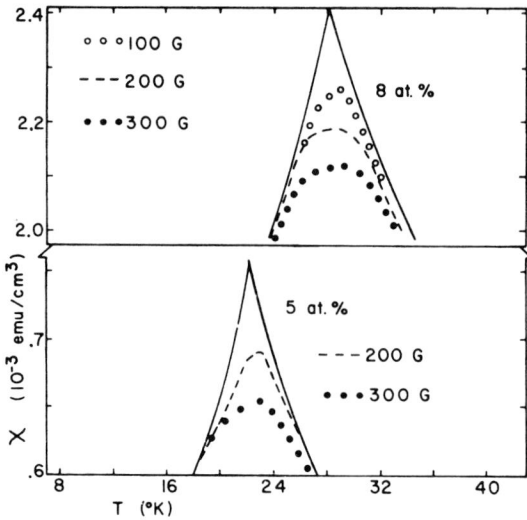

Fig. 3.13 Susceptibility data for AuFe with $x = 5$ and 8 at. %, showing the curves for zero field, and for various applied fields; from Cannella and Mydosh (1972).

gauss fields typically employed in static magnetization measurements. The field effect is surprising, for, we have a T_f of ≈ 25 K being strongly affected by a field of ≈ 1000 gauss, although $k_B T_f \gg \mu_{eff} H$. (Assuming $\mu_{eff} = 10\mu_B$, then 0.1 T corresponds to a temperature of ≈ 0.67 K.) This means the spin-glass state has a peculiar sensitivity even to a small external field.

Certainly from looking at the sharpness of the ac-susceptibility cusp we could exclaim 'phase transition' and immediately delve into the statistical mechanics of critical phenomenon. But, be careful, and let us blow up the χ-T region around the ac-cusp and vary the frequency ω – something which took about ten years to accomplish with sufficient accuracy. Figure 3.14 presents one such set of data on CuMn. Over the wide-temperature range the cusp is clearly seen, however, on the fine scale of the inset there is a slight rounding of the cusp into a peak. Yet more important the peak is shifted downwards in temperature with decreasing frequency of measurement. For the frequency variation of $2\frac{1}{2}$ decades in Fig. 3.14, T_f is reduced by about 1%. Other similar metallic spin-glasses have roughly the same or a slightly larger frequency shift, a few per cent. For the insulating spin glasses the frequency dependence is even larger (see below). A quantitative measure of the frequency shift is obtained from $(\Delta T_f / T_f)$ per decade ω. Table 3.1 shows this quantity for a number of spin glasses and also for a known superparamagnet a-$(Ho_2O_3)(B_2O_3)$. Notice the difference in magnitude and how really little the canonical spin glasses are affected by frequency. This frequency shift in T_f offers a good criterion for distinguishing a canonical spin glass from a spin-glass-like material from a superparamagnet. A similar frequency dependence of T_C (Curie temperature) for a ferromagnet or T_N

3.2 Experiments spanning T_f

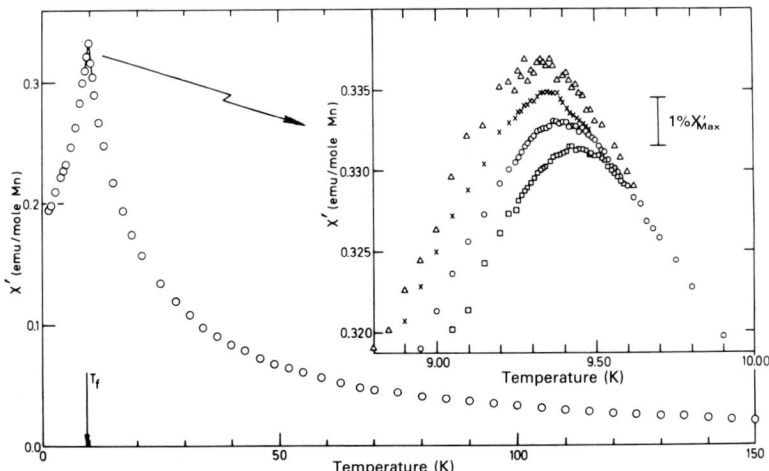

Fig. 3.14 Zero-field susceptibility χ' as a function of temperature for sample IIc (*Cu*Mn (0.94 at. %) powder), measuring frequencies: □, 1.33 kHz; ○, 234 Hz; ×, 10.4 Hz; and △, 2.6 Hz. From Mulder *et al.* (1981).

Table 3.1 Frequency shifts of T_f (defined as the maximum in the ac-susceptibility, χ') for various spin glasses

System	$\Delta T_f/[T_f \Delta(\log \omega)]$
*Cu*Mn	0.005
*Au*Mn	0.0045
*Ag*Mn	0.006
*Pd*Mn	0.013
*Ni*Mn	0.018
*Au*Fe	0.010
(LaGd)Al$_2$	0.06
(EuSr)S	0.06
(FeMg)Cl$_2$	0.08
a-CoO·Al$_2$O$_3$SiO$_2$	0.06
a-(Ho$_2$O$_3$)(B$_2$O$_3$)	0.28

(Néel temperature) for an antiferromagnet has *not* been observed at such low ω-values. Here one needs very high frequencies (mega to giga Hz) to find $T_C(\omega)$ or $T_N(\omega)$ dependencies in these long-range ordered materials. We shall return below to the significance and interpretation of this frequency dependence.

Since we are using an ac technique, the susceptibility, as mentioned above, will have two components a real part $\chi'(\omega)$, the dispersion, and an imaginary part $\chi''(\omega)$, the absorption. They are usually related through a relaxation time by a set of equations called the Casimir–du Pré equations which we shall discuss later. For a spin glass there is a sudden onset of the imaginary component near T_f. We depict this behaviour in Fig. 3.15 for (EuSr)S, an insulating spin glass. Since χ'' is so small it is difficult to measure accurately and in a metal it will be distorted because of eddy currents induced by the alternating field h. This limits χ''-measurements to about a thousand hertz even on powdered conducting samples. As with χ' the frequency shift of χ'' is downwards in temperature with decreasing frequency. Now, however, the peak in $\chi'(T)$ corresponds to the maximum slope in $\chi''(T)$, $(d\chi''/dT)_{max}$. The appearance of an imaginary component means relaxation processes are affecting the measurement and by decoupling the spins from the lattice they cause the absorption. Such effects are usually not found at magnetic transitions except when hysteresis arises as in a ferromagnet. So with an analysis of the complex ac susceptibility we have a handle on the low-frequency or intermediate $(10^{-1}$–10^{-5} s) relaxation

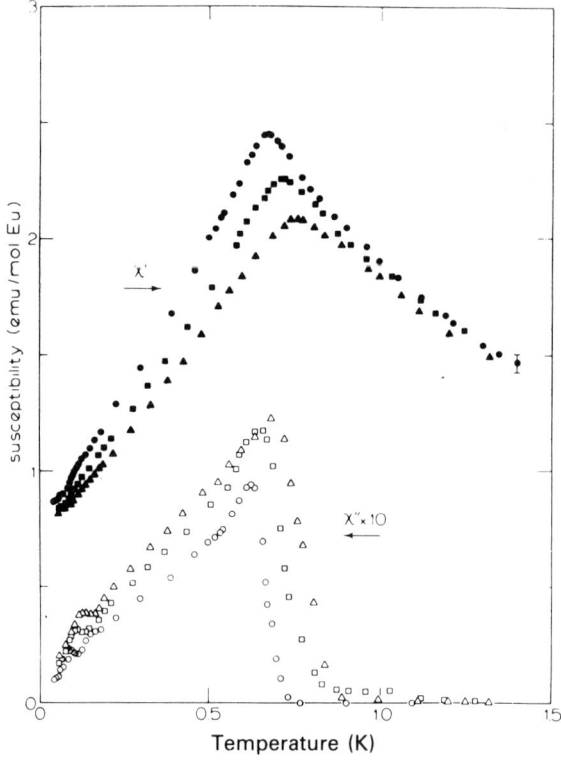

Fig. 3.15 Temperature dependence of the dispersion χ' (solid symbols) and absorption χ'' (open symbols) for $Eu_{0.2}Sr_{0.8}S$: ●○, 10.9 Hz; ■□, 261 Hz; ▲△, 1969 Hz (applied ac field $h \approx 0.1$ Oe); from Hüser et al. (1983).

3.2 Experiments spanning T_f

times. Remember that the neutron scattering tells us about the fast relaxation processes (10^{-8} s and shorter).

Finally, with the newly developed SQUID techniques a highly sensitive method exists for determining the magnetization in a very small applied field. As $H \to 0$ there must be a proportionality $\chi_{dc} = M/H$, and a quasi-static (governed by the time of the measurement at a given T, e.g. ≈ 100 s) delineation of the susceptibility is gained. It is remarkable how many decades of time or frequency may be probed using the various experimental techniques, for, with the SQUID we have times available which can approach 10^5 s. There are two distinct ways to measure the susceptibility with a small dc field. The first is to apply the field far above T_f and cool the sample in a field to $T \ll T_f$ all the while recording the magnetization. This method we call field cooling (FC). Secondly, we can cool the sample in zero field to $T \ll T_f$ and at this low temperature apply the field. Then we heat the sample while measuring M to $T \gg T_f$ with the field constant. Here we use the term zero-field cooling (ZFC). Figure 3.16 illustrates the different temperature dependences (FC versus ZFC) for CuMn (1 and 2 at. %) with a field of 6 gauss. Notice what happens at T_f. The FC-χ becomes constant in value and to a great extent independent of time: if we stop and wait at a given $T(< T_f)$, χ remains unchanged. (Only at extreme sensitivities and very long waiting times is there a drift in χ.) Furthermore, the process FC followed by field warming is reversible. If we cycle the temperature back and forth (with H = const.) the FC-susceptibility traces

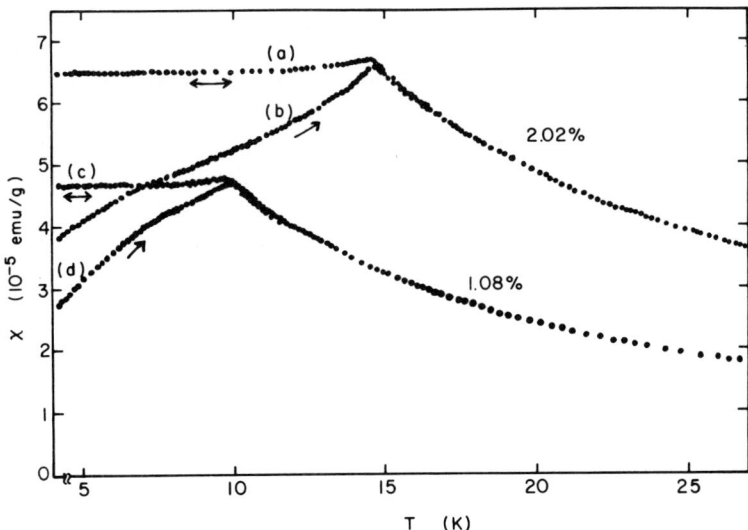

Fig. 3.16 Field cooled [(a), (c)] and zero-field cooled [(b), (d)] magnetizations ($\chi \equiv M/6$ gauss) for CuMn (1 and 2 at. %) as a function of temperature; from Nagata et al. (1979).

the same path (see double-tipped arrows in Fig. 3.16). On the other hand, the ZFC susceptibility is zero until the field is applied, in Fig. 3.16 at ≈ 5 K. Then with H on and T constant, the ZFC-χ jumps to a value comparable with that found from χ_{ac}. However, now there is a slow, clear drift upwards which continues over many decades in times. If we wait long enough (and no one has yet achieved this – the average lifetime of a graduate student is only about 10^8 s), χ_{ZFC} $(t \to \infty) \approx \chi_{FC}$ at the given $T < T_f$. So, when we proceed to increase the temperature at some reasonable rate dT/dt, we arrive at the single-arrow curves in Fig. 3.16. These traces are irreversible, χ_{ZFC} is always drifting upwards. Therefore, the actual curve or T-dependence is a function of dT/dt. Such does not occur with χ_{ac} – it is fully reversible. If we continue to increase the temperature above T_f and then cool the sample backdown, there is first the reversible 'paramagnetic' regime above T_f and then the susceptibility will become the reversible χ_{FC} (double-tipped arrows) with subsequent cooling. In short, the dc field (albeit small), if applied below T_f creates a metastable, irreversible state, and T_f is nicely defined by the onset of these irreversibilities, i.e., difference between FC and ZFC curves. More about the low-temperature effects in the next section.

How can we better analyse and interpret the data near T_f? Let us begin with the frequency shift of T_f. First of all we could try the Arrhenius law for thermal activation already used for a superparamagnet

$$\tau = \tau_o \exp[E_a/k_B T] \qquad (3.29)$$

which we can rewrite as

$$\omega = \omega_o \exp[-E_a/k_B T_f] \qquad (3.30)$$

Here ω is the driving frequency of our χ_{ac}-measurement and T_f its peak. By plotting ln (ω/ω_o) versus $1/T_f$, we can from the slope and value of the logarithm determine E_a and ω_o. For the canonical spin glasses using the data in Fig. 3.14, best-fitting to the Arrhenius law gives completely unphysical values for ω_o and E_a, e.g. $\omega_o \approx 10^{200}$ Hz and $E_a = 4400$ K. This nonsense is clearly due to the very small changes in T_f (maximum in χ_{ac}), and it distinguishes a spin glass from a superparamagnet where the Arrhenius law does indeed hold. There is much more than simple energy-barrier blocking and thermal activation in the transition of a spin glass.

A second method of analysis is ferreted out from the literature on real glasses. It is the empirical law which describes the viscosity of supercooled liquids, namely, the Vogel–Fulcher law. Written for use in describing T_f shifts this becomes

$$\omega = \omega_o \exp[-E_a/k_B(T_f - T_o)] \qquad (3.31)$$

where T_o is a new parameter (for real glasses it is called the 'ideal glass' temperature). With three fitting parameters (ω_o, E_a and T_o) the agreement is naturally much better, except perhaps at low frequencies, using a

3.2 Experiments spanning T_f

more physical set of parameter values. Typically for CuMn (4.6 at. %) $\omega_o = 1.6 \times 10^8$ Hz, $E_a = 11.8$ K and $T_o(< T_f = 27.5$ K$) = 26.9$ K. The problem here is the precise physical meaning of the large T_o. Some attempts have been made to relate it to the interaction strengths between the clusters in a spin glass. Since a spin glass is *not* a non-interacting collection of clusters something must be added to take into account the inter-cluster couplings. T_o might also be related to the true critical temperature of an underlying phase transition for which T_f is only a dynamic manifestation. Thus far, a deeper understanding of T_o and its connection to the longer-range interactions is lacking.

Finally, a third approach draws on the standard theory for dynamical scaling near a phase transition at T_c. The conventional result of dynamical scaling relates the critical relaxation time τ to the correlation length ξ as $\tau \sim \xi^z$. Since ξ diverges with temperature as $\xi \sim [T/(T - T_c)]^\nu$, we have the power-law divergence

$$\tau = \tau_o \left(\frac{T}{T - T_c}\right)^{z\nu} \tag{3.32}$$

where the product exponent $z\nu$ is called the dynamical exponent. In the language of spin glasses we can write

$$\tau_{AV} = \tau_o \left(\frac{T_c - T_f}{T_f}\right)^{-z\nu} \tag{3.33}$$

where T_f is the frequency-dependent freezing temperature determined by the maximum in χ_{ac}. Once again with three parameters (τ_o, T_c and $z\nu$) an experimental extraction is attempted and now reasonable, and perhaps not unphysical, values are obtained. As a typical example CuMn (4.6 at. %) has $\tau_o = 10^{-12}$ s, $z\nu = 5.5$ and $T_c = 27.5$ K which should correspond to the dc (equilibrium) value of T_f ($\omega \to 0$). The trouble here is that $z\nu$ varies between about 4 and 12 for the different spin glasses. For conventional phase transitions, $z\nu$ usually is around 2. This latter (power-law) approach seems to be the acceptable one with which to compare to Monte Carlo simulations of spin glasses – more of this when we get to the theory.

For the sake of completeness we include another relation for τ based on the dynamical-scaling exponent ν. This expression was proposed for the limit of a zero temperature [$T_c = 0$ or $T_f(\omega \to 0) = 0$] phase transition

$$\tau = \tau^* \exp\left[E_a^*/k_B T\right)^{z\nu}] \tag{3.34}$$

Assuming the critical temperature is zero the T_f (maximum in χ_{ac}) represents a relaxational effect not related to a phase transition at finite temperature. An attempt was made by fitting some real spin glasses to obtain the three parameters (τ^*, E_a and $z\nu$), but the results were totally meaningless and a $T_c = 0$ phase transition for a 3D spin glass can be eliminated.

We now consider the real and imaginary susceptibilities and their frequency and temperature dependences. Let us start with the Casimir–du Pré equations

$$\chi' = \chi_s + \frac{\chi_T - \chi_s}{1 + \omega^2 \tau^2} \quad (3.35)$$

and

$$\chi'' = \omega\tau \left(\frac{\chi_T - \chi_s}{1 + \omega^2 \tau^2} \right) \quad (3.36)$$

where τ is a single relaxation time, χ_T is the isothermal susceptibility in the limit $\omega \to 0$, and χ_s is the adiabatic susceptibility in the limit $\omega \to \infty$. If $\chi'' = 0$, then there are two possibilities: either $\omega\tau \to 0$ and we measure an equilibrium isothermal χ_T, or $\omega\tau \to \infty$ and we measure the, totally decoupled from the lattice, nonequilibrium adiabatic χ_s. At $\omega = 1/\tau$ the dispersion $\chi'(\omega)$ will have an inflection point, whereas the absorption $\chi''(\omega)$ will show a maximum. Thus, this maximum provides a method for determining an average relaxation time τ_{av} for each temperature. Also according to equation (3.36), the absorption χ'' should follow a sech (ln $\omega\tau$) functional dependence for single τ and be considerably broadened if a distribution of relaxation times $g(\tau)$ is present. Figure 3.17 exemplifies this behaviour for the $(Eu_{0.8}Sr_{0.2})S$ sample shown previously in Fig. 3.15. No clear maximum in χ'' is observable over the frequency range investigated. Nevertheless, the absorption at the lowest frequencies dramatically increases in the temperature interval from 700 to 600 mK which spans $T_f(10 \text{ Hz}) = 640$ mK. For $T < 600$ mK, the absorption is essentially flat with all isotherms remaining

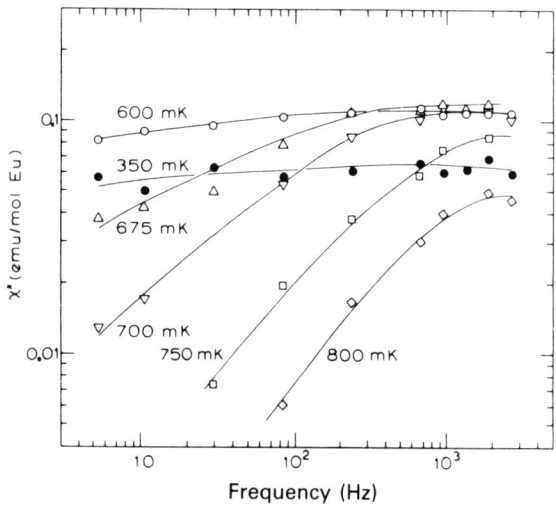

Fig. 3.17 Absorption χ'' as a function of frequency for different temperatures in $(Eu_{0.8} SR_{0.2})S$. The solid lines are visual guides; from Hüser et al. (1983).

3.2 Experiments spanning T_f

parallel to one another. In order to illuminate further this relaxation-time distribution, the susceptibility data can be plotted in the complex plane as χ' versus χ'', where each frequency represents a point. These so-called **Argand diagrams** are illustrated in Fig. 3.18 for the same (EuSr)S sample. Note now that the expected half-circle for a single τ, which is given by the maximum where $\omega\tau = 1$, is squashed into an arc length whose flatness signifies a large distribution of relaxation times. Hence, the deviations from a full half-circle are an estimate for the width of the distribution function. A mathematical method exists to treat these deviations, if the frequency range of measurements can be made sufficiently large. Such analysis we shall reserve until our chapter on ideal spin glasses.

As for the present situation we must modify the Casimir–du Pré equations to account for this broad distribution of relaxation times existing between τ_{min} and τ_{max}. Consequently, equations (3.35) and (3.36) become

$$\chi'(\omega) = \frac{1}{h} \int_{\tau_{min}}^{\tau_{max}} \frac{1}{1 + (\omega\tau)^2} m_o(\tau) g(\tau) \, d(\ln \tau) \tag{3.37}$$

and

$$\chi''(\omega) = \frac{1}{h} \int_{\tau_{min}}^{\tau_{max}} \frac{\omega\tau}{1 + (\omega\tau)^2} m_o(\tau) g(\tau) \, d(\ln \tau) \tag{3.38}$$

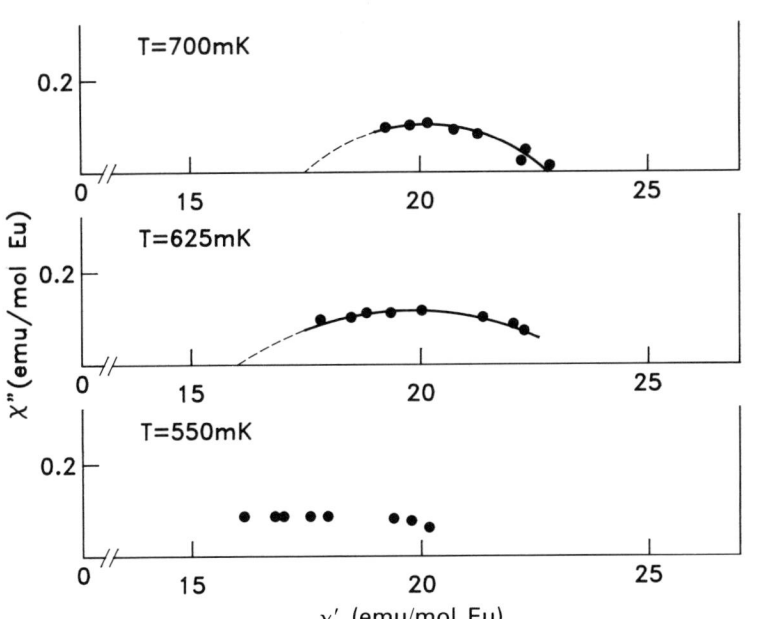

Fig. 3.18 Argand diagrams for three different temperatures. The frequency of the ac driving field h increases from right (10.9 Hz) to left (1969 Hz). The lines are computer fits to the data points assuming a symmetric diagram (see text). From Hüser et al. (1983).

where $m_o(\tau)$ is the time variation of cluster moment m_o which responds to a driving field h and which becomes blocked at its individual blocking temperature $T_B(\tau)$. We remark in passing that there is a simple relationship between χ' and χ'' for frequencies ω in the middle of a flat distribution, namely

$$\chi'' \approx -\frac{\pi}{2}\frac{\partial \chi'}{\partial \ln \omega} \approx -\frac{\pi}{2}\frac{m_o(\tau_m)}{h}g(\tau_m) \qquad (3.39)$$

where $\tau_m = 1/\omega$ is this middle relaxation time.

While the above χ' and χ'' equations work reasonably well in extracting $g(\tau)$ for superparamagnets and even for some insulating spin-glass-like materials (e.g. cobalt aluminosilicate glass) and a temperature evolution of $g(\tau, T)$ may be calculated, they do not do as well for the canonical (RKKY) spin glasses. These latter systems seem to possess such a wide range of relaxation times 10^{-14} s to ∞, so that experiments cannot deliver the many decades of frequency for χ' and χ'' which are necessary to determine $g(\tau, T)$. A combination of various techniques is needed. Additionally, there are indications here that $g(\tau, T)$ exhibits a sudden shift to longer times as T is reduced to and crosses T_f. Does this mean a phase transition? In any case, for a metallic spin glass we can schematically sketch the distribution function and its evolution with temperature in Fig. 3.19. The true paramagnetic (noninteracting isolated spins) regime ($T \gtrsim 10\, T_f$) is distinguished by a nearly single Korringa relaxation time at $\approx 10^{-12}$ s. As T is lowered, the spatial correlations begin to form and this perturbs the distribution function.

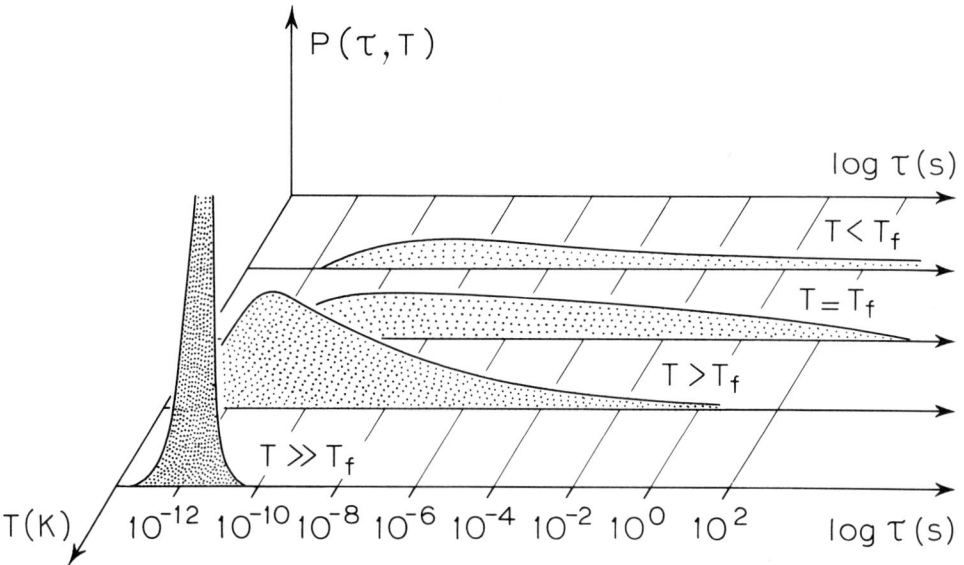

Fig. 3.19 Schematic representation of the probability distribution for spin relaxation times with its evolution as a function of temperature.

3.2 Experiments spanning T_f

Then, as $T \to T_f$ there is a rapid shift of $g(\tau)$ to slower time scales due to the long-range competing interactions. Finally, at T_f a very long-time tail suddenly appears and for all practical purposes the system is frozen or 'transformed' on a laboratory time scale. The static times continue to grow below T_f, but what do they mean if we cannot measure them? Remember the lifetime of a graduate student.

Before completing this subsection on susceptibility a final word must be said about the non-linear susceptibility χ_{nl}. This is an important new variable which according to theory exhibits the critical susceptibility divergence and exponent of a spin glass. Such a quantity may be defined in two different ways:

(i)
$$\chi_{nl} = 1 - \frac{M}{\chi_o H} \qquad (3.40)$$

where χ_o is the linear susceptibility M/H in the limit of $H \to 0$. Since we can expand M in odd powers of the field

$$M = \chi_o H - a_3(\chi_o H)^3 + a_5(\chi_o H)^5 - \cdots \qquad (3.41)$$

where the as are a function of temperature. Then

$$\chi_{nl} = +a_3(\chi_o H)^2 - a_5(\chi_o H)^4 + \cdots \qquad (3.42)$$

Hopefully, our power series rapidly converges so we are left with only one or two leading terms.

(ii) Alternatively, by applying an ac driving field, h, at frequency ω, we perform a similar expansion, but now as a function of odd frequency harmonics $3\omega, 5\omega, \ldots$.

$$M(\omega) = \sum_{k=\text{odd}} [\Theta'_k \cos k\omega t + \Theta''_k \sin k\omega t] \qquad (3.43)$$

where

$$\Theta'_1 = \chi'_1 h + \frac{3}{4}\chi'_3 h^3 + \frac{5}{8}\chi'_5 h^5 + \cdots \qquad (3.44)$$

$$\Theta'_3 = \frac{1}{4}\chi'_3 h^3 + \frac{5}{16}\chi'_5 h^5 + \cdots \qquad (3.45)$$

$$\Theta'_5 = \frac{1}{16}\chi'_5 h^5 + \cdots \qquad (3.46)$$

and similarly for the imaginary part Θ''. If we assume that the driving field is very small ($h \ll 1$ in our expansion), then

$$M(\omega) = \chi'_1 h \cos \omega t + \chi''_1 h \sin \omega t +$$
$$+ \frac{1}{4}\chi'_3 h^3 \cos 3\omega t + \frac{1}{4}\chi''_3 h^3 \sin 3\omega t \qquad (3.47)$$
$$+ \frac{1}{16}\chi'_5 h^5 \cos 5\omega t + \frac{1}{16}\chi''_5 h^5 \sin 5\omega t$$

Thus, we can measure the various coefficients of the different harmonics. The higher order or non-linear ones (3ω, 5ω, etc.) through scaling theory should reflect the critical divergences characterized by a set of critical exponents (γ and β). Neglecting the imaginary part, the non-linear susceptibility becomes

$$\frac{M}{h} - \chi_1' = X_3 h^2 \left(\frac{T - T_c}{T_c}\right)^{-\gamma} + X_5 h^4 \left(\frac{T - T_c}{T_c}\right)^{-(\beta + 2\gamma)} + \cdots \quad (3.48)$$

where the Xs are the amplitudes of the divergences.

Highly sophisticated experimental techniques, e.g. fast Fourier transforms, have been brought to bear on the determination of χ. However, difficulties exist with each definition (i) and (ii) of χ_{nl}. For the field expansion (i), the rather large applied fields will definitely affect the spin-glass transition smearing out the critical phenomenon. For the harmonic expansion (ii), the relaxation times will soon become larger than ω^{-1} and thereby influence or broaden the critical behaviour. For this reason only in the limit $\omega \to 0$ should the static transition be accessible. Nevertheless, χ_{nl}, and *not* χ_{ac}, is crucial to establish the properties of a phase transition in spin glasses. By way of illustrating the typical behaviour of χ_{nl}, we reproduce the data at $\omega = 10^{-2}$ Hz for an AgMn (0.5 at. %) sample in Fig. 3.20. The plot of χ_3' versus reduced temperature, $(T - T_g)/T_g$ is made on a log-log scale in order for the slope to give the (negative) critical exponent. Here T_g (= 2.95 K) is the critical temperature fixed as the peak in χ_1' with $\omega = 0.001$ Hz. The results do show what looks like a power law behaviour with $\gamma = 2.1$ for at least a decade of reduced temperature. But take note of the rounding beginning below 2×10^{-2} in reduced temperature. This means we cannot approach too close to T_g before the transition becomes smeared out, most probably due to these relaxation time effect, even a 100 s measurement time is not long enough at $2 \times 10^{-2} T_g$. It would seem that the spin glass begins by following a critical divergence. However, the distribution of relaxation times becomes too great and the system drifts out of equilibrium. For, in usual phase transitions, these divergences can be tracked much closer to T_c, say 10^{-3} or even 10^{-4}. Well then, the spin glasses certainly do not display an ordinary phase transition.

3.2.2 Specific heat

After all this action around T_f for the susceptibility, the specific heat is a big disappointment. Already from Fig. 3.5 there is little to be seen at T_f. As a matter of fact without further analysis T_f does not even seem to exist in a C_m versus T plot.

What can we do further to reanalyse or blow-up the specific-heat behaviour? One suggestion for extracting T_f was to make plots for the most

3.2 Experiments spanning T_f

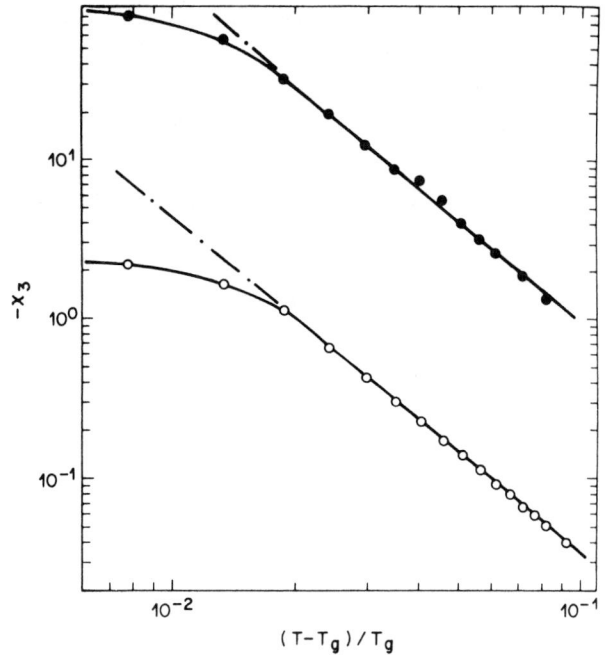

Fig. 3.20 Temperature dependence of $-\chi_3$ above T_g ($=T_c$) measured at 10^{-2} Hz in static fields of 0 (open circles) and 90 G (solid circles) as a function of reduced temperature. The same slope ($-\gamma$) indicates that the divergence of χ_3 in dc fields less than 100 G remains well described by the same exponent ($\gamma = 2.3$) as in zero field. From Lévy (1988).

accurate specific-heat data of C_m/T versus T. The hope was to establish a break or a rapid fall in the rate of change of entropy $dS_m/dT = C_m/T$ at T_f. By examining the meticulous results shown in Fig. 3.21 for three samples of CuMn (0.08, 0.43 and 0.88 at. %) we conclude the following: while there is some indication for a drop, it is certainly not sharp enough to relate to a distinct and well-defined T_f. Consequently, on the basis of the best-available data, no peak or singularity can be found in C_m, in contrast to the cusps and incipient divergences in χ_{ac} and χ_{nl}, respectively.

We still need a reason for the absence of a clear effect in C_m at T_f. Perhaps the critical phenomenon is very subtle. A large negative critical exponent, α, for the specific heat would produce a broad maximum centred at T_f. And, if this is superimposed on a flat background of a non-critical specific-heat contribution, it could be quite difficult to discern the critical part. Perhaps the critical region for C_m is so narrow that the finite temperature steps ΔT, which are necessary for the specific-heat measurement (remember experimentally $C \equiv \Delta Q/\Delta T$), are enough to wipe out the critical fine-peak structure. What we learned from our high-T specific-heat analysis

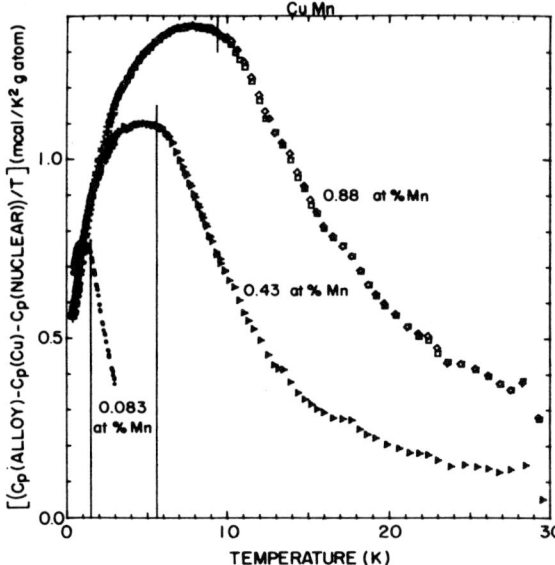

Fig. 3.21 Spin-glass specific heat divided by temperature plotted against temperature. The vertical bars intersecting the plots mark the estimate of T_f for each composition. From Martin (1980).

(section 3.1.2) was that most of the entropy ($\approx 80\%$) is already frozen out in short-range order above T_f. So conceivably there is simply not enough entropy remaining to produce a significant peak at T_f. Recall how our cluster configuration model nicely spanned the C_m data through T_f (Fig. 3.7) whereas it could not deal with the susceptibility peak.

Such conjectures are trying to preserve a phase-transition explanation for C_m. However, we must live with the empirical fact that little happens in the specific heat. Either the phase transition is unconventional or maybe we should mimic the real glasses and their freezing transitions (fall out of equilibrium) as a viable model for the spin glasses. Before we choose let's look at some other experiments.

3.2.3 Resistivity

For usual phase transitions the critical behaviour of the electrical resistivity is reflected in its temperature coefficient or derivative $d(\Delta\rho)/dT$. Theory has even related this derivative to the specific heat and its critical divergence for certain metallic cases. So it is not surprising that little can be found in $d(\Delta\rho)/dT$ at T_f. Figure 3.22 illustrates the point by plotting the T-coefficient $d(\Delta\rho)/dT$ for some CuMn alloys as a function of T. The arrows represent

3.2 Experiments spanning T_f

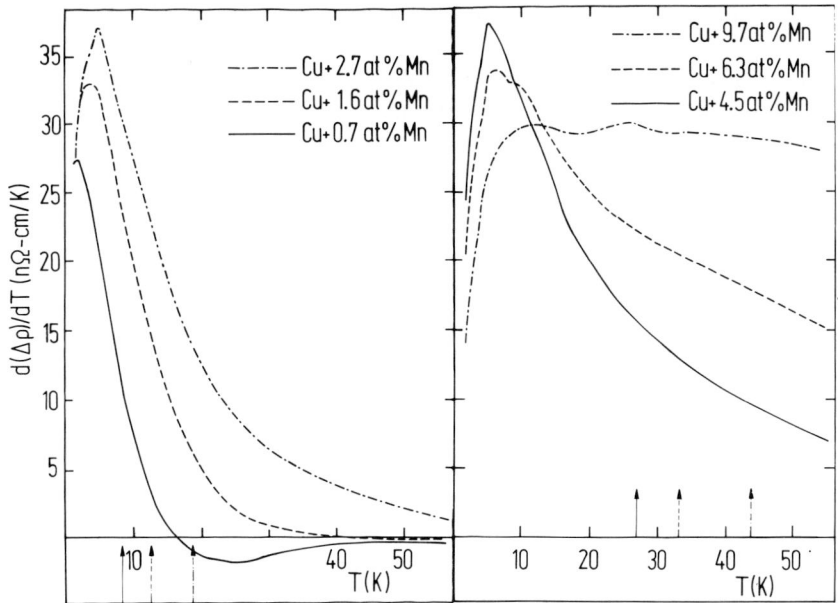

Fig. 3.22 Temperature dependence of the temperature derivative of the impurity resistivity d$(\Delta\rho)$dT (nΩ cm/K) for the CuMn alloys. The arrows represent T_f as determined from the susceptibility measurements. From Ford and Mydosh (1976).

temperatures of the sharp cusps observed in χ_{ac}. There is no correlation with T_f only a broad maximum at a temperature significantly below T_f.

Once again we have the same smeared behaviour as C_m. The transport properties, really a scattering of conduction electrons from magnetic impurities, exhibit no transition, only a gradual transformation. For, certainly, the spin-disorder scattering is steadily reduced upon lowering the temperature through T_f. A comparable situation occurs for the thermoelectric power which is more complicated due to a variety of contributions – nothing striking happens at T_f. Why is there no effect? Well, one suggestion is that it is to do with the lack of a *periodically* ordered state, and thus, no long-range spatial correlation function. The frozen state is random and frozen disorder will still make a significant contribution to the scattering. Unquestionably the time correlations are progressively becoming slower, but even a rapid change to longer time scales near T_f is not affecting the much shorter local-moment/conduction-electron scattering times which determine the transport properties. There are always some fast relaxation times available in our $g(\tau)$ distribution which will continue to scatter the conduction electrons. It seems exclusively when there is the formation of a periodic spin lattice that the transport properties will reflect the critical behaviour. This periodicity may even create distortions of the Fermi surface, which, of course, will be seen in d$\Delta\rho$/dT and the thermopower. So our job of experimentally characterizing the freezing process T_f is a most difficult task without a periodic spin order. The intrinsic randomness of the freezing masks many of the key experimental effects which might be related to critical phenomenon, and we certainly have an unusual type of phase transition.

3.2.4 Neutron scattering

Let us return to our structure-factor separation into static and dynamic parts (section 3.1.4). We recall that the total scattering cross-section is proportional to the sum of these two parts which, within our relaxation time approximation, we can write as equations (3.24) and (3.26). The limits of the integrals depend upon the instrumental resolution of our neutron spectrometer. So we just repeat the measurements of the scattering cross-section at a fixed value of momentum transfer q on as many different spectrometers as are available. (Fig. 3.10 was limited to only one spectrometer.) Easier said than done! Another hindrance is the exact separation of elastic from quasi-elastic. But this is a fitting 'detail' disentangled by the experts.

For an arbitrary value of $q = 0.13$ Å^{-1} and a CuMn (3 at. %) sample we show the results in Fig. 3.23 for two elastic scattering cross-sections measured with two different spectrometer resolutions, 1.5 μeV ($\approx 10^{-9}$ s) and 230 μeV ($\approx 10^{-11}$ s). This means that only relaxation times longer than 10^{-9} s (or 10^{-11} s) will be seen in the elastic window. Notice how the total cross-section given by the squares in Fig. 3.23 remains relatively constant over the entire temperature range. Below 75 K the fast scattering rate is being shifted to slow scattering and this appears as 'elastic' intensity – it contributes to a continuously increasing central peak. T_{sg} is the freezing temperature (T_f) determined via χ_{ac}. Here the 'a' length or amount of scattering cross-section represents the very fast quasi-elastic range ($< 10^{-11}$ s), the 'b' amount is the scattering with relaxation times in the interval between the two spectrometers (10^{-11}–10^{-9} s), and 'c' is the slow scattering ($> 10^{-9}$ s) all lumped together. The next step is to construct a schematic form for the distribution of relaxation times $N(\tau)$. Figure 3.24 presents a sketch of $N(\log \tau)$ versus $\log \tau$ as expected from the data in Fig. 3.23. Note the sudden and asymmetric shift of the distribution function to much longer times at T_f. Surprising even for me is the remarkable similarity between this distribution function and that independently obtained from the $\chi'(\omega)$ and $\chi''(\omega)$ results (Fig. 3.19). While both are qualitative, merely schematic, the same basic idea is at hand, namely the sudden shift in the distribution of relaxation times to much longer values at T_f.

We could continue with our neutron scattering analysis and examine the q-dependence of the scattering as a function of T. A bit of this was already introduced in Fig. 3.11. But let us postpone this discussion until the low

3.2 Experiments spanning T_f

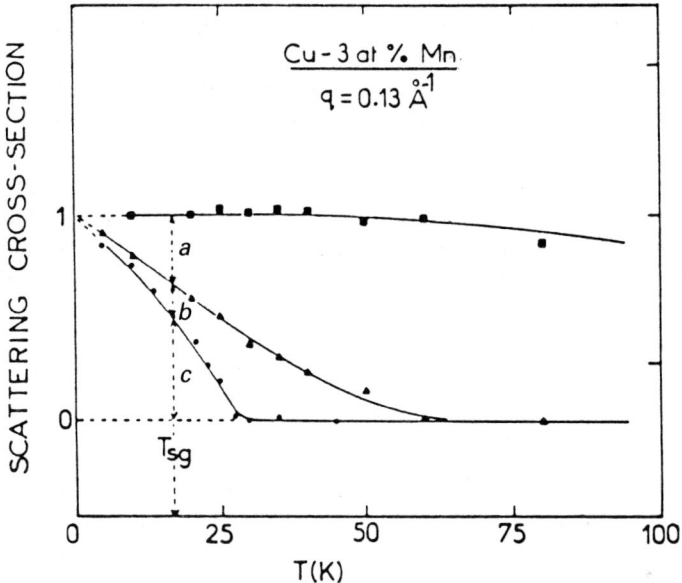

Fig. 3.23 The 'elastic' cross-sections measured with two different energy resolutions along with the quasistatic part of the total cross-section (as defined in the text) for the $CuMn$ (3 at. %) alloy at $q = 0.13$ Å^{-1}. The dots, ●, represent the 'elastic' cross-section measured with energy resolution of 1.5 μeV. The triangles, ▲, represent the elastic cross-section measured with energy resolution of 230 μeV. The quasistatic part of the total cross-section is represented by the squares, ■. From Murani (1981).

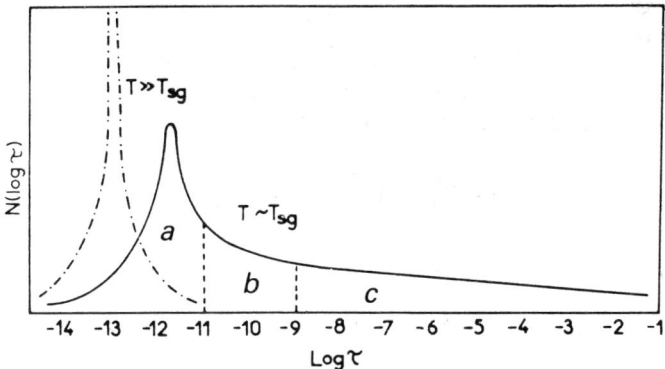

Fig. 3.24 The possible forms of the density of relaxation times $N(\tau)$ per unit logarithmic interval of τ, for the $CuMn$ (3 at. %) alloy at (i) $T \gg T_{sg}$ ($=T_f$) and (ii) $T \approx T_{sg}$. The regions marked a, b and c correspond to the lengths a, b and c in Fig. 3.23 representing the spectral weights in the various time zones. From Murani (1981).

temperature ($T \ll T_f$) section (section 3.3.4), and instead, briefly consider a novel neutron scattering technique which directly measures the time-correlation function at some average small q-value for longer time scales. The method is called neutron spin-echo and essentially it enables us to detect small inelastic scattering components by comparing the Larmor-precession phase of a spin-polarized neutron beam before and after scattering with a sample. Since this is essentially a resonance process, for which slight deviations in the polarization phase are measurable, it can extend the time detection range down to 10^{-8} s. The formula for the average polarization along a given axis is

$$\langle P \rangle = P_o \frac{\int S_T(q, \omega) \cos\theta \, d\omega}{\int S_T(q, \omega) \, d\omega} \tag{3.49}$$

where θ is the phase angle. According to our original definition of $S_T(q, \omega)$ in equation (3.13), $\langle P \rangle$ is related to the time correlation function which for a spin glass is $\langle P \rangle \equiv \xi(q, t) \propto \langle S_i(0)S_i(t) \rangle$. In the limit of $q \to 0$ and $F^2(q \to 0) = 1$, and substituting for $S_T(q, \omega)$ via equation (3.25), we have

$$\xi(t) = \frac{2k_B T}{g^2 \mu_B^2} \frac{1}{\pi} \int_{-\infty}^{+\infty} \int_0^\infty g(\Gamma) \frac{\Gamma}{\Gamma^2 + \omega^2} \cos \omega t \, d\omega \, d\Gamma \tag{3.50}$$

$$\xi(t) = \frac{2k_B T}{g^2 \mu_B^2} \int_0^\infty g(\Gamma) \exp(-\Gamma t) \, d\Gamma \tag{3.51}$$

At $t = 0$ the last expression reduces to

$$\xi(0) = \frac{2k_B T}{g^2 \mu_B^2} \int_0^\infty g(\Gamma) \, d\Gamma \tag{3.52}$$

which is nothing but our old friend S_T or equation (3.26) in the limits $q \to 0$ and $\tau \to \infty$. If $g(\Gamma)$ is a slowly varying function, then we can define a correlation function at relaxation time τ as

$$\xi(\tau) = \frac{2k_B T}{g^2 \mu_B^2} \int_0^{1/\tau} g(\Gamma) \, d\Gamma \equiv S_S^\tau (q = 0) \tag{3.53}$$

See equation (3.24). Therefore,

$$\xi(0) - \xi(\tau) = S_T(q = 0) - S_S^\tau (q = 0) = S_D^\tau (q = 0) \tag{3.54}$$

where

$$S_D^\tau (q = 0) = \frac{2k_B T}{g^2 \mu_B^2} \chi(q = 0, \tau) \propto \chi_{ac}(\omega) \tag{3.55}$$

So we can collect these results and plot the measured $\xi(t)/\xi(0)$ against log-

3.2 Experiments spanning T_f

time. Figure 3.25 shows how the neutron-spin-echo determined correlation function (at $\langle q \rangle = 0.09$ Å^{-1}) decreases at various temperatures for times $\approx 10^{12}$–10^{-8} s. Again the alloy is $CuMn$ (5 at. %) with $T_f = 27.4$ K. Clearly one can discern the temperature evolution from exponential decrease $\exp(-\Gamma t)$ ($\Gamma = 0.5$ meV) via power-law fall off to approximately $\log(t)$.

However, the neutron technique is not suited for really long time scales $> 10^{-8}$ s. So what we have also included in Fig. 3.25 are the $\xi(t)$ data (i) for μSR (muon spin relaxation), discussed in the next section (3.2.5), which tracks the intermediate times $\approx 10^{-7}$ s, and (ii) from the $\chi_{ac}(\omega)$ according to equation (3.55). Using equation (3.54) we can write

$$1 - \xi(\tau) = S_S^\tau(q=0) = \frac{2k_BT}{g^2\mu_B^2}\chi(\tau) \tag{3.56}$$

where we have assumed a phase transition and $\xi(0) = 1$. Note how the correlation function jumps up at T_f and develops its long-time tail (right-hand side of Fig. 3.25). Once more there is the same type of experimental behaviour and phenomenological model. At least we have an empirical picture of the freezing process that is *unlike* that of the standard phase transitions which exhibit a rapid critical slowing down and disappearance of the fluctuations, a small critical regime where all this happens, and a becoming infinite correlation length (long-range order). Using these various techniques we have traversed almost ten decades of time – something necessary for the spin glasses. Yet needed are probes of the really long times which characterize the frozen state ($T \ll T_f$). Section 3.3 will handle this situation, as for now we persevere with two further experiments which are useful for 'around T_f' behaviour.

3.2.5 μSR (muon spin relaxation)

Here a spin-polarized beam of positive muons μ^+ ($m = 250m_e$, $Q = +e$ and $S = \frac{1}{2}$), produced by an accelerator, is implanted in our spin glass. The

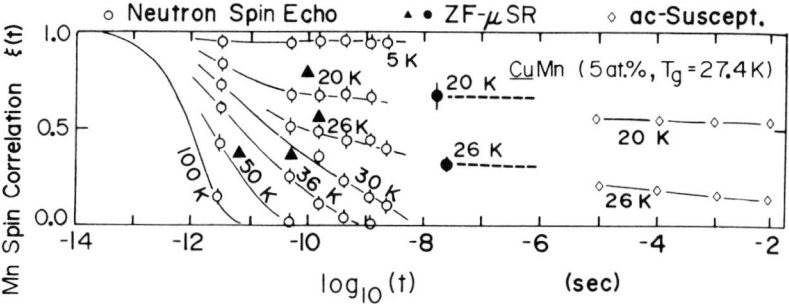

Fig. 3.25 Comparison of the time correlation $\xi(t)$ of the Mn moment in $CuMn$ (5 at. %) measured by neutron-spin echo (NSE), zero-field μSR (ZF μSR), and ac susceptibility (χ_{ac}); from Uemura et al. (1985).

μ⁺ will process about any *local* magnetic fields (internal or external) felt at its rest site. After 2.2 μs the muon decays by emitting a positron along the instantaneous direction of its spin **S**. One measures the asymmetry in the direction of the emitted positrons. Although various configurations of fields and detectors are possible, we, for simplicity, treat one important case for the spin glasses, namely, that of a weak (LF-μSR), or even no (ZF-μSR), longitudinal field, H_L, applied parallel to the incoming beam and thus $\mathbf{H}_L \| \mathbf{\mu}_L$. The difference between forward emitted positrons and backward emission is related to a spin-relaxation function $G_z(t)$.

We assume for a spin glass that there exists a lorentzian distribution of random internal fields which fluctuate as a markovian process e^{-vt} (v is the rate of fluctuation). This model introduces two parameters: 'a' the half-width at half-maximum of the lorentzian field distribution, and v which is taken as the inverse average spin-correlation time reflecting the fluctuations of the local field, $v = 1/\tau_c$. For $v/a \ll 1$ (slow modulations or quasi-static local fields)

$$G_z(t) = \frac{1}{3}\exp\left(-\frac{2}{3}vt\right) + \frac{2}{3}(1-at)\exp(-at) \tag{3.57}$$

In the opposite limit $v/a \gg 1$ (fast modulation or rapidly fluctuating magnetic moments)

$$G_z(t) = \exp\left(-\sqrt{4a^2t/v}\right) \tag{3.58}$$

Hence, the LF-μSR experiments can determine the two important parameters, a as a spatial distribution of local fields, and τ_c as a sort of spin-relaxation time. An advantage of this technique is that it can be performed in zero external field. We remark that the time window available in μSR to study relaxation processes is $\approx 10^{-10}$–10^{-6} s.

With the above oversimplified parametrization we obtain results for $G_z(t)$ in the different limits. Then τ_c can be extracted at each temperature and plotted versus $1 - T_f/T$, as is shown in Fig. 3.26 for various alloy systems and concentrations. This figure illustrates the rapid upturn in τ_c as T_f is approached from above followed by a saturation very close to T_f. The effective correlation time appears to move through the μSR-time window with a power law dependence. Best fitting of the data shows that

$$\tau_c = \tau_o \left(\frac{T-T_f}{T_f}\right)^{-n} \tag{3.59}$$

where $\tau_o \approx 10^{-12}$ s and $n = 2.6$–2.9. Note that rounding does seem to occur for reduced temperatures less than 5×10^{-2}. Analogously, we have a consistent picture with our average or effective relaxation time slowing down, but not diverging, as T_f is spanned.

Regarding the spatial distribution of local fields (parametrized in 'a') which are sampled by the muons, two extreme cases have been considered.

3.2 Experiments spanning T_f

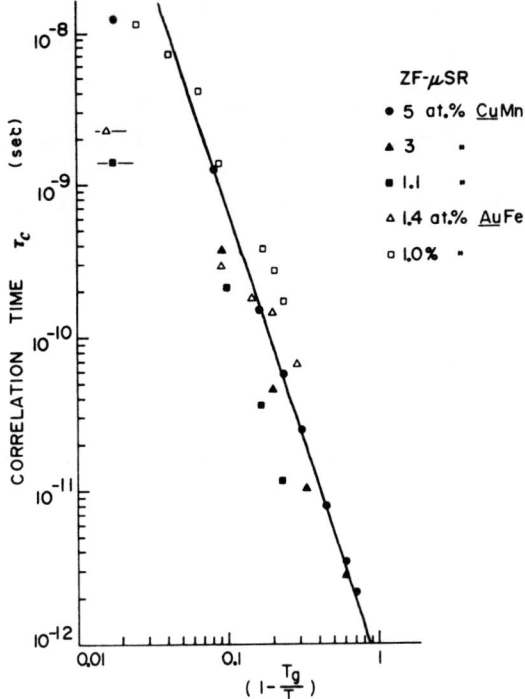

Fig. 3.26 Correlation time τ_c of Mn (or Fe) moments determined by zero-field μSR from the dynamic depolarization rate observed above T_f. The straight line corresponds to the power law of critical slowing-down. From Uemura *et al.* (1985).

(i) Spatially inhomogeneous dynamics, where various sized clusters independently fluctuate with the particular correlation time of the given volume. In other words, each (rigid) cluster has its own chracteristic τ_c and there are many different cluster sizes and thus τ_cs.

(ii) Spatially homogeneous dynamics, the local field at a given site is a superposition of various contributions from spins belonging to different clusters with different τ_cs. Said otherwise, the separate clusters are really coupled via the long-range interactions and each site feels a superposition of these many fields which in turn lead to a similar fluctuation spectrum of the local field at different sites. Hence there is no spatial inhomogeneity.

Since we are not dealing with a superparamagnet we would expect the latter (ii) model. After a lengthy and systematic analysis of the muon-spin-relaxation function, (ii) was confirmed around and slightly above T_f. Therefore, we have gained new and corroborative information from the local probe of μSR about our experimental model of interacting cluster out of which the spin-glass state is formed.

Another achievement of the μSR technique is to relate its measured relaxation time to the spin correlation function determined from neutron spin echo and χ_{ac}. If we set $\xi(t = \tau_c) = 1/e$ (e is the base of the natural logarithms), then we obtain the triangular points in Fig. 3.25. On the other hand, if we bring an order parameter Q, as some sort of static component (with linewidth a_s), into the μSR random-field distribution via $Q = (a_s/a)^2$, then $\xi(t = \tau_c) = (1 - Q)/e + Q$. And these are represented by the two solid dots in Fig. 3.25. Hence, we have a few agreeing points in the intermediate time range spanned by the muons.

It is the combination of spatial (a local probe) and temporal (10^{-10}–10^{-6} s) resolution that has made this very sophisticated method so powerful for the spin glasses. μSR seems to have been the contact point with a fractal cluster model (see section 5.10) which will give us a little more quantitative insight into our phenomenological picture. What does μSR say about the question of a phase transition? First of all, the static linewidth a_s remains finite at T_f, i.e., all the spins are not statically frozen. Secondly, the $\tau_c \propto (T - T_f)/T_f$ does not really diverge in the allowed time window, and even when one can extract or estimate a critical exponent, it is much different from those values obtained using other techniques. At the risk of repeating oneself, if there is a phase transition, it is certainly unconventional: the usual theories and intuitions are not instructive.

3.2.6 Mössbauer effect

Another microscopic or local-probe technique is the Mössbauer effect which involves the resonance absorption of a 'recoil-less' emitted gamma rate from a radioactive nucleus. ^{57}Fe turns out to be one such Mössbauer nucleus. If a local magnetic field is present, then there will be an induced hyperfine field which splits the ^{57}Fe nuclear spin, I, into a doublet ground state ($I = \frac{1}{2}$) and a quartet ($I = \frac{3}{2}$) excited state. It follows that the emission of a γ-ray from this latter state with the selection rule $\Delta m_I = \pm 1$ or 0 has six paths. Accordingly, the single-line emission in zero field becomes six lines in a field. Both the amount of hyperfine splitting and the intensity ratios give valuable information about the magnetic state of the sample in which the ^{57}Fe is embedded.

This hyperfine field can be related to the local spontaneous magnetization. For, say, ferromagnetic (pure) iron below its $T_c = 1040$ K, the full spontaneous magnetization and thus the Mössbauer splitting obeys the standard (order-parameter) temperature dependence. What about a spin glass? First we need to know and include the time scale of the Mössbauer effect. The decay processes have an intrinsic lifetime of about 10^{-7} s. So if a spin in a AuFe alloy is frozen on a time scale longer than $\approx 10^{-7}$ s, the hyperfine splitting will appear, regardless of the orientation of the frozen spin. The technique is a zero-field one and measures the local freezing

3.2 Experiments spanning T_f

without determining the various alignments. In Fig. 3.27 we show a very early measurement (Violet and Borg 1966) of the hyperfine field versus the reduced temperature for a series of AuFe spin glasses. While there is a lack of data near T_o, now our T_f, the solid line through the data has a form which does not differ much from that of pure Fe. This was one of the first clues that perhaps a phase transition was occurring.

By adding an external field to the Mössbauer effect, the changes in the intensity ratios can be used to determine the spin alignment. Then it was called a peculiar type of antiferromagnetic (weak, canted) ordering. Further analysis pointed out the existence of a random distribution of magnetic moments and various probability distributions for the internal field could be established. The randomness of the local-spin arrangements was greatly affected by the external field – a ferromagnetic component could be induced. We shall have much more to say about these low-temperature properties in the next section.

Modern Mössbauer experiments have focused on the smearing of the transition and relaxation effects near T_f. There is indeed a finite temperature interval of some degrees which signifies the Mössbauer transition. The best estimate of a Mössbauer 'transition temperature' is systematically higher ($\approx 20\%$) than T_f from the susceptibility cusps. Local-moment relaxation times may be deduced from the paramagnetic Mössbauer (lorentzian) linewidths. For a typical AuFe (3 at. %) spin glass τ_M, the average relaxation time, begins to deviate from its paramagnetic value (10^{-12} s) far above T_f.

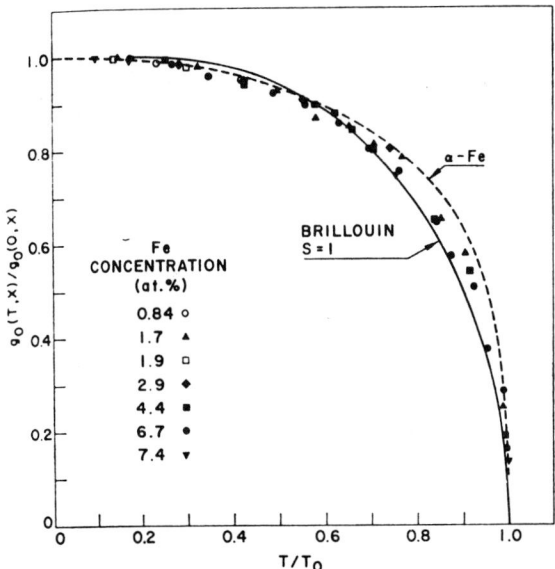

Fig. 3.27 Normalized Mössbauer-effect splitting versus T/T_f for a series of AuFe alloys. From Violet and Borg (1966).

As $T \to T_f$, τ_M rapidly increases to $\approx 5 \times 10^{-9}$ s and then the single-line spectrum evolves into the six-finger one of the 'static' field state, i.e., slower than 10^{-7} s. Our analysis and interpretation follow the previous lines of thought, namely, a distribution of relaxation times which progressively broadens and shifts via an 'average' time to slower and slower values. In addition, a careful examination of the six-finger pattern and its evolution suggests an inhomogeneous 'static' broadening related to non-zero average spins $\langle \vec{S_i} \rangle$ with a distribution of magnitudes $|\langle S_i \rangle|$. This conclusion which is a direct result of local-probe or microscopic measurement technique, is certainly connected to our collection ferromagnetic clusters of different size proposed previously. It is interesting to note that the inhomogeneous static broadening from the Mössbauer effect (not related to the spatially inhomogeneous dynamics proposed and rejected for the μSR) goes through a maximum very close to T_f from the susceptibility cusp. Such behaviour is what we would expect at a percolation transition so once again we have another, now Mössbauer effect, confirmation of our spatial cluster and lengthening relaxation-time model.

3.3 LOW-TEMPERATURE ($T \ll T_f$) EXPERIMENTS

Now we must confront the experimental consequences of our frozen state. Certainly there is a new and unique magnetic phase, but how can we succinctly characterize its basic properties? In this section we select a few simple, yet representative, experiments that can maintain the thread of continuity from the previous temperature regimes, and use them to delineate the ground state of a spin glass. We fully realize that there is no long-range periodic order, so conventional ordering behaviour and elementary excitations will be transmogrified or non-existent. Dynamical effects persisting down into the frozen state govern many properties, especially how they are modified with an external field – the key variable of the new phase. Our choice of measurement techniques will be to persevere with our four basic experiments: magnetization will take the place of susceptibility, since the proportionality $M \propto \chi$ no longer exists, specific heat and resistivity our two standard thermodynamic/transport properties, and neutron scattering, where we consider some recent results and their iconoclastic interpretation. Finally, two new techniques will be introduced, namely, torque and electron paramagnetic resonance (EPR) which have clarified the particular importance of anisotropy but requires rather large field sweeps. As always we try to treat only generic behaviour, i.e., we carefully select our examples to correspond to the most basic and general characteristics of the canonical spin glasses.

3.3 Low-temperature ($T \ll T_f$) experiments

3.3.1 Magnetization

Let us use one of the many magnetization techniques and zero-field cool (ZFC) the spin-glass sample to $T \ll T_f$. We begin with sweeping the applied field slow enough that the measurement is effectively dc, however, the sweep rate is sufficiently rapid to avoid large relaxation or creep effects mentioned previously. A generic sketch is given in Fig. 3.28. The small initial slope near the origin is comparable with our ac-susceptibility, limit $(H \to 0)$ $M(ZFC)/H \approx \chi_{ac}$, and before discernible time dependences occur a small finite H is required. These usually appear around the start of the S-shaped form in the virgin M versus H curve. Thus, already in small field, the magnetization slowly drifts upwards roughly following a log-time scale. The inset of Fig. 3.28 illustrates the upward creep for increasing field and a downward one for decreasing field by a thick line.

As we reach very large fields (≈ 40 T), now shown in Fig. 3.29, there is a tendency towards saturation, but full saturation $M_{sat} = g\mu_B S$ (≈ 4 μ_B for CuMn) is never attained even at the lowest temperatures and concentrations. Notice also how the magnetization at constant field decreases with increasing concentrations (Fig. 3.29). What does this all mean? Well we certainly do not have a superparamagnet which is easy to polarize! Instead, not only

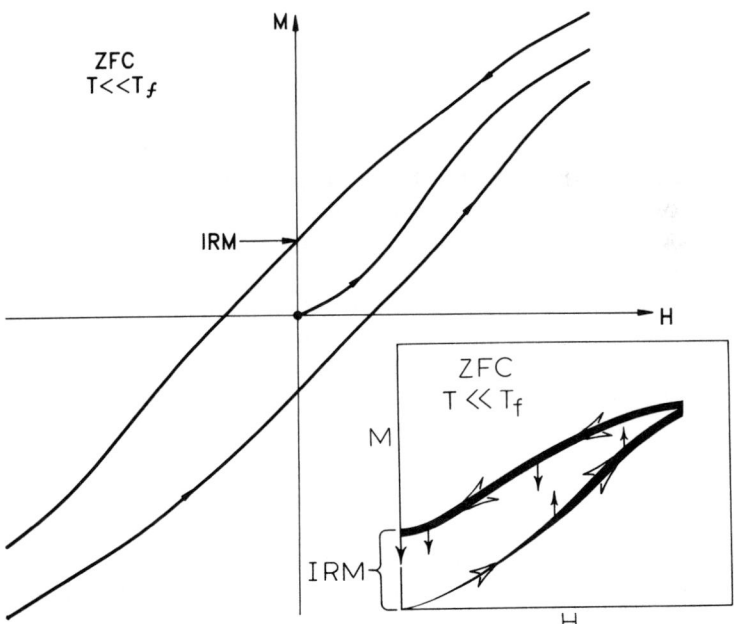

Fig. 3.28 Schematic of zero-field cooled hysteresis loop for $T \ll T_f$ illustrating the S-shape, the lack of saturation, and the isothermal remanent magnetization (IRM). The time dependences are depicted in the inset.

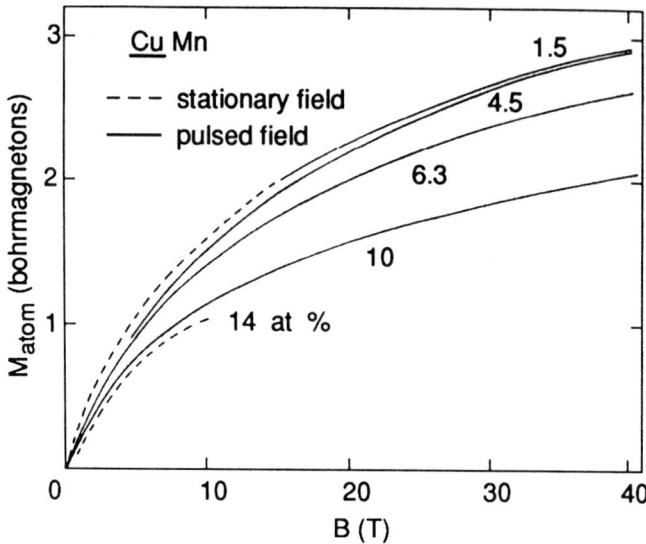

Fig. 3.29 Magnetization per Mn-atom for the various concentrations of CuMn studied at 4.2 K; from Smith *et al.* (1979).

the long-range RKKY interactions, but also the peculiar anisotropy associated with CuMn play a major role and prevent the external field from fully rotating the frozen moments into its direction. One could try to calculate the various exchange interactions based upon a superposition of three Brillouin functions fitted to a universal (concentration scaled) magnetization curve. However, for CuMn too much antiferromagnetic coupling is required for a reasonable fit. The results totally disagree with our cluster configuration model's estimates of the J_is from section 3.1.1 and Fig. 3.4. Since the present magnetization versus field behaviour is a low-temperature $T \ll T_f$ measurement, there must be a *new* aspect to the frozen state. And this is the random anisotropy which is formed below T_f and is particularly strong in CuMn. The external field has to overcome an array of local anisotropy axes, randomly oriented, before the various clusters can point along the field direction and fully align, i.e., saturate, the spins. It takes field energy to rotate the cluster moment away from its anisotropy-pinned 'frozen' orientation. Such an energy barrier for rotation was already introduced for a superparamagnet, but now for a frozen spin glass the barrier orientations are randomly distributed and the barrier heights, also a distribution, are temperature dependent disappearing above T_f. We shall delve more closely into this anisotropy when we consider the torque and ESR experiments (section 3.3.5).

The AuFe system is somewhat different. The magnetization at constant field increases with increasing concentration and saturation, while not completely reached, appears a bit closer, here $M_{sat} = g\mu_B S \approx 3\ \mu_B$, while

3.3 Low-temperature ($T \ll T_f$) experiments

$M(40\ T) \approx 2\text{--}2.5\ \mu_B$. In addition, the Brillouin-function fitting gives a more reasonable estimate for the exchange parameters J_1 and J_2 which compare favourably with that found from the cluster configuration model. The random anisotropy also is much weaker and field cooling (FC) does not induce a unidirectional component (see below).

If we FC CuMn to $T \ll T_f$, a most unusual 'displaced' hysteresis loop occurs upon reducing the field through zero to negative values and back, thereby tracing out the loop. Figure 3.30 presents this behaviour for CuMn (0.5 at. %) after a few kgauss FC to 1.35 K $\ll T_f$ = 6.5 K. Metastable with respect to time, point A represents the thermal-remanent magnetization (TRM) – a field-cooling effect where, after the temperature change, the FC field has been reduced to zero. Note what happens when the applied field is increased in the negative direction. The magnetization remains positive. At point B (or B' dependent on the field sweep in different runs) there is a sharp switching to negative magnetization – a jump in M to point C (C'). Upon proceeding to point D and then returning the field back through zero, the reversal $-M$ to $+M$ of remanent magnetization, E to F, (at $H \gtrsim 0$) is approximately equal to the initial value of the remanence (neglecting the small time dependence). Therefore, we have a displaced hysteresis loop or memory effect whose *width* is determined by the weak

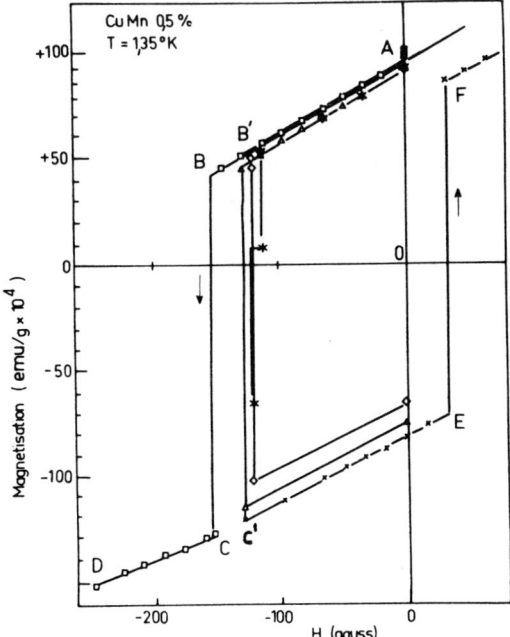

Fig. 3.30 Hysteresis for field-cooled CuMn with saturated remanence magnetization. Different symbols are used for different runs. From Monod *et al.* (1979).

uniaxial anisotropy and whose *displacement* (position of the loop centre with respect to the origin) by the *unidirectional* anisotropy.

A simple physical picture here is that the FC induces a preferred direction along which a small net ferromagnetic moment is frozen. This causes the TRM. However, a small negative perturbing field can flip the preferred direction and thus the TRM from $+$ to $-$. We visualize an external field-dependent energy barrier which has its lowest lying minimum along the field direction (Fig. 3.31 a). When the field is reversed ($H < 0$), so is this the minimum and now a thermal fluctuation or 'noise' will push the TRM over the small energy hump into this opposite-spin minimum: $-M$ (Fig. 3.31 b). Finally, when the field is reduced towards zero, the return process takes place (Fig. 3.31 c) and we have generated a hysteresis loop. The value of the return field in Fig. 3.30 is slightly positive and depends, in general, on the value of the uniaxial anisotropy, which determines the width, and the unidirection anisotropy which gives the displacement. The asymmetry in our naïve potential-barrier sketch is created by the unidirectional component. The entity which spin flips between $+M$ and $-M$ is our small TRM. Since these switches of magnetization are so sharp, there must be a co-operative nature to the frozen spin-glass state, i.e., a long correlation length between the frozen spins such that they can collectively turn the TRM back and forth on a macroscopic scale. For AuFe there is simply not enough unidirectional anisotropic which can be induced by field cooling. By including a third element, e.g. Pt, to increase the spin-orbit coupling of the alloy system, a displaced hysteresis loop can be established.

Before we proceed to the time dependence of the various magnetizations, let us classify the two types of remanent magnetizations which occur in a spin glass. As already mentioned, the thermoremanent magnetization (TRM) is a field-cooling effect for which the sample temperature is reduced from $T > T_f$ to $T < T_f$ in an FC field H. At this lower temperature, H is set

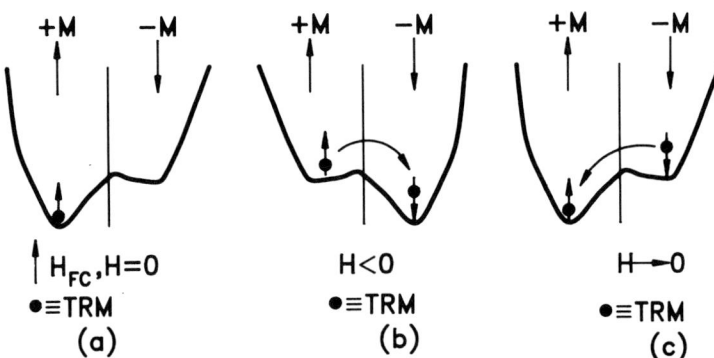

Fig. 3.31 Memory effect as represented by energy-barrier deformations for the field-cooled magnetization of CuMn.

3.3 Low-temperature ($T \ll T_f$) experiments

equal to zero and the magnetization measured. Here saturation of the TRM is quickly reached for low values of H. Figure 3.32 illustrates this behaviour for a canonical spin glass AuFe (0.5 at. %). For the TRM an irreversible susceptibility may be defined as $\chi_{ir} = \text{TRM}(H)/H$. By adding χ_{ir} to $\chi \equiv \lim_{H \to 0}(M/H)$, called the reversible susceptibility, χ_{FC} may be indirectly obtained. From this procedure χ_{FC} is found to be constant for $T \leq T_f$ – the result agrees nicely with the direct method of simply FC a spin glass in field H. The second type of remanence is called isothermal remanent magnetization (IRM) and is obtained by cooling the spin glass in zero field (ZFC), followed by cycling an external field $0 \to H \to 0$, and then measuring the magnetization. A more difficult saturation of the IRM is observed for sufficiently large H values. Figure 3.32 also shows the IRM(H) with its slower approach to reaching the already saturated TRM.

We have frequently mentioned relaxation and time dependences of the magnetization within the frozen state. Let us now examine these in some detail. With the presently available, highly sensitive SQUID magnetometer, it is quite easy to detect very small changes (10–100 ppm) in magnetization. What is more difficult is to prepare the sample magnetically for a meaningful and reproducible experiment. We treat here a few simple well-defined examples, which have created a great deal of interest and introduced a totally new concept, viz., the waiting or ageing time, t_w.

Magnetization of a spin glass will react in time to any change of external field while in the frozen state. There are various routes to prepare a spin glass for systematic measurements of these long-time relaxational effects. Since the behaviour is irreversible and complicated by ageing processes (see

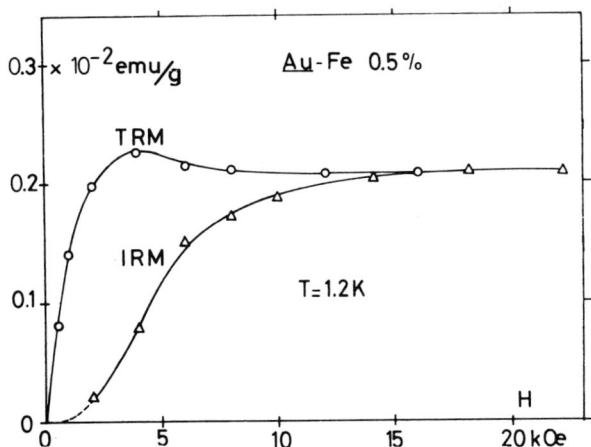

Fig. 3.32 Field dependences of the thermoremanent magnetization (TRM) obtained after cooling from T_i ($\gg T_f = 5.4$ K) to $T = 1.2$ K in a field H, and of the isothermal remanent magnetization (IRM) obtained when a field H applied at 1.2 K is suppressed; from Tholence and Tournier (1974).

below), it is imperative to employ a well-defined H–T cycling procedure. Let us consider two such preparation recipes as sketched in Fig. 3.33. The first utilizes field cooling through T_f in a weak field, and after the FC, one waits a time t_w before reducing the field to zero and measuring $M(t)$. Alternatively, we can ZFC (or T-quench) the spin glass and wait t_w before applying a small field, see Fig. 3.33 b. Afterwards $M(t)$ is tracked over many decades of time for different t_w. Note in both cases that the waiting time begins running after the sample is cooled to its final temperature $T < T_f$ and stops at the change of field when the measurement time t starts. There is also a field-quench route whereby a very strong field is applied in the frozen state, then switched off; and after t_w, a weak field is restored and $M(t)$ measured. Here the behaviour is similar to the ZFC or T-quench method.

Figure 3.34 shows some typical data for CuMn (10 at. %) after the ZFC procedure of Fig. 3.33 b (H is usually limited to fields less than 20 gauss). The base line of the figure is arbitrary and the relative scale is given in units of the FC-magnetization divided by H. It is interesting how $M(t)$

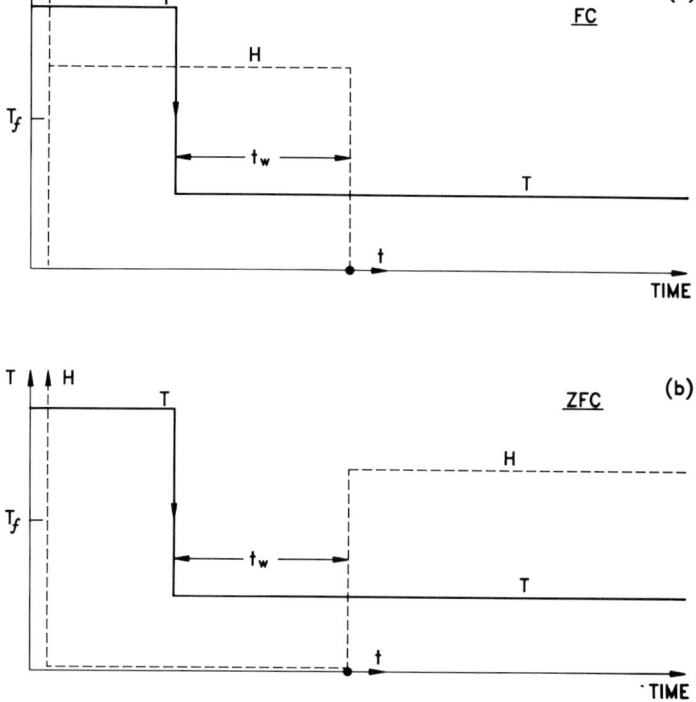

Fig. 3.33 (*a*) FC-TRM $M_R(t)$ measurement: solid line T-profile, dashed line H-profile; (*b*) ZFC or T-quench $M(t)$ measurement: solid line T-profile; dashed line H-profile.

3.3 *Low-temperature ($T \ll T_f$) experiments* 95

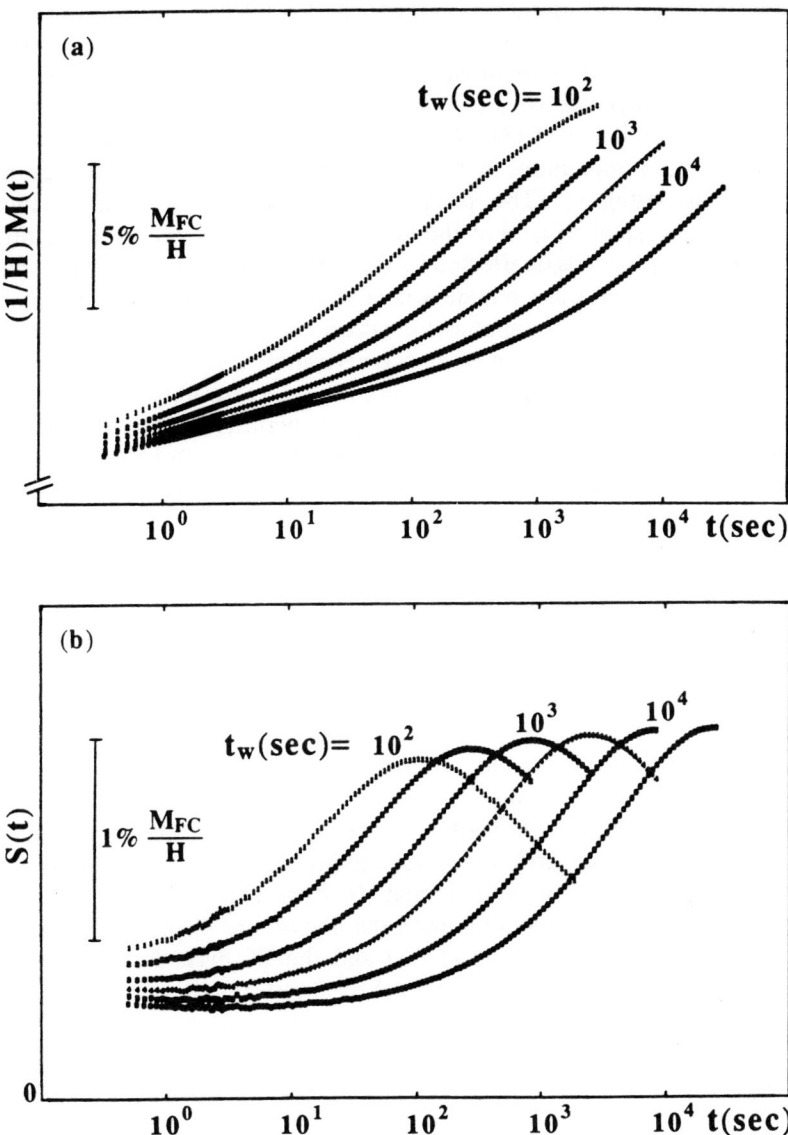

Fig. 3.34 ZFC susceptibility $(1/H)M(t)$ and the corresponding relaxation rate $S(t)$ at different wait times, $t_w = 10^2$, 3×10^2, 10^3, 3×10^3, 10^4 and 3×10^4 sec plotted versus log t: (*a*) $(1/H)M(t)$, 5% of $(1/H)M_{FC}$ is indicated; and (*b*) $S(t)$, 1% of $(1/H)M_{FC}$ is indicated. *Cu*Mn (10 at. %), $T/T_f = 0.91$, $T_f = 45.3$ K. From Sandlund *et al.* (1988).

always increases, has an inflection point, and is a strong function of t_w. The latter behaviour is due to the ageing processes of the spin glass below T_f which are continuing even though $H = 0$. These are then reflected in the different time dependences of $M(t)$ after each t_w. The lower plot gives the relaxation rate $S(t) = (1/H)\partial M/\partial \ln t$. Here the inflection point of $M(t)$ corresponds to a maximum in $S(t)$ which shifts to longer observation times (t) with increasing t_w. The peak in $S(t)$ occurs at a time t which is approximately equal to t_w. For the small fields used in the experiments the ageing process proceeds unaffected by the field. Also for small H ($<$ 20 gauss) there is a linear-response regime where the data scales as $M(t)/H$ and the principle of superposition applies for different field-cycling procedures. For example, the isothermal remanent magnetization (IRM) may be measured by ZFC, applying a field after a waiting time t_w, returning the field to zero at t'_w and then tracking IRM(t). The same quantity can be calculated by applying a $+H$ at t_w and superimposing a $-H$ at t'_w, and then subtracting the two measured $M(t)$s. Needless to say the agreement for the decay of IRM(t) from direct measurement and this subtraction is excellent.

Various functional forms have been proposed to describe the magnetization as a function of observation time and waiting time. One of the most popular relations is the stretched exponential

$$M(t) = M_o \exp[-(t/t_p)^{1-n}] \qquad (3.60)$$

where M_o and t_p (the time constant) depend upon T and t_w, while n is only a function of T. If $n = 0$, we have the Debye, single time-constant, exponential relaxation. On the other hand, for $n = 1$, $M(t)$ is constant. So the value of n will critically govern the exact relaxation rate from very strong to none at all. In order to verify such dependences sensitive experiments must be carried out over many decades of observation time and waiting time for a well-defined H–T procedure. Some illustrative measurements, which have attempted the above, are displayed in Fig. 3.35. Once again we have $CuMn$ (5 at. %) and the FC-cycle has been used to create a TRM. In fact, $M_R(t)$ is tracked for different waiting times and temperatures over four decades of observation time (1–10^4 s). This is indicated by the bold solid line in the figure. Next an attempt is made to accommodate the data with a stretched exponential using a two parameter (t_p and n) fit of the above equation. The results of the best-fit are denoted by the fine solid lines. While the agreement is good at short times, clear deviations (see Fig. 3.35) commence as the time interval is extended. In all four cases the stretched exponential function is insufficient to match the data over a broad range. Something more is needed and this has been provided by the Uppsala group, Nordblad et al. (1986). Superimposed upon the stretched-exponential contribution is a purely logarithmic-decay term

$$M'_R = SH \ln t \qquad (3.61)$$

where S is the relaxation rate in 'dynamical equilibrium' which only weakly

3.3 Low-temperature ($T \ll T_f$) experiments

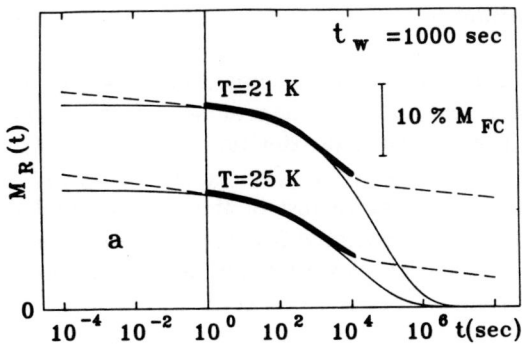

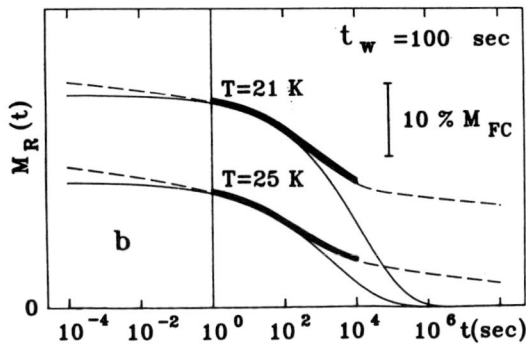

Fig. 3.35 Relaxation of the remanent magnetization (M_R) in the time interval $10^{-4} < t < 10^8$. Thick full lines are experimental data. Dashed lines are the relaxation according to power-law plus stretched-exponential dependences. Thin full lines represent the stretched exponent alone: (a) $t_w = 10^3$ sec., temperatures 21 K (top) and 25 K (bottom; and (b) $t_w = 10^2$ sec., temperatures 21 K (top) and 25 K (bottom); from Nordblad et al. (1986).

depends upon the time and waiting time. With the addition of equation (3.61) to equation (3.60) the fitting is greatly improved as illustrated by the dashed line in Fig. 3.35. Therefore, the time decay of $M_R(t)$ is logarithmic for $t \ll t_w$ and $t \gg t_w$, and the ageing process superimposes a stretched-exponential term on the relaxation around $t \approx t_w$. At this point a bit of theoretical interpretation is required to distinguish the quasi-equilibrium dynamics represented by the $\log(t)$ decay at very long or very short time scales from the non-equilibrium dynamics causing an ageing-time dependent, stretched-exponential behaviour spanning $t \approx t_w$. We shall defer such explanations until the theoretical chapter (section 5.9) when we will introduce a domain-growth model.

3.3.2 Specific heat

Here we can return to the low-temperature portion of our previously discussed measurements on the magnetic specific heat, e.g., Figs. 3.5, 3.6 or 3.21. Below T_f to a first approximation the temperature dependence of C_m in these figures seems linear. This means that C_m/T versus T should stay constant in a more severe test of the experimental data. For CuMn and (EuSr)S there is a substantial low-temperature regime where C_m/T is indeed constant. Yet at the lowest temperatures definite deviations occur particularly in CuMn. A better example of the linear dependence is AuFe (Fig. 3.36) which shows a 1 at. % sample with only a very slight deflection around 0.4 K from the line extending down from $T_f = 8.5$. A positive curvature must take place to force C_m to zero as $T \to 0$ (3rd law of thermodynamics). In this particular case less than half the entropy $S_m = cR \ln (2S + 1)$, where $S = \frac{3}{2}$ for Fe, appears between $T = 0$ and T_f. Such is consistent with our high-temperature C_m conclusion that most of the available entropy is already lost above T_f.

This strong (recall there will always be a weak γT-term for a normal metal) linear T-dependence of the *magnetic* specific heat at low temperatures (known already since the early 1960s) marked the starting point of the spin glasses in the early 1970s. For, then it was discovered that real glasses or amorphous solids also had an excess specific heat with respect to their

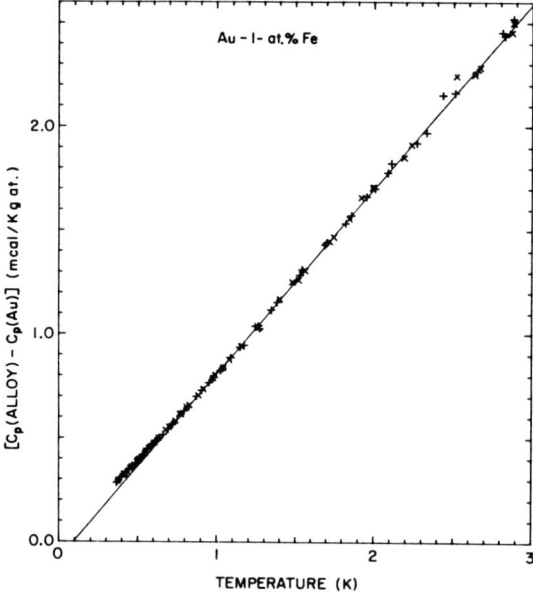

Fig. 3.36 Plot of the spin-glass specific-heat data fro AuFe (1 at. %). Note that the linear part extrapolates to a positive temperature intercept. From Martin (1980).

3.3 Low-temperature ($T \ll T_f$) experiments

crystalline counterparts that was directly proportional to the temperature. And this led to the now-famous two-level tunnelling model. At low temperatures, where thermal activation is unimportant, various positional or spin configurations in an amorphous material or spin glass are separated in phase space by energy barriers with similar energy minima. Figure 3.37 gives a sketch of a two-level energy diagram with a tunnelling barrier separating the two nearly equal energy minima. Also shown is the constant density of states for the two-level excitations. These configurations exist in a spin glass because of the frustrated, degenerate nature of the ground state. Hence, quantum-mechanical tunnelling *through* the barrier may occur between the nearly similar spin clusters – a small rearrangement or partial reorientation of some spins modifies the configuration by a tunnelling process which is temperature independent. The two slightly different configurations represent the two levels, and assuming a constant density of level states (see Fig. 3.37), a linear proportionality at low T naturally arises

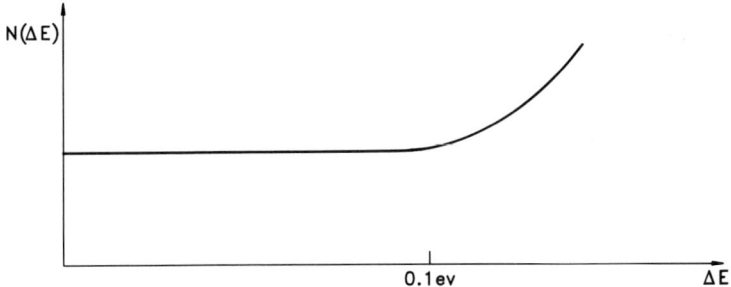

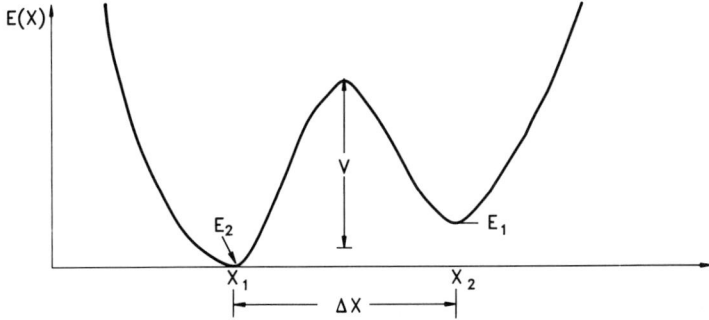

Fig. 3.37 Two-level energy-barrier construction and density of states at very low energies $\Delta E = E_1 - E_2$.

for the excess 'amorphous' specific heat or for the randomly frozen spin-glass C_m. It was this connection with the real glasses via the similar specific-heat behaviours that caused the word 'glass' to be attached to the random magnetic systems whether metallic or insulating. So an approximately linear specific heat is the low temperature signature of a spin glass.

Since the specific heat is in principle a rather difficult measurement, it has not been fully used to explore the dynamic of spin glasses by introducing a frequency or time of measurement into the technique. A related experiment can be performed to determine the energy flux (or heat flow) dQ/dt for $T \ll T_f$ as the external field is changed. Here the spin-glass sample is weakly connected to a reservoir by a constant heat leak. For both FC and then setting $H = 0$ or ZFC and then turning on a field H, heat flows *out* of the sample with a long-time relaxation (similar to the magnetization). At very low T and for a given ΔH, the energy flux $dQ/dt \propto 1/t$ and the energy $Q = $ (const) ln t. One can picture this behaviour as a lowering of the spin-glass energy by a change of field, either removal after FC or application after ZFC. Such alterations drive the system slowly towards a new equilibrium state. A more complete interpretation may be gained from the two-level tunnelling system with a distribution of both energy-barrier heights P(V) as in our Néel superparamagnetic model and splittings of the double-well (nonequal depths) P(ΔE) caused by the field change ΔH (see Fig. 3.37). Note that the tunnelling processes are essential at these low temperatures ($T \ll T_f$), since thermal activation (or surmounting the barrier) is mainly ineffective. Therefore, we maintain our nexus among relaxation, energy barriers and tunnelling, all participating in the quest for equilibrium.

3.3.3 Resistivity

At low T the magnetic resistivity of a metallic spin glass, $\Delta\rho$, has a large residual ($T = 0$) component – a combination of the disorder of the alloy due to random A and B site occupancies, and the spin disorder of the randomly frozen state. We call this contribution $x\Delta\rho_o$ since these disorders are roughly proportional to x, the concentration. As the temperature is increased

$$\Delta\rho(T, x) = x\Delta\rho_o + A(x)T^{\frac{3}{2}} \tag{3.62}$$

where $A(x)$ decreases very slowly with increasing concentration. We illustrate this behaviour, which seems rather general, in Fig. 3.38 for a series of CuMn alloys.

Some evidence has been offered for the appearance of a limiting T^2 dependence at the very lowest temperatures ($T < 0.3$ K). However, this must quickly crossover to the $T^{\frac{3}{2}}$ behaviour for a larger T-range approximating T_f. The interpretation of these $A(x)$ and $T^{\frac{3}{2}}$ dependences has been discussed in terms of long-wavelength elementary excitations which are diffusive in

3.3 Low-temperature ($T \ll T_f$) experiments

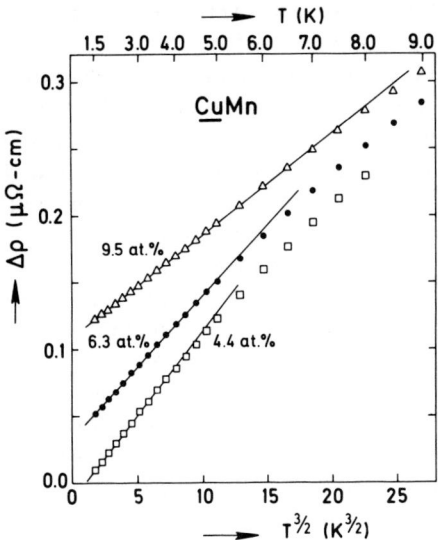

Fig. 3.38 The temperature variation of the resistivity for 4.4, 6.3 and 9.5 at. % Mn in Cu versus $T^{\frac{3}{2}}$. The lines through the data represent a fit to this temperature dependence. From Mydosh and Ford (1974).

character. The excitations are thought to be highly damped, non-coherent (independent), localized spin fluctuations which scatter the conduction electrons. Such a model is against propagating excitations, e.g. spin waves which occur in long-range order magnets. Once again as in the two-level system, we resort to a localized description for our frozen state. The state may be a co-operative one with correlation length, ξ_{SG}, large, but because of the disorder the excitations cannot propagate.

3.3.4 Neutron scattering

What does inelastic neutron scatter say about the above conjecture? Usually this technique provides a powerful probe for the study of collective spin excitations. If we recall our dynamic structure factor $S_D(\mathbf{q}, \omega)$ given by equation (3.17), then we could look for propagating spin waves, namely, two inelastic lines at $\omega = \pm \omega_q$. Hence,

$$\text{Im} \chi(\mathbf{q}, \omega) = \frac{\omega \Gamma}{(\omega - \omega_q)^2 + \Gamma^2} + \frac{\omega \Gamma}{(\omega + \omega_q)^2 + \Gamma^2} \quad (3.63)$$

where Γ is the inverse lifetime of the spin wave which has a typical dispersion relation $\omega_q = Dq^2$ (with D the spin-wave stiffness constant).

Much experimental effort has gone into searching for these excitations in the metallic spin glasses – all in vain. There are no sharp inelastic lines to be found, meaning the modes are overdamped, thus localized. A similar conclusion was reached from our resistivity measurements. For the insulating spin glasses, e.g. (EuSr)S, some indication was found of broad quasi-elastic to inelastic scattering intensity as weak shoulders to the elastic central peak. The broadness implies Γ is very large, distinctive of short lifetime or highly-damped excitations. Here the experimental and analytic difficulties become so great that we can only conclude that propagating spin waves do not, in general, exist in the canonical spin glasses.

More recently a quite radical interpretation has been proposed for the various neutron scattering results in the noble-metal-based spin glasses (e.g. CuMn, AuFe, etc.). We alluded to this explanation in section 3.1.4 and Fig. 3.11, where single-crystal samples must be used to determine the exact place in reciprocal lattice (**q**) space from which the scattering occurs. So let us consider in more detail the two relatively sharp peaks in Fig. 3.11 which emerge at low T out of the broad in q-space magnetic scattering. These peaks, called satellites, appear at **q**-values of $(1, 0, \frac{1}{2} \pm \delta)$ and other symmetry-related positions. In Fig. 3.11 δ takes the value of 0.19. The position indicates that there is a modulation, λ, of spin correlations which has a period longer than the unit-cell spacing ($a_o = 3.6$ Å for Cu), namely $\lambda = a_o/(\frac{1}{2} - \delta) = 11.6$ Å. The halfwidth Γ of the satellite peaks is not resolution limited – they are not true Bragg peaks. This means the modulation of spin correlations is not a long-range propagating sinusoidal wave with period λ. In other words, its correlation length ξ is finite and may be estimated from Γ. For the case of q-dependent peaks a gaussian fit is usually made to determine the halfwidth in units of $2\pi/a_o$. Hence $\xi = 2\sqrt{\ln 2}/\Gamma \approx 6a_o$ or 22 Å dependent on the concentration x of Mn or Fe. This corresponds to a domain of roughly 2000 atoms of which the fraction x are magnetic impurities.

We call this incommensurate (having a different periodicity λ than that of the lattice which is an integer times a_o) modulation of spin correlations a spin density wave (SDW). A standard example of an SDW material is the antiferromagnetism of Cr, others are the helical ordering of certain rare-earth elements or alloys. SDW represents a collective, excited state of the conduction-electron gas with an energy greater than the ground-state energy. This higher energy can preclude its formation in a pure non-magnetic metal, e.g. Cu or Au. However, in a dilute magnetic alloy at low T, the orientation of the solute spins in the effective field of the SDW provides an interaction energy which may more than compensate the higher energy of the SDW. Here the spin polarization of the electron gas **s** varies with position **R** as

$$\mathbf{s} = Nb\boldsymbol{\epsilon} \cos(\mathbf{q} \cdot \mathbf{R}) \tag{3.64}$$

where N is the number of atoms per cm^3, b is the order-parameter amplitude of the SDW, $\boldsymbol{\epsilon}$ its polarization and **q** its wave vector. Paramagnetic solute spins, e.g. Mn or Fe, will interact with the SDW via s-d (or s-f for the rare

3.3 Low-temperature ($T \ll T_f$) experiments

earths) exchange interactions. Hence, a solute spin $\mathbf{S}_j$ at $\mathbf{R}_j$ will experience an effective field

$$\mathbf{H}_j = Gb\epsilon \cos(\mathbf{q} \cdot \mathbf{R}_j) \qquad (3.65)$$

where G is related to the exchange interaction between conduction electrons and solute spins. The interaction energy is $-\mathbf{H}_j \cdot \mathbf{S}_j$. It tries to orient the solute spin along the effective field, and thereby compensate for the positive energy necessary to form the SDW. Upon removing the thermal disorder by going down to low T, b, the SDW order parameter, will become non-zero at an 'antiferromagnetic' critical temperature T_N.

This phase transition had already been derived by Overhauser in the late 1950s and within the mean-field approximation T_N and the various thermodynamic quantities may be calculated. The Overhauser theory leads to long-range antiferromagnetic order via a second order phase transition of the randomly distributed solute spins according to the SDW polarization $s(\mathbf{R})$. The ordering wave vector $\mathbf{q}$ is usually incommensurate with the lattice, i.e., unequal to the reciprocal lattice vectors $\mathbf{G}$, since a SDW is a Fermi-surface effect, most favourable when $q = 2k_F$ (Fermi momentum) unrelated to the lattice.

Now, what are the neutron-scattering results telling us about the possible SDW in CuMn et al.? First of all, the ordering $\mathbf{q}$-wave vectors have been determined by the satellite peaks which clearly emerge below T_f. However, the linewidths Γ of these peaks remain too broad for a long-range propagation of the SDW. In CuMn, our example in Fig. 3.11, the sinusoidal propagation extends only about 20 Å (its correlation range) then breaks off and with different amplitude and direction a new SDW begins and traverses ≈ 20 Å before the process is repeated. We sketch this short-range (SR) SDW in Fig. 3.39 (lower part) where there is an additional complication due to the formation of very short-range ferromagnetic clusters embedded within the SR-SDW. That the chemical or atomic short-range order leads to these ferromagnetic clusters has also been established from the polarized neutron-scattering experiments and its separation into nuclear and magnetic scattering, so, the 'chunks' of ferromagnetic spins simply follow the SDWs sinusoidal polarization of the electron gas.

As a basis for comparison we have shown the results for a typical long-range (LR) SDW (linearly polarized) in Fig. 3.39 (top part). The alloy YGd is similar regarding concentration, but it has a preferred SDW propagation direction along the c-axis of its hexagonal crystal structure. Here the correlation range $\xi \to \infty$ and $\lambda \approx 20$ Å with an incommensurate periodicity. The Gd 7 μ_B moments align themselves in the a–b plane according to the modulation of the SDW.

This picture of a SR-SDW is entirely consistent with the arbitrary $\mathbf{q}$-value scattering on polycrystalline samples discussed in section 3.2.4. Then the model was a distribution of relaxation times that comes from the fluctuations of the ferromagnetic clusters which are a direct result of the chemical SR

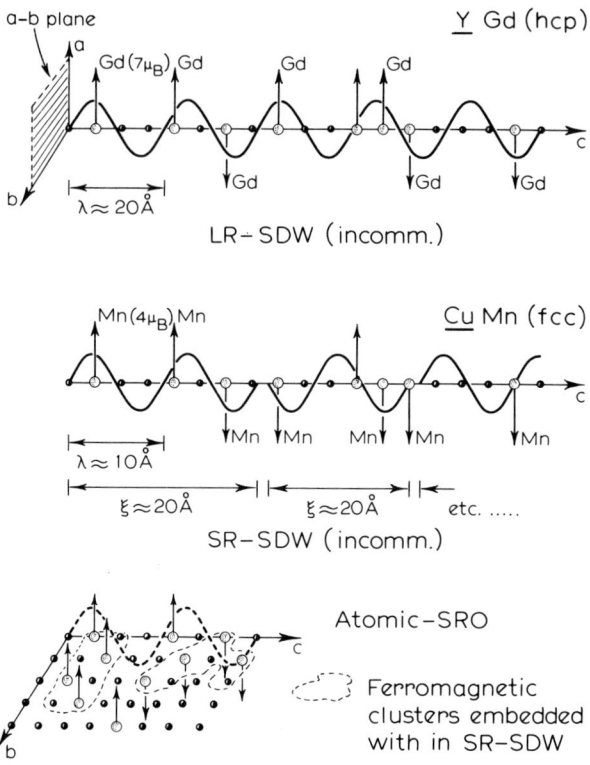

Fig. 3.39 Schematic of linear SDW propagation, upper part: long-range for YGd and lower part: short-range for CuMn. From Mydosh (1988).

order. The satellite magnetic peaks can only be observed on single-crystal samples with a separation of nuclear and magnetic scattering. And indeed our clusters do slow down and become static as the SDW forms below T_f. We conclude by mentioning that the thermodynamic and transport properties of our LR-SDW (YGd) have all the 'sharp' behaviours of a good second-order, long-range phase transition. There are *none* of the time dependences, irreversibilities, remanences, etc. characteristic to the spin-glass state. Why a LR-SDW does not form in CuMn, AuFe etc. has to do with a number of specific reasons, such as the weak polarizability of the noble-metal matrix, the more itinerant 3D moments, the strengths of the RKKY interaction between the moments and the cubic crystal structure which by symmetry allows twelve different propagation directions each associated with a specific **q**-vector. For these reasons the spin-glasses become unique because they are disordered at long range and this makes them so very difficult to describe, even if we realize an incipient SDW is starting to form.

3.3 Low-temperature ($T \ll T_f$) experiments

3.3.5 Torque and ESR

According to our experimental model, a random anisotropy should appear in the frozen spin-glass state. For metallic systems the origin is thought to be the Dzyaloshinsky–Moriya interaction, while for the insulators the dipole interaction seems to provide the random anisotropy (see section 1.4). In any case, the problem with a random (both in magnitude and direction) and local anisotropy is to measure it, since on a macroscopic scale it would average out to zero. As was seen from the magnetization studies, we can via FC or ZFC, then a field, induce a remanent magnetization M_r along the field direction. The field, in whatever direction it is applied, will probe the local random anisotropy for each frozen spin by trying to rotate it. If the spin-glass state is a cooperative, collective one, i.e., all spins locked into a 'rigid body', then the rotation of this *stiff* spin system will be governed by a single macroscopic anisotropy parameter K. But how can we probe the 'stiffness to rotation' of our rigid collection of randomly frozen spins? The trick is to use the induced remanence M_r as a handle fixed in our spin system and then to turn M_r with the external field. Yet M_r and K are independent quantities. Usually a combination of unidirectional (K_1) and uniaxial (K_2) anisotropies are required to describe the magnetization results for a displaced, broad hysteresis loop. Thus, in general, the total anisotropy energy may be written for rotation Θ as

$$E_A(\Theta) = K_1(1 - \cos \Theta) + \tfrac{1}{2} K_2 \sin^2 \Theta \tag{3.66}$$

A very effective way of investigating the anisotropy is through torque measurements. The torque Γ is defined as the vector product of magnetization times external field, or for field-prepared spin glasses

$$\Gamma = M_r H \sin(\Theta_H - \Theta) \tag{3.67}$$

where Θ_H is the rotation of H, and Θ is the rotation of M_r from their original (FC) direction. With respect to the energy

$$\Gamma = \frac{dE_A}{d\Theta} \tag{3.68}$$

The measurement of rotation or torque may be carried out very accurately by determining the corresponding displacements of two metal plates using a capacitance bridge. For simplicity, we choose a sample (large Mn concentration in Cu) and a field-cooling procedure such that K_2, the uniaxial component, is negligible. With our applied field along the z-axis, the FC direction, we have induced a value of M_r, which at the very low temperatures of the measurement remains roughly time-independent, and created a macroscopic value for K_1 by our rigidly frozen spin system. Next we rotate the field **H** about the x-axis, this in turn drags along the M_r at some delay angle and we can measure the torque. This process is sketched in Fig. 3.40

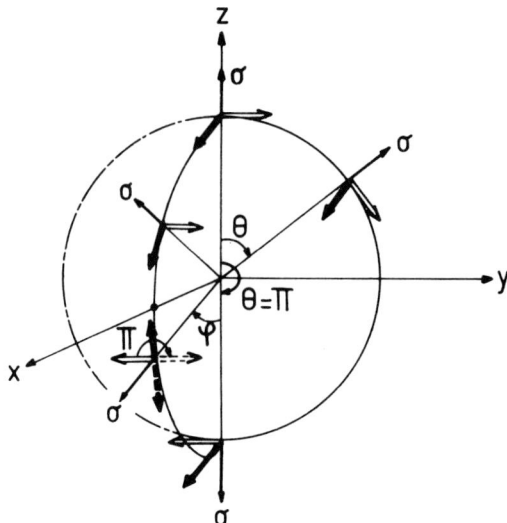

Fig. 3.40 Rotation of the anisotropy triad during a torque experiment. A rigid rotation of the spin system is induced from $\Theta = 0$ to $\Theta = \pi$ when H is rotated in the y–z plane from $\Theta_H = 0$ to $\Theta_H = \pi$. When a subsequent rotation φ is induced in the x–y plane, the total rotation angle of the triad is still π with respect to the $\Theta = 0$ starting position. However, a non-controlled irreversibility π rotation of the anisotropy triad around σ brings back the rotation angle to $\pi - \varphi$. From Alloul (1983).

where σ is used to represent M_r, and Θ is the angle of rotation of M_r with respect to the z-axis. The field angle Θ_H is somewhat larger due to the 'pinning' of M_r by the unidirection anisotropy K_1.

Figure 3.41 shows the measured torque Γ_x versus the field-rotation Θ_H in curve (a), while the calculated torque derived from equations (3.66) to (3.68) is also plotted as a dashed line. Notice how the two curves nicely agree until about $\Theta_H \approx 50°$ (or $\Theta \approx 40°$). The measured torque is here reversible and independent of time. This means a rigid rotation of the entire spin system is occurring with M_r. However, at larger angles small rearrangements are taking place with the corresponding irreversibilities and time dependences. Nevertheless, the deviations of the experiment from the calculated curves are not too bad (see Fig. 3.41) for the particular system, and the remanent magnetization turns over for a field rotation of 180° and comes back into phase with **H**.

Now what happens, if we rotate the applied field back to its original $+z$-axis, not via a return rotation about the x-axis, but instead via a rotation ϕ about the y-axis? This results in a Γ_y, which is measurable and shown in Fig. 3.41 as curve d. Here there is a large difference in the Γ_y values compared to Γ_x at least for the first 20° of rotation. Therefore, a return rotation about the y-axis after an x-axis rotation is non-identical to a direct

3.3 Low-temperature ($T \ll T_f$) experiments

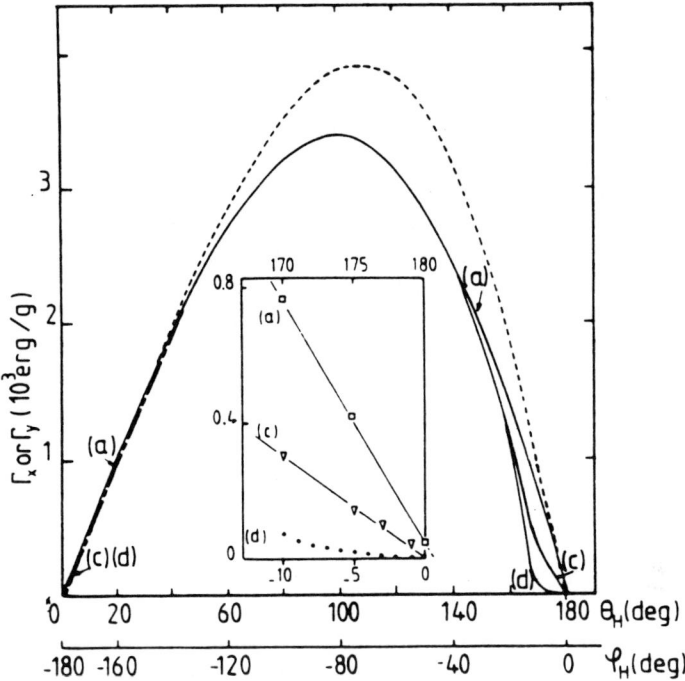

Fig. 3.41 Torque measurements in a CuMn 20% sample at 1.5 K. The torque Γ_x and Γ_y are measured after a π rotation of σ in the y–z plane. These two responses are quite different and point out the triad character of the anisotropy. Γ_y is found to be very weak for small φ as the total anisotropy energy is then φ independent. From Fert and Hippert (1982).

rotation from the FC z-axis about the y-axis and back. For, we must attach a threefold orthonormal coordinate system or *triad* to our M_r (or σ) as is done in Fig. 3.40. Note the dissimilarity between an x-axis followed by a y-axis rotation and a simple y-axis, back and forth, rotation. Although σ always points in the same direction after these rotations the two non-σ triad axes are oriented in opposite directions along the direct and return (after x) y-axis rotations. This coordinate-system difference or triad property is manifested in the two unlike values of Γ_y measured around 180°.

Our external field controls the rotation of the remanent magnetization M_r, but we cannot control the rotations of the collective rigid spin system around M_r – slippage will occur at large angles and the spin system will revert back to its direct-rotation and lower-energy form and $\Gamma_y = \Gamma_x$. Since the random freezing is isotropic, our spin system comprises all orientations in the 3D space, i.e., is non-collinear or Heisenberg-like. This means the random local anisotropy is also distributed over all directions as an 'isotropic anisotropy'. When we probe the macroscopic anisotropy via a FC-field, we establish the 'memory' direction $\hat{N}$ of our fixed triad independent of the crystal axes. Moreover, we have also induced M_r and its triad vector $\hat{n}$ which rotates with M_r as we turn $M_r\hat{n}$ away from the FC orientation by applying and rotating the external field $\mathbf{H}$. With the proper choice of system, temperature and fields, we can only hope that the randomly frozen spin system remains rigid and moves bodily and in phase with M_r which is being controlled by $\mathbf{H}$. This seems to be roughly true for the x-axis rotation in Fig. 3.41, but, if followed by a y-axis rotation, the spin system readjusts or inverts two of its triad axis and returns to the torque and lower energy of the original x-axis rotation. All these contrasting processes are illustrated in Fig. 3.40 using the triad coordinate that turns with $\mathbf{M}_r$ (or $\boldsymbol{\sigma}$). Note the striking difference with a collinear (Ising) spin system where all of the randomly frozen spins point either 'up' or 'down' along a given unit-vector axis. For this 'vector model', unidirectional anisotropy only, there is no difference in the spin system after the various axes rotations. It remains completely the same no matter which path is followed. Thus, we would expect to see a dramatic difference in the rotational properties between a Heisenberg and an Ising (collinear) spin glass.

Another powerful experimental technique which can be used to study the consequences of the spin-glass anisotropy is electron spin resonance (ESR). Here one detects the spin resonance frequency of the Mn atoms. It has long been known (late 1950s) that Mn impurities in a noble metal satisfy the condition for ESR and many interesting physical properties can be derived from such experiments in the dilute limit. But now we wish to investigate the co-operative behaviour of the frozen Mn spin system for $T \ll T_f$. For this purpose we can apply a hydrodynamic approach and construct an appropriate free energy. Hydrodynamic theory represents a collective approximation at low frequency and long wave-length using macroscopic variables, and as a result it neglects the microscopic details of the systems under consideration. For a canonical spin glass we have a non-collinear, isotropic, frozen, spin structure with its magnetization, $\mathbf{M}(= \chi\mathbf{H})$, a possible remanent magnetization M_r and the unidirectional (K_1) and uniaxial (K_2) anisotropies. Accordingly, the free energy (FE) becomes

$$\text{FE} = \frac{1}{2\chi}(\mathbf{M} - M_r\hat{n})^2 - \mathbf{M} \cdot \mathbf{H} - K_1 \cos\Theta - \tfrac{1}{2}K_2 \cos^2\Theta \qquad (3.69)$$

where the notation is similar to before. The triad coordinates must be used with $\hat{N}$ the FC direction and $\hat{n}$ the rotated M_r direction. One can check this FE by minimizing it with respect to Θ and obtain the equilibrium rotation angle Θ_o after field cooling H_c and rotating the external field $\mathbf{H}$. Substituting $\mathbf{M}_o = \chi\mathbf{H} + M_r\hat{n}$, as shown in Fig. 3.42, we arrive at the equilibrium condition with $K_2 = 0$:

$$\sin\Theta_o = \frac{M_r H}{K_1}\sin(\Theta_H - \Theta_o) \qquad (3.70)$$

which is identical to that used in the torque experiments. Figure 3.42

3.3 Low-temperature ($T \ll T_f$) experiments

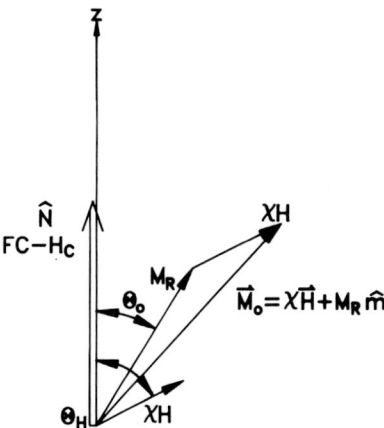

Fig. 3.42 Equilibrium rotation angle Θ_o for a field-cooled spin glass followed by a rotation of the field **H** to an angle Θ_H.

sketches the various vectors and their equilibrium angles. These conditions have been verified by several magnetization and torque measurements as long as the frozen spin system remains rigid; however, often such is not the case.

For ESR experiments more is needed. Equations of motion must be derived for the dynamical variables **M** and $\hat{n}$. This is usually carried out by forming Poisson brackets of these variables with the free energy. Let us consider a special case, namely, $M_r\hat{n}$ is parallel to **H** which is parallel to $\mathbf{H_c}$. Now three resonance frequencies occur because of the three orthonormal triad vectors. For a collinear (Ising) spin system there would be only two resonances. The solutions of the eigenvalue equations for the (3) frequencies are

$$\frac{\omega_L}{\gamma} = \sqrt{\frac{K_1}{\chi}} \quad \text{and} \quad \frac{\omega_\pm}{\gamma} = \pm \tfrac{1}{2}\left(H - \frac{M_r}{\chi}\right) + \tfrac{1}{2}\left[\left(H + \frac{M_r}{\chi}\right)^2 + 4\left(\frac{K_1}{\chi}\right)\right]^{\frac{1}{2}} \quad (3.71)$$

where γ is the gyromagnetic ratio. Taking the simplest case of $M_r = 0$, i.e. ZFC, the two $\omega_\pm$ equations reduce to

$$\frac{\omega_\pm}{\gamma} = \pm \tfrac{1}{2}H + \tfrac{1}{2}\left(H^2 + 4\frac{K_1}{\chi}\right)^{\frac{1}{2}} \approx \left(\frac{K_1}{\chi}\right)^{\frac{1}{2}} \pm \tfrac{1}{2}H \quad (3.72)$$

if $H \ll K_1/\chi$. The first ESR experiments, shown for this special case in Fig. 3.43, gives good agreement for the ω_+ mode (linear slope $\approx \tfrac{1}{2}$ and expected intercept) even with resonance fields as low as 300–400 gauss. The predicted situation for the three modes develops according to Fig. 3.44, where the longitudinal mode ω_L remains at a fixed frequency (no dispersion), thereby making it difficult to find. This mode was finally detected indirectly

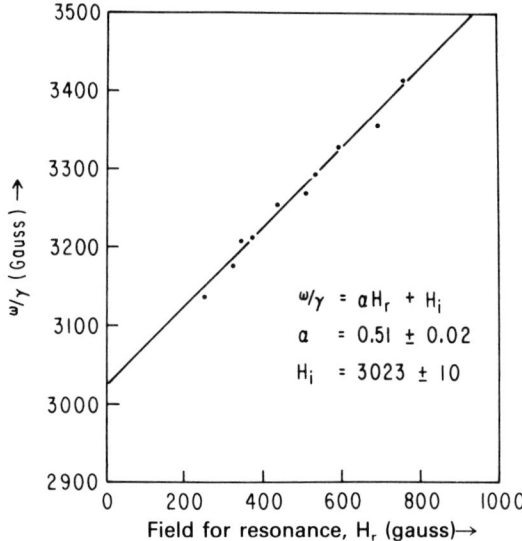

Fig. 3.43 ESR (spectrometer frequency versus field for resonance) for a ZFC spin glass, the resonance corresponds to the ω_+ mode according to the equation given in the text. From Schultz et al. (1980).

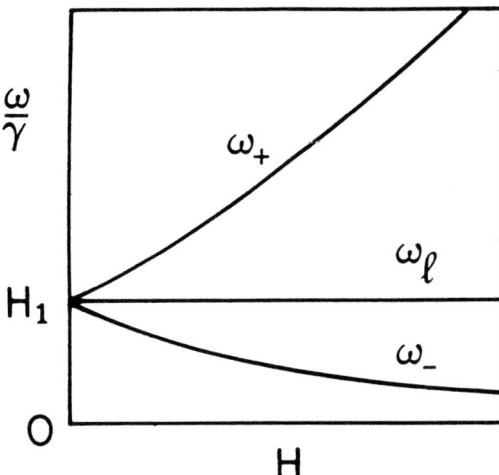

Fig. 3.44 A plot of the three expected resonance modes as a function of H for zero remanence.

by using the mode-crossing repulsion phenomenon which divaricates two distinct frequencies for a given field-for-resonance at a particular rotation angle of M_r. There were several experimental attempts to verify the various angular dependences of the different ESR models. Reasonable results in

3.4 Spin glasses in a field

accord with hydrodynamic theory were mostly gathered, provided the spin system rotated bodily, that is, stayed rigid. At large angles without a proper preparation the *Cu*Mn spin system usually broke apart and along with it the hydrodynamic description. The crowning achievement of all this low-temperature effort was to demonstrate the co-operative nature of the spin-glass phase. Clearly a new and unique state exists in the frozen spin glasses, so we can conclude that there must be a type of phase transition which transforms the system from the paramagnetic state at high temperatures to the collective entity at $T \ll T_f$. And this state cannot be a mere assortment of superparamagnetic particles or domains.

3.4 SPIN GLASSES IN A FIELD

Now that we have surveyed many of the important experimental results in the three temperature regimes of a spin glass, we should briefly explore what happens to a spin glass in a large applied field. The only regime of real interest is that spanning T_f, since not much will be different at high temperatures – only more deviations from paramagnetism and we have already considered in the previous section the effects of a large field on the frozen state. While there are a certain group of theoretical predictions which give an H–T_f phase diagram, let us first view the problem experimentally.

We know already from the ac-susceptibility that the cusp is smeared out even in a small field. The broad maximum of the specific heat above T_f is made even less visible by the field-generated spreading. The various relaxation times just below T_f seem to increase and become faster as a larger field is applied and the system seeks its new equilibrium state. One conclusion of these experiments would be that the field removes the criticality of the phase transition, yet it does not fully prevent the formation of the frozen state, although with a useful handle – the remanent magnetization. This means there are strong driving forces, e.g. competing exchange and random anisotropy, whose average energies are greater than that created by the field. Nevertheless, when the phase transition is trying to take place at T_f the field can have its potent effect.

In order to avoid all the complications with relaxation times and irreversibilities, a good experiment to perform is that of FC. We saw in section 3.2.1 that the FC magnetization displayed a clear kink in the plateau at T_f and time dependences were minimal. For certain of the canonical spin glass even a small peak develops with FC at T_f. In Fig. 3.45 a plot is given of the *inverse* magnetization divided by FC field (H/M) versus the temperature. At small fields < 100 gauss the inverted cusp denotes T_f (similar value to that obtained by χ_{ac}). However, note how the reverse peak rapidly disappears with increasing field. Now we must employ another criterion to determine T_f, for example, the onset of the plateau, as

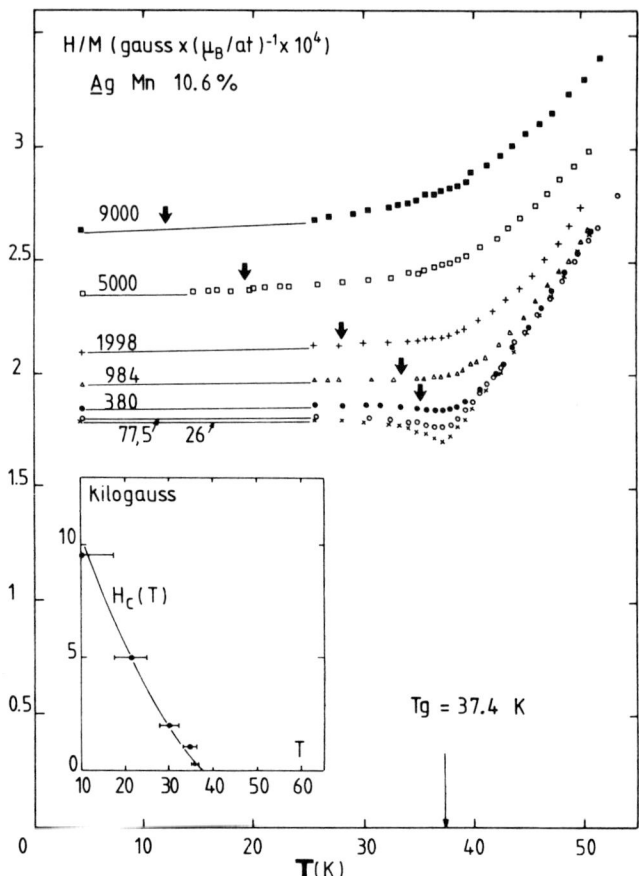

Fig. 3.45 Inverse of the FC-susceptibility (H/M) for AgMn (10.6 at. %) as a function of temperature for various magnetic fields (indicated on each curve in gauss). Data were obtained by slow cooling in constant applied field. The onset of the 'plateau' (marked by arrows) is taken arbitrarily as the point of the $M(T)$ curve departing by 3% from its low-temperature value thereby defining T_f. The resulting boundary of the spin-glass phase $H_c(T)$ is shown as an inset. From Monod and Bouchiat (1982).

represented by the arrows in Fig. 3.45. These not only shift downward in T with increasing H, but become greatly smeared, an estimate of which is given in the figure's inset where an H–T phase diagram is attempted, albeit with large error bars. At larger fields T_f is simply not well defined.

If we use other measurement techniques, first of all there are significant differences in establishing an $H(T_f)$ line between the methods and analytic procedures. And secondly, the time scale of the measurement enters and gives different forms for $H(T_f)$ depending on the experimental time window. Hence, the dynamics of the transition are playing the more important role

3.5 Experimentalist's (intuitive) picture

and a static $H(T_f)$ phase diagram seems impossible to generate based upon general experimental criterion. When the strong influence of dynamics is coupled with the large smearing effects on a so-called critical temperature, the whole meaning of a line of phase transitions created by a field becomes moot.

3.5 EXPERIMENTALIST'S (INTUITIVE) PICTURE

We summarize this lengthy chapter with an experimentalist's model. Or, in other words, how a spin-glass researcher, who has never had any contact with the theory, would interpret his measurements. Two important concepts have emerged and been repeatedly employed in our discussions of the spin-glass experiments. The first is the local correlations or 'clusters' which are detected at high temperatures $(T > T_f)$. These represent the 'building blocks' of spin for the spin-glass state. The second is our need for not just a single relaxation time, but a broad distribution of such times to describe the measurements. Both of these concepts go hand-in-hand, for, big clusters have long relaxation times.

Our simple intuitive picture of a spin glass is as follows. There exists compelling experimental evidence that for $T > T_f$ short-range magnetic interactions are present. Thus we divide the system up into dynamical, evolving magnetic clusters (the building blocks) which develop out of the high-temperature paramagnetic collection of spins. A competition occurs between the short-range exchanges $J(r_{ij})$ (where r_{ij} is the separation between two spins) and the disordering effect of temperature $k_B T$. When $J(r_{ij}) > k_B T$ for a given group of spins a cluster is formed. The clusters need not be fully ferromagnetic entities, they can be mostly random with a $\sqrt{n}$ small net moment. Yet, they possess a correlation length ξ_{SG} which rigidly couples together two or more spins at r_{ij} of arbitrary orientation. As T is decreased the clusters will grow in size (ξ_{SG} increases) and take on diverse shapes based upon the distribution of competitive interactions among the various spins in its immediate neighbourhood. As an illustration of this cluster behaviour, the results of a computer simulation using the $J(r_{ij})$ versus $k_B T$ model are shown in Fig. 3.46. Note the complex structures which are built via the various range contours. There is even an isolated spin that does not belong to any cluster in the lower right of Fig. 3.46.

Concomitant to this *spatial* magnetic clustering will be the *temporal* relaxation rate $1/\tau$. We can easily describe a collection of isolated non-interacting spins in terms of a single Debye lifetime. But our spin-glass situation is more complex, the various sized clusters all have different relaxation times. And clearly the temperature evolution will be towards longer and longer time scales as T is reduced to T_f. Nevertheless, our isolated spin in Fig. 3.46 will still keep its orientation fluctuating at a rapid rate, only when this spin finally joins a cluster does its turning rate slow

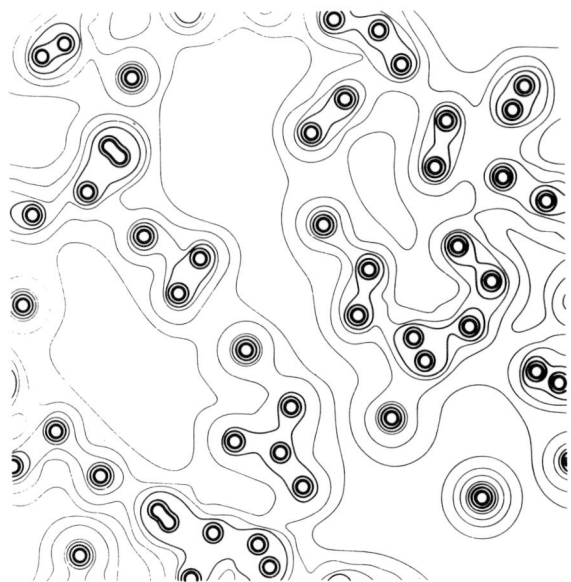

Fig. 3.46 Computer simulation for a random lattice of points (spins) with the contours and their overlap representing the different ranges of spin–spin interactions. From Verbeek *et al.* (1980).

down. Hence, there is a natural emphasis on the dynamics especially near T_f which results in a wide distribution of relaxation times shifting with temperature.

Since these time dependences are so intimately connected to the spatially rigidity, the experimental model requires both and the particular measurement techniques will highlight one or the other of the correlations. As $T \to T_f$, the random anisotropy seems to take hold and preferred orientations are established throughout the crystal. This anisotropy creates the random-freezing directions. At T_f a sort of percolation ensues which generates an infinite cluster of rigidly frozen spins. The infinite cluster comprises many, many much smaller clusters that are randomly frozen in orientations. Spins inside the infinite cluster are correlated, but they have their local direction governed by the random shape or exchange anisotropy of the smaller clusters. Therefore, the small clusters retain their original identity, however, they are no longer able to react to an external field, since they are firmly embedded or *fixed* within the infinite cluster. In general, for a percolation transition the order parameter is represented by the fraction of spins belonging to the infinite cluster. Consequently, at $T < T_f$ there are still many free spins or small clusters which behave superparamagnetically. These latter entities contribute a fast-relaxation component and cause the wide distributions of relaxation times to be also

3.5 Experimentalist's (intuitive) picture

observed below T_f. So the freezing process in our depiction is a percolation-like one, but the infinite cluster is composed of many randomly frozen smaller clusters. Quite an unusual 'mosaic-fractal' picture is required for the spin glasses in order to elucidate the preceding experimental results.

Such a phenomenological description may be carried down to the lowest temperature and used to explain the collective and co-operative properties where the anisotropy is all important. In addition, the various irreversibilities and metastabilities can occur in an inherent way within the inhomogeneously percolated state. Here thermal activation is sufficient to flip some special clusters or possibly tunnelling processes reorient other cluster configurations. Energy barriers are a natural consequence of our model and this concept nicely describes the remanence, hysteresis and long-time relaxations.

Spatial inhomogeneities and a vast spectrum of relaxation times both over a wide temperature range are the experimental conclusions. Our special percolation model seems to accommodate these deductions in a reasonable qualitative description. In such a model the freezing temperature is a function of the time or frequency of measurement. Since the infinite cluster is a dynamical entity consisting of weak and strong links with many fluctuations and excitations, it will appear percolated according to the particular time window in which it is viewed. Thus, with a very rapid measurement time the slowly occurring breaks in the ∞-cluster will not be observed and T_f will be discerned at a higher temperature than by slower measurements. We can represent this schematically according to Fig. 3.47. In the limit of very long times of observation, a single T_{fo} or critical temperature T_c should be measured without any time dependences.

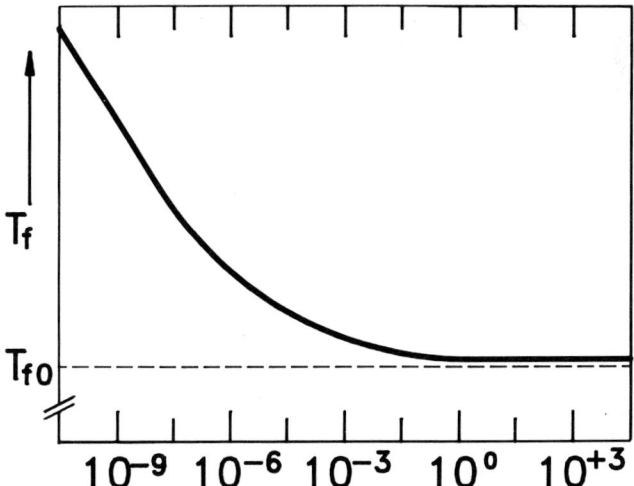

Fig. 3.47 Schematic representation of the freezing temperature as determined by different measurement techniques having different 'time constants' of measurement.

This then would indicate that an equilibrium state has been reached. Experimentally, it is very difficult to know if T_{fo} has been attained, one can always try for another decade of time, and see if there is a small change in a measurable quantity. Most likely, there is and this then makes the question of an equilibrium phase transition unresolvable. How can we rid ourselves of these dynamical processes to determine the basic properties of the underlying phase transition? At present an answer has not yet appeared.

REFERENCES

Alloul, H. (1983) *Heidelberg colloquium on spin glasses*, Vol. 192 Lecture Notes in Physics (eds J. L. van Hemmen and I. Morgenstern), Springer, Berlin, 18.
Brodale, G. E., Fisher, R. A., Fogle, W. E., Philips, N. E. and van Curen, J. (1983) *J. Magn. Magn. Mater.*, **31–34**, 1331.
Cable, J. W., Werner, S. A., Felcher, G. P. and Wakabayshi, N. (1984) *Phys. Rev. B*, **29**, 1268.
Cannella, V. and Mydosh, J. A. (1972) *Phys. Rev. B.* **6**, 4220.
Fert, A. and Hippert, F. (1982) *Phys. Rev. Lett.*, **49**, 1508.
Ford, P. J. and Mydosh, J. A. (1976) *Phys. Rev. B*, **14**, 2057.
Hüser, D., Wenger, L. E., van Duyneveldt, A. J. and Mydosh, J. A. (1983) *Phys. Rev. B*, **27**, 3100.
Levy, L. P. (1988) *Phys. Rev B*, **38**, 4983.
Martin, D. L. (1980) *Phys. Rev. B*, **21**, 1902.
Meschede, O., Steglich, F., Felsch, W., Maletta, H. and Zinn, W. (1980) *Phys. Rev. Lett.*, **44**, 102.
Monod, P., Préjean, J. J. and Tissier, B. (1979) *J. Applied Phys.*, **50**, 7324.
Monod, P. and Bouchiat, H. (1982) *J. Phys (Paris) Lett.*, **43**, 145.
Morgownik, A. F. J. and Mydosh, J. A. (1982) *Physica*, **107B**, 305.
Morgownik, A. F. J. and Mydosh, J. A. (1981) *Phys. Rev. B*, **24**, 5277.
Morgownik, A. F. J. and Mydosh, J. A. (1983) *Solid State Commun.*, **47**, 321.
Mulder, C. A. M., van Duyneveldt, A. J. and Mydosh, J. A. (1981) *Phys. Rev. B*, **23**, 1384.
Murani, A. P. (1978 a) *Phys. Rev. Lett.*, **41**, 1406.
Murani, A. P. (1978 b) *J. Appl. Phys.*, **49**, 1607.
Murani, A. P. (1981) *J. Magn. Magn. Mater.*, **22**, 271.
Mydosh, J. A. and Ford, P. J. (1974) *Phys. Lett.*, **49A**, 189.
Mydosh, J. A. (1988) *J. Magn. Magn. Mater.*, **73**, 247.
Nagata, S., Keesom, P. H. and Harrison, H. R. (1979) *Phys. Rev. B*, **19**, 1633.
Nordblad, P., Svedlindh, P., Lundgren, L. and Sandlund, L. (1986) *Phys. Rev. B*, **33**, 645.
Sandlund, L., Svedlindh, P., Granberg, P., Nordblad, P. and Lundgren, L. (1988) *J. Appl. Phys.*, **64**, 5616.
Schultz, S., Gullikson, E. M., Fredkin, D. R. and Tovar, M. (1980) *Phys. Rev. Lett.*, **45**, 1508.
Smit, J. J., Nieuwenhuys, G. J. and de Jongh, L. J. (1979) *Solid State Commun.*, **31**, 265.
Tholence, J. L. and Tournier, R. (1974) *J. Phys. (Paris)*, **35**, C4-229.
Uemura, Y. U., Yamazaki, T., Harshman, D. R., Senba, M. and Ansaldo, E. J. (1985) *Phys. Rev. B*, **31**, 546.

References

Verbeek, B. H., Nieuwenhuys, G. J., Mydosh, J. A., van Dijk, C. and Rainford, R. D. (1980) *Phys. Rev. B*, **22**, 5426.
Violet, C. E. and Borg, R. J. (1966) *Phys. Rev.*, **149**, 545.
Wenger, L. E. and Keesom, P. H. (1976) *Phys. Rev. B*, **13**, 4053.

4
Systems of spin glasses

For the sake of generality and completeness we should try at this stage to collect our various systems of spin glasses. The previous chapters have, in their generic approach, dwelt on a few very specific materials, e.g. CuMn, AuFe and to a lesser extent (EuSr)S. Yet the spin-glass phase is, as stated previously, a very general phenomenon. Over 500 different systems have been claimed to be spin glasses, i.e., exhibit at least some (even one) of the experimental characteristics mentioned before for the canonical (RKKY) spin glasses. After ferro- and antiferromagnetism, spin-glass freezing is the third type of magnetic 'order' or, better put for the latter case, *co-operative* magnetic state.

While it is not the purpose of this chapter to consider all 500-plus materials, nevertheless, we wish to list the most important spin-glass categories and to distinguish their class according to the strength of the magnetic interactions which are present. Remember the conduction-electron-mediated RKKY interaction of the transition-metal impurities have the strongest coupling. Oppositely, superparamagnets (rock magnets or CoO particles) have little or no coupling and are, therefore, not spin glasses.

We begin with the transition-metal-solute alloys, work our way through the rare-earth binary or pseudo-binary alloys and add some of the vast number of multi-component, amorphous magnetic alloys. Then we move onto the semiconducting materials for which there is an enhanced superexchange. Such a coupling may be sturdy enough to create a good spin glass at a reasonable temperature. And finally, we reach the magnetic insulators where the superexchange is weak and, thus, the concentration must be increased to promote a possible spin-glass state. Or, if the coupling is insufficient, we return to our (super) paramagnetic, at least, down to the lowest available temperature of measurement (see section 2.10). Here the blocking of individual spins or clusters at T_B would take precedent over a cooperative and collectively frozen ground state which might begin to form at $T_f < T_B$.

4.1 TRANSITION-METAL SOLUTES

4.1.1 Noble metals with transition-metal impurities

We start with the well-studied noble-metal/3d transition metal alloys, our canonical spin glasses. Since 4d or 5d transition metals are non-magnetic, i.e., do not form local moments, they cannot be used as impurities in a noble metal to form a spin glass. Two criteria can be invoked to find the simplest spin-glass behaviour:

(a) 'good' moment systems, meaning that the Kondo temperature should be less than ≈ 1 K, so that no complications are encountered with weakening of the local moments at low temperatures; and
(b) a favourable solubility such that at least 10 at. % of the 3d metal may be dissolved in the noble-metal host.

This latter criterion, in conjunction with a proper homogenization process, provides for a random distribution of impurities and eliminates difficulties with chemical clustering. In Table 4.1 the various combinations of noble-metal solvent 3d solute are given. Besides the archetypal examples of CuMn and AuFe, there are only three other uncomplicated spin-glass systems, denoted by a 'good' in Table 4.1: AuCr, AuMn and AgMn. By referring to the table, problems are encountered for many of the other combinations, either with the Kondo temperature or especially with the solubility limit, represented by XT and XS, respectively.

However, a number of special cases exist as, for example, with CuFe which possesses a rather poor solubility and a $T_K \approx 30$ K but, nonetheless, shows the magnetization characteristics of a mictomagnetic. The high T_K means that isolated or single Fe atoms cannot participate in the magnetic interactions, yet Fe pairs or triplets may. The deviations from randomness (precipitation of Fe for $x \geq 1$ at. %) are reflected in the low-field susceptibility by the expected sharp peaks at T_f changing into broad maxima. Also

Table 4.1 Spin glass combinations

Host	Impurity:	Noble metal-transition metal					
		V	Cr	Mn	Fe	Co	Ni
Cu		XS	XS	GOOD	XS+T	XS	XT
Ag		XS	XS	GOOD	XS	XS	XS
Au		XT	GOOD	GOOD	GOOD	XS+T	XT

'GOOD' represents the most favourable combinations, XS or XT means that the spin glass behaviour is limited by lack of solubility or too high a Kondo (fluctuation) temperature, respectively.

in this special subset we should include *Au*Co for which the Co solubility is poor, but not as bad as *Cu*Fe. The difficulty here is with the very large, few hundred kelvin, Kondo (or spin-fluctuation) temperature. As before T_K is strongly dependent upon the local environment, i.e., the number of Co nearest neighbours (nn). In order to reduce T_K to less than 1 K, a Co triplet is required and then modified spin-glass behaviour is observed. For this system an effective magnetic concentration of Co–nn triplets should be used instead of the actual atomic percent.

A few non-noble-metal host systems may be grouped with the above collection. Examples of which include *Zn*Mn and *Cd*Mn, whose crystal structure, being hexagonal, adds a crystalline single-site anisotropy to restrict the orientations of the freezing spins. This breaks the isotropic distribution of frozen-spin directions discussed in section 3.3.5 and creates Ising (up/down) or x–y (spins confined to lie in a plane) types of spin glasses. Nevertheless, for these systems the solubility of Mn in Zn or Cd is rather limited, in the thousand ppm range. A final extreme example of a local-environment spin glass is $Fe_{0.3}Al_{0.7}$ where the very large Fe concentration is necessary to make the Fe magnetic, since Fe in Al has a very high spin-fluctuation temperature of some thousands of degrees. Note that in all of the above systems the competing ferro-/antiferromagnetic RKKY interaction is active and results in a strong and oscillating exchange coupling.

4.1.2 Transition-metal/transition-metal combinations

By using the previous criteria of good moments and high solubility, binary combinations of two transition metals may be used to create a spin glass. That is, if a giant-moment system does not sweep away the random freezing and leave a dilute, but long-range ordered, ferromagnet. Table 4.2 offers a collection of the various possibilities. We begin with the giant-moment alloys whose formation requires an exchange-enhanced host (see section 2.3). This means that, firstly, there is an abundant density of itinerant (d-

Table 4.2 Transition metal-transition metal spin glass/giant moment combinations

Host	Impurity:	Cr	Mn	Fe	Co	
Mo		XT	SG	SG	XS+T	simple
Rh		XT	SG	XT	XT	
Pd		XT	GM+SG	GM	GM	exchange
Pt		XT	SG	GM	SG	enhanced

SG and GM represent favourable combinations for spin glass or giant moment behaviour. Note the strong, mixed behaviour of *Pd*Mn, XS or XT means solubility or too high a Kondo (fluctuation) temperature limits the appearance of both the spin glass or giant moment states.

4.1 Transition-metal solutes

states near the Fermi energy $N(E_F)$ such that the Pauli susceptibility is large: $\chi_p = 2\mu_B^2 N(E_F)$. And secondly, an intra-atomic exchange, $\bar{I}$, should be present on the host sites. This results in an exchanged-enhanced susceptibility $\chi = \chi_p[1 - \bar{I}N(E_F)]^{-1}$ where the factor $[1 - \bar{I}N(E_F)]^{-1} \equiv \Theta$ is known as the Stoner enhancement factor. For Pd, $\Theta = 10$, while for Pt, $\Theta = 3$. Consequently, Pd or Pt hosts alone will produce the giant-moment polarization. As mentioned previously (section 2.3) only in the very dilute limit of Fe impurities (ppm) in Pd will a spin-glass state form at very low temperatures (< 1 mK).

An especially interesting situation is with *Pd*Mn where for $x < 3$ at. % Mn, the giant-moment ferromagnetism prevails. However, upon further increasing the Mn concentration ($x > 4$ at. %), the probability of having two Mn atoms as first or second nearest neighbours increases. This then supplies the essential element of competing exchange for the appearance of the spin-glass phase. Since Mn–nn couple antiparallel, they thereby produce the antiferromagnetic coupling which mixes into the longer-range ferromagnetic, giant-moment polarization. By making a ternary alloy *Pd*MnFe and utilizing different concentrations of Mn and Fe we can independently vary the ratio of ferromagnetic to antiferromagnetic exchange. This procedure has become very important in comparing with theory and studying the re-entry spin glasses of section 2.9.

Cr impurities, when introduced into the Pd or Pt host, locally 'blot-out' the uniform exchange enhancement, but in this process the Cr loses its magnetic moment and becomes a weak moment with a high T_K. Only when the Cr concentration is large enough to provide a suitable Cr local environment does a stable (or good) Cr moment appear, now however, in a non-enhanced Pd or Pt matrix. Accordingly, the RKKY-oscillating interaction can occur leading to a spin-glass state at low temperature. Here there is no giant-moment phase, if anything, the term 'dwarf' or 'destroyed' moment is more applicable. *Cr*Fe, a combination of two magnetic elements, represents another exceptional case. Fe when diluted into the Cr–bcc host first eradicates the antiferromagnetic (itinerant) spin-density wave. Yet, the Fe moments remain intact, and again, due to RKKY interaction a spin-glass phase appears around 15 at. % Fe. At slightly lower concentrations, there is a re-entry transition back to the spin-density-wave antiferromagnetism. At slightly higher Fe concentrations, a series of transitions occur: paramagnetic → ferromagnetic → (re-entrant) spin glass.

At this point in our survey of binary, crystalline alloys we should also mention the various Ni-solute systems. Either in noble-metal or transition-metal matrix, Ni moments require a local environment of other Ni nearest neighbours. For *Pd*Ni three or more nearest neighbours are needed to produce a giant-moment ferromagnet. For *Cu*Ni at least seven or eight nearest neighbours give a narrow spin-glass regime before ferromagnetism takes over. The all-important concept, used time and time again for moment formation, is the local-environment model that a particular number of weak-moment atoms are present as nn to produce a sufficiently strong magnetic surrounding and thereby generate the local magnetic moment which is *per se* a magnetic cluster.

4.1.3 Collection of characteristic temperatures for *Au*Fe and *Cu*Mn

We now return to our two most-popular canonical spin glasses, *Au*Fe and *Cu*Mn. There has been an enormous amount of effort put into studying these materials and it seems appropriate to gather the various experimental results in a temperature-concentration phase diagram. According to Larsen (1978), different characteristic temperatures are defined in the following way:

(a) the freezing or ordering temperature usually defined by a susceptibility cusp or for a ferromagnet the 'demagnetizing-limit' plateau;
(b) the temperature T_m of the maximum in the magnetic resistivity $\Delta\rho$, see equation (3.11);
(c) freezing or ordering temperatures derived from other experimental techniques with different time windows such as Mössbauer spectroscopy and neutron scattering; and
(d) a 'noise' temperature Δ_c calculated from $\Delta_c = T_m Q[\ln(T_m/T_K)]$ where T_K is the Kondo temperature and $Q[\cdot]$ is the noise function derived by theory.

Figures 4.1 and 4.2 display the many data points collected over the years for *Au*Fe and *Cu*Mn, respectively. The labelling corresponds to the above definitions while the solid lines represent Larsen's theoretical calculation of the freezing $T_f(x)$. Here the model relates T_f to the root-mean-square RKKY interaction energy. Exponential damping of the RKKY interaction is included to take into account a finite mean-free path via a parameter $r(x)$. Some salient features of these data, which extend over five decades in concentration, are listed below.

1. The magnetic-resistivity maximum always occurs at a significantly larger temperature than the freezing temperature.
2. Within the limits of a log–log plot there is a reasonable agreement between the various experimental determinations of critical (freezing or ordering) temperatures.
3. The noise model can nicely relate the temperature of the resistivity maximum to the freezing temperature, except in the 1–10 at. % concentration regime of *Au*Fe.
4. While *Cu*Mn continues to follow a smooth curve over its entire concentration range, *Au*Fe has the above-mentioned anomaly, followed at larger x by an upward jump in the ordering temperature beginning

4.1 Transition-metal solutes

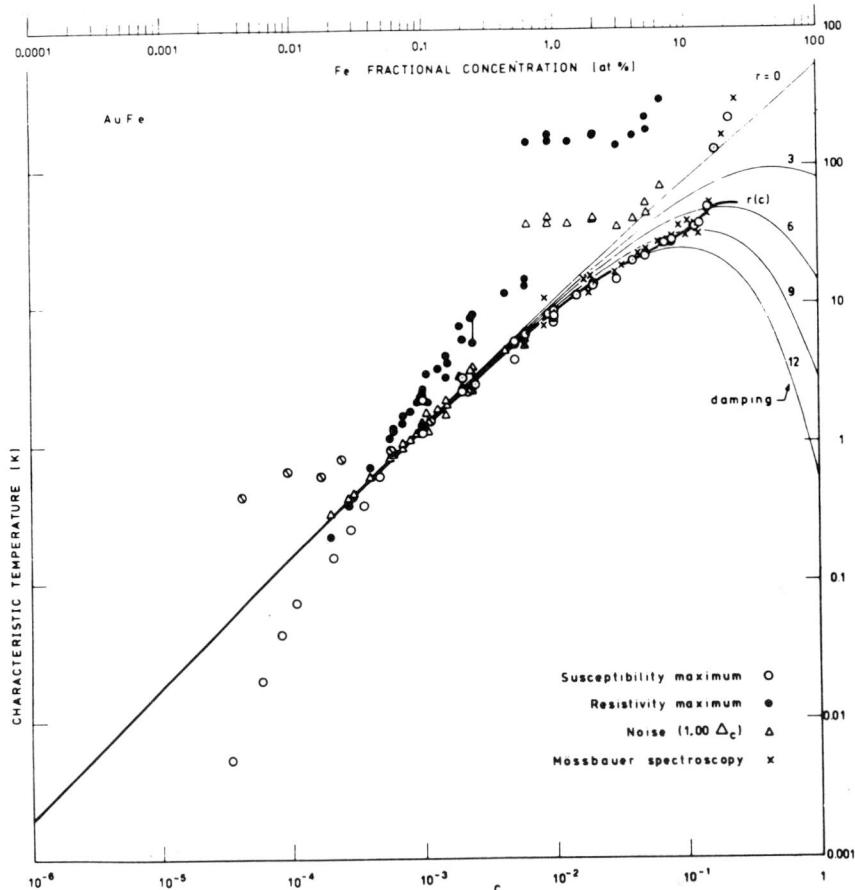

Fig. 4.1 Collection of characteristic temperatures for AuFe alloys over the complete concentration range; from Larsen (1978).

at about 15 at. %. This deviation corresponds to the onset of percolation ferromagnetism. The precursor effect is probably due to the $\Delta\rho_{max}$ being shifted to much larger temperatures by the formation of big ferromagnetic clusters.

5 The theoretical calculation of T_f from the RKKY energy seems to work well with the proper choice $r(x)$, the damping parameter. It looks like this parameter must be made a function of the concentration above about 10 at. % for both systems in order to maintain the fit.

The important conclusion of all this data plotting and theoretical analysis is that the RKKY-interaction is the driving force behind the spin-glass state of these many transition-metal alloys. The coupling strength is large, its spatial oscillations create the mixed couplings of the randomly distributed

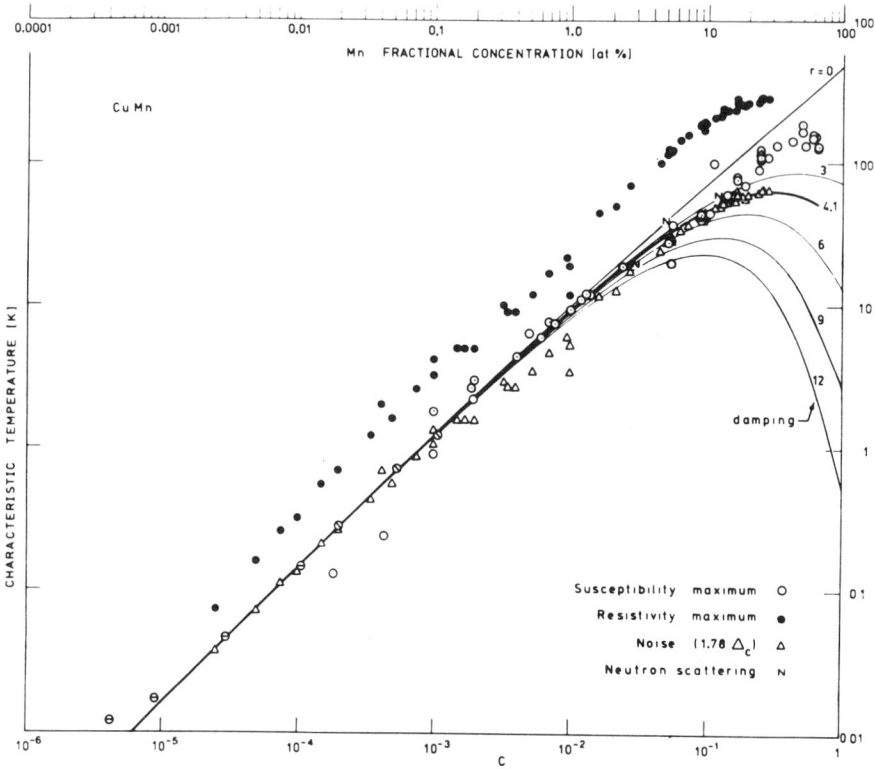

Fig. 4.2 Collection of characteristic temperatures for *Cu*Mn alloys over the complete concentration range; from Larsen (1978).

impurities, and frustration is a natural consequence. So we have a simple recipe for creating a new spin glass: take a transition metal, alloy it into a non-magnetic metal, and, with the caveat of insufficient solubility or too high a Kondo temperature, we will discover another spin-glass material. But what have we really learned by adding another spin glass to our long list? Answer: not much about the essence of the phase transition or transformation, how and out of what the frozen state is formed, and what governs its metastable properties. Such crucial points have not been touched. For we must work (measure everything) on a few ideal systems which closely correspond to the theoretical models. Only then can we use the microscopic theory to explain and delve deeper into the nature of the spin glasses. And experiments on these sparsely chosen materials will point the

4.2 SOME RARE-EARTH COMBINATIONS

The same game can be played with magnetic rare-earth elements: dilute them into a non-magnetic host metal and let the RKKY interaction perform its coupling. Nevertheless, there are some advantages and certain disadvantages of the rare earths when compared to the transition-metal solutes. First the good news, for most rare earths, with the notable exception of Ce, there is no Kondo effect, because the coupling between local moment and conduction electrons is a positive (or ferromagnetic) one. The rare-earth elements, which do exhibit a Kondo effect (antiparallel coupling), e.g. Ce, usually have a very low T_K.

And now the bad news. In many cases limited solubility is a problem one cannot simply alloy over a large concentration range, e.g. *La*Gd exists as a solid solution for only a few percent. A way of overcoming this difficulty is to use a pseudo-binary intermetallic compound, e.g. $La_{1-x}Gd_xAl_2$. Here there are a vast number of such combinations, a few of which are listed in Table 4.3. Further bad news is that the strength of the RKKY, i.e., its coefficient J_o in equation (1.2), becomes much weaker for rare-earth impurities. This smaller polarization of the conduction electrons results in much lower T_fs even at increased concentration of impurities. A final complication with the rare earths is the crystalline-electric-field splittings which are rather low in energy ≈ 100 K. This means that the effective moment will change as a function of temperature and Schottky anomalies will influence the various experimental behaviours. On the positive side, the single-ion crystal field may be used to generate a crystalline anisotropy. Such, in turn, can create Ising (up-down) and *x-y* (spins confined to a plane) types of spin glasses which should have different critical phenomenon

Table 4.3 Some good rare-earth spin-glass combinations

Binary	*La*Gd, *La*Eu *Sc*Gd, *Sc*Tb, *Y*Gd, *Y*Tb and *Pr*Tb
Pseudo-binary	$La_{1-x}Gd_xAl_2$, $La_{1-x}CeAl_2$ $La_{3-x}Gd_xIn$, $La_{3-x}Ce_xIn$ $La_{1-x}Gd_xB_b$, $La_{1-x}Ce_xB_6$ $La_{1-x}Ce_xRu_2$, $Th_{1-x}Gd_xRu_2$ $Ce_{1-x}Gd_xRu_2$, $Ce_{1-x}Tb_xRu_2$ $Th_{1-x}Nd_xRh_2$

associated with their freezing behaviour than our isotropic (Heisenberg), transition-metal spin glasses.

After all these pluses and minuses of the rare earths are taken into account, we have essentially similar ingredients and behaviours as with the transition-metal systems. The common property here is the conducting host and the RKKY interaction connecting the site-random impurities. We might have to measure at lower temperatures or higher concentrations (renormalized T/x), but the basic effects will be the same. And we can include via the above recipe many new systems to our ever expanding collection of spin glasses.

A very interesting question arises concerning the mutual coexistence of *magnetism* and *superconductivity*. This query has been the subject of much effort for the past 30 years. The conventional wisdom is such that superconductivity can coexist with all the various types of antiferromagnetic long-range order (even spiral sorts with rather long periods). But ferromagnetic order destroy the superconductivity. So what about a spin glass? After a long search, finally a suitable rare-earth system, $(Th_{1-x}Nd_x)Ru_2$, was fabricated which was a good ($T_c^S = 4$ K) superconductor at $x = 0$ and a ferromagnet ($T_c^F = 22$ K) at $x = 1$. For intermediate concentrations $x \approx 0.35$ a superconducting/spin glass developed. Studies of the critical fields for superconductivity, and the freezing behaviour and remanence of the frozen state proved that there was indeed a coexistence of these two states. However, the spin-glass transition did weaken the superconductivity by depressing T_c^S and reducing the critical fields. If $T_f \leq T_c^S$, a remarkable quenching of the superconductivity occurs at T_f followed by a recovery or 're-entry superconductivity' at very low temperatures. This unique behaviour results in one spin-glass (or more accurately a cluster-glass) and three superconducting-normal transitions at a single concentration ($x = 0.35$) value. Figure 4.3 illustrates the exceptional phase diagram of $Th_{1-x}Nd_xRu_2$ where the coexistence regime is shaded. Thus, we conclude that a weakened superconducting state can coexist within a frozen spin-glass state.

4.3 AMORPHOUS (METALLIC) SPIN GLASSES

As already mentioned in section 1.2 an amorphous solid, viz., an alloy or an intermetallic compound without crystallographic order, can also be a spin glass. In the cases considered below all the samples are metallic with a reduced conductivity due to the disordered matrix. We represent such materials with an 'a' before their elemental symbols. There are many combinations of these systems starting with a single magnetic species of a transition metal, e.g. a-$Fe_xPd_{80-x}P_{20}$ or a-Fe_xSn_{1-x}. Table 4.4 collects a few additional examples. Notice how complex the element manipulation and compositions can become in order to 'fine tune' the material. Then we can proceed to more complicated double transition-metal glasses such as

4.3 Amorphous (metallic) spin glasses

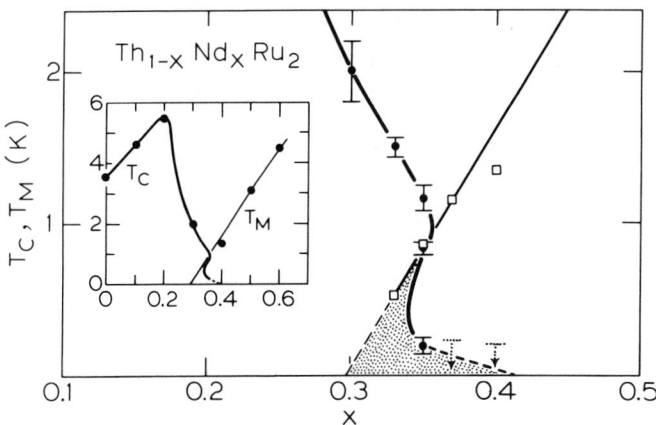

Fig. 4.3 Superconducting (T_c) and magnetic ($T_m = T_f$) phase diagram for $Th_{1-x}Nd_xRu_2$. The lines are a visual guide; from Hüser et al. (1983).

Table 4.4 Some good amorphous metallic spin glasses

Binary	a-Fe_xB_{1-x}, a-Fe_xSn_{1-x}; a-Fe_xZr_{1-x} a-Mn_xSi_{1-x}, a-Mn_xGe_{1-x} a-Fe_xY_{1-x}, a-Ni_xY_{1-x} a-$Gd_{0.37}Al_{0.63}$
Pseudo-binary	a-$(Fe_xPd_{1-x})_{80}P_{20}$, a-$(Fe_xCr_{1-x})_{80}B_{20}$ a-$(Fe_xNi_{1-x})_{77}Si_{10}B_{13}$, a-$(Fe_xNi_{1-x})_{79}P_{13}B_8$ a-$(Co_xNi_{1-x})_{78}P_{14}B_8$, a-$(Mn_xN_{1-x})_{78}P_{14}B_8$ a-$(Fe_xNi_{1-x})_{75}P_{16}B_6Al_3$, a-$(Co_xNi_{1-x})_{75}P_{16}B_6Al_3$ a-$(Fe_xMn_{1-x})_{75}P_{16}B_6Al_3$, a-$(Fe_xCr_{1-x})_{75}P_{16}B_6Al_3$ a-$(La_{1-x}Gd_x)_{80}Au_{20}$

a-$(Fe_xNi_{1-x})_{79}P_{13}B_8$ or even a-$(Fe_xNi_{1-x})_{75}P_{16}B_6Al_3$. These multi-element compounds (see Table 4.4) formed by melt-quenching can be tailor-made to give the desired amounts of ferro- and antiferromagnetic exchange, in both ratio and absolute magnitudes. Hence, the spin-glass freezing temperature can be adjusted to a convenient range and re-entry (ferromagnetism → spin glass) behaviour can be extended over a wide composition region.

In addition, amorphous materials can be fashioned out of the rare-earth elements: a-$Gd_{37}Al_{63}$ and a-$La_{80-x}Gd_xAu_{20}$ are good examples. Particularly with the latter system, a wide x-range of composition (a few percent up to 50 at. %) is available to trace the spin-glass phase with T_fs spanning $\approx$ 2–50 K. We could easily employ melt-quenching or splat-cooling and sputtering techniques to fabricate, at will, other rare-earth spin glasses. The same caveats as for the crystalline alloys apply for the amorphous ones.

A special property of the amorphous spin glasses is their higher resistivity which will dampen the range of the RKKY-interaction. Therefore, large concentrations of the magnetic elements are required to form a spin glass. This reduced interaction range places such system intermediate between the short-range insulators and the long-range undamped RKKY alloys. From the sharp, well-defined freezing temperatures of these amorphous compounds, the range of the competing interactions does not seem to affect the freezing characteristics.

In general, the spin-glass aspects of these amorphous magnetic materials are mainly similar to those of their crystalline counterparts. One advantage of using them is the vast region of magnetic concentration available in an amorphous structure. This permits an examination of cluster glass and re-entry (ferromagnet → spin glass) behaviours. Another benefit comes from the damped RKKY interaction which allows the coupling strength and range to be varied. A third is that any crystalline or single-ion anisotropy will be averaged to zero, due to the disordered lattice, leaving a fully isotropic (Heisenberg) spin glass. In summary, we would gather these amorphous alloys into our universal class of metallic, competing exchange, random-site spin glasses.

4.4 SEMICONDUCTING SPIN GLASSES

We commence this section with a most important family of magnetic semiconductors, namely the rare-earth monochalcogenides. As discussed earlier, $Eu_xSr_{1-x}S$ is a prime example of a short-range (only first and second nearest-neighbour exchanges are important) Heisenberg spin glass. Let us briefly consider the phase-diagram of this system which is reproduced in Fig. 4.4. The J_1 and J_2 exchange couplings are competing, ferro- and antiferromagnetic, respectively, with ratio $J_2/J_1 = -0.5$. Since the Eu sites occupy an fcc lattice, there will be a two neighbour percolation threshold at $x = 0.13$. According to the phase diagram in Fig. 4.4, this is exactly where the spin-glass phase ends. For $x < 0.13$ a sort of superparamagnetic cluster blocking begins without any of the collectivity or co-operativeness of inter-cluster coupling – a prerequisite for the spin-glass transition. This behaviour nicely illustrates the need for mixed interactions and the disadvantage of a short-range random site system which cuts off the spin glass at the percolation limit.

Other members of this family include $Eu_xSr_{1-x}Te$ and $Eu_xSr_{1-x}Se$ which, also due to the competing $\pm J_i$ exchanges, show spin-glass-like characteristics. An exception to this trend is $Eu_xSr_{1-x}O$ where both J_1 and J_2 are ferromagnetic. Hence no spin-glass phase occurs, only a ferromagnetic one down to the percolation threshold. We should at this point mention the superexchange mechanism in these chalcogenides. Theoretical work has proposed a novel f-d overlap of wave functions between two Eu-ions, which are either first or second nearest neighbour. For the closest Eu–Eu nearest

4.4 Semiconducting spin glasses

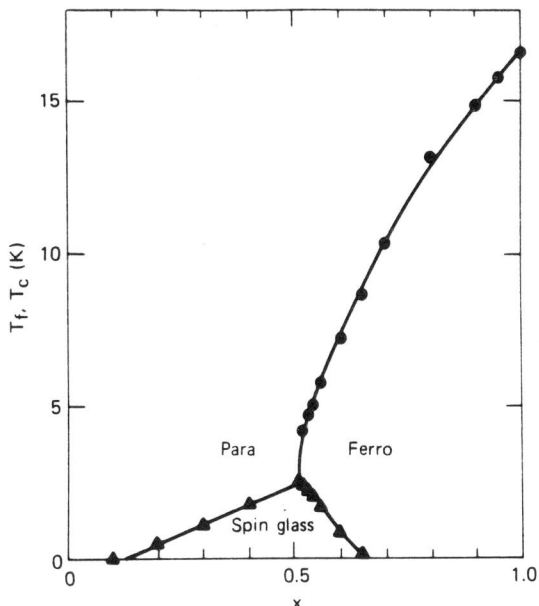

Fig. 4.4 Magnetic phase diagram of Eu_xSr_{1-x}; from Maletta (1982).

neighbours a virtual transfer of a 4f electron to the 5d-t_{2g} excited state of its neighbour results in a strong ferromagnetic coupling. In contrast, an indirect (or more typical) superexchange via the ligand (see section 1.3.3) creates the next-nearest neighbour J_2 which involves the antibonding 5d-e_g orbitals of the Eu and the p-orbitals of the ligand. Here the coupling is either ferro- or antiferromagnetic depending on the distance.

Another interesting, but more troublesome, class of 'dilute' magnetic semiconductors (DMS) are the II–VI group compounds (the II and VI refer to columns of the periodic table). These include the sulfides, tellurides, sellurides and arsenides of Zn, Cd, Hg, Ge, Pb, and Sn, the latter sites being doped with Mn (or Fe) as the magnetic impurities. A prime example is $Cd_{1-x}Mn_xTe$, others are listed in Table 4.5. So we have a random-site problem and must consider the exchange coupling between the Mn^{2+} ions.

Again theoretical work has established that superexchange is the main source of the magnetic coupling. Accordingly, the filled valence band of the semiconductor exchanges electrons with the half-filled 3d band of the Mn. With such 2-electron (or hole) processes the interaction between the Mn will always lead to an antiparallel (AF) orientation and this seems to be true for both first and second nearest neighbours. Although, the second neighbour J_2 is usually much weaker in magnitude, since $J(r)$ falls off either as $\exp(-r^2)$ or $1/r^7$. Experiment seems to support these conclusions concerning the exchange coupling. Therefore, we must pose the question

Table 4.5 Some dilute magnetic semi-conductors (DMS) as possible spin glass combinations

---
$Zn_{1-x}Mn_xS$, $Zn_{1-x}Mn_xSe$; $Zn_{1-x}Mn_xTe$
$Cd_{1-x}Mn_xS$, $Cd_{1-x}Mn_xSe$; $Cd_{1-x}Mn_xTe$
$Hg_{1-x}Mn_xSe$; $Hg_{1-x}Mn_xTe$
$Pb_{1-x}Mn_xS$, $Pb_{1-x}Mn_xSe$; $Pb_{1-x}Mn_xTe$
$(Zn_{1-x}Mn_x)_3As_2$; $(Cd_{1-x}Mn_x)_3As_2$
$Zn_{1-x}Fe_xSe$, $Cd_{1-x}Fe_xSe$; $Hg_{1-x}Fe_xSe$
$Sn_{1-x}Mn_xTe$; $Ge_{1-x}Mn_xTe$
---

with these II–VI DMSs: can an *all* antiferromagnetic, random-site systems produce a good spin glass? We let experiment be our guide.

Many measurements have been performed on the DMS systems listed in Table 4.5. The following characteristics are generally found. A Curie–Weiss (CW) law is obeyed at high temperatures with a paramagnetic temperature indicative of the predominantly antiferromagnetic coupling. Deviations from CW occur at low temperatures depending on the Mn concentration usually resulting in smooth kink or rounded plateau. But there is no cusp or sharp maximum except at large Mn–x values ($x \geq 0.4$). A difference in FC versus ZFC magnetization is also observed, however, this is very gradual without the sharp features of the canonical (RKKY) spin glasses. The magnetic specific heat exhibits a broad maximum shifting to higher temperatures with x. Attempts at scaling the static and dynamical behaviour of such DMSs via $\chi'(\omega)$ and $\chi''(\omega)$ measurements give roughly similar results and exponents with respect to the canonical or even ideal (Chapter 6) spin glasses. In spite of this there remain the troublesome features referred to earlier, viz., the lack of sharpness in the ac-susceptibility and magnetization. It would seem that the freezing behaviour is non-co-operative at low x-values. A significant portion of the spins remains independent and only joins the randomly frozen, infinite cluster at a much lower temperature than the 'kink temperature'. This absence of finely-defined freezing temperature is probably caused by the short-range nature of the exchange interaction $J(r)$ and the enormous frustration present. Too much frustration appears not to be good for a co-operative spin-glass freezing. With most two or three nearest neighbours participating in the coupling scheme, the percolation cut-off should be around 10%. For $x < 0.1$ only superparamagnetic clusters would become blocked as in the case $(Eu_xSr_{1-x})S$. However, for the DMS this seems not to occur around 10% but around 40%. Is the frustration taking its toll? Furthermore, the exclusively antiferromagnetic alignment makes the susceptibility and magnetization very small. With such a weak magnetic response, experiment will have difficulty tracking the subtleties of freezing or blocking. Thus, a major problem may be unresolvable, i.e., how to distinguish an all-antiferromagnetic spin glass from its two possible counterparts: a $\pm J$ spin glass or a non-interacting 'super'-paramagnet. More work is needed to sort out these nuances and to arrive at the fine (or significant?) distinctions between a competing or mixed-interaction spin glass and an only antiferromagnetic one. As we shall soon see theory only treats the mixed-interacting spin glasses.

4.5 INSULATING SPIN GLASSES

Chemical compounds with one magnetic element have long been a testing ground for magnetic critical phenomenon. The chemists can almost at will produce a material which has the desired spatial (1-, 2- and 3D) and order parameter (Ising, x–y and Heisenberg) dimensionalities. Exchange interactions can be varied (ferro, mixed and antiferromagnetic) with superexchange and sometimes dipolar mechanisms. In the 1960s and 1970s such *pure* compounds dominated the subject of magnetic phase transitions.

For the chemist there is no great difficulty in randomly replacing a magnetic element by a non-magnetic one. Thus, in principle, we can create multitude spin glasses. However, what is to be gained by discovering yet another spin glass? There must be something particularly interesting or 'ideal' about the given material. As a case in point let us consider the mixed compound $Fe_xMg_{1-x}Cl_2$. Pure $FeCl_2$ is a hexagonal-lattice layered compound which has ferromagnetic a-b planes (with a triangular lattice) that stack antiferromagnetically along the c-axis; for this reason we have a long-range ordered antiferromagnet. A strong uniaxial anisotropy aligns the moments along the c-axis so the material is Ising-like. The in-plane interactions are ferromagnetic for nearest neighbours, but antiferromagnetic for next-nearest neighbours (nnn) with $J_2/J_1 = -0.13$. In diluted $Fe_xMg_{1-x}Cl_2$ one expects for $x \leq 0.4$ a random site, 3D, Ising spin glass which should evolve out of an ordered antiferromagnetic structure. Very interesting here is the *coexistence* (at intermediate $x \approx 0.45$), proven by neutron-scattering experiments, of an infinite antiferromagnetic network with a collection of randomly frozen spins in the various a-b planes. At lower x, until two-neighbour percolation cuts off the spin-glass phase, the usual competing-interaction spin-glass behaviour is observed, but now in an Ising system, i.e., predominantly along the c-axis are there magnetic responses (freezing). Hence, this material is a good candidate for studying the dynamics of an Ising random-site spin-glass.

Another example is the $Fe_xZn_{1-x}F_2$, also an antiferromagnetic Ising system, which for $x \geq 0.5$ represents a prototypical 3D random-exchange Ising model (REIM). In the presence of a field applied along the Ising axis it exhibits crossover phenomenon to random-field Ising-model (RFIM) behaviour. The above two models (REIM and RFIM) represent unusual critical phenomenon and crossovers of different types of phase transitions, albeit both to a long-range-ordered state which will be discussed in Chapter 7. Now what happens when the concentration is reduced? Before attempting to answer this question we must first consider the superexchange present in the body-centre-tetragonal (rutile) FeF_2 structure. Best determinations for these parameters give $z_1J_1 = +0.14$ K, $z_2J_2 = -42$ K, and $z_3J_3 = -1.15$ K where z_i is the coordination number: for the ith nn, $z_1 = 2$, $z_2 = 8$ and $z_3 = 4$. If we neglect the tiny ferromagnetic J_1, then the ratio of z_2J_2/z_3J_3 is about 40 and this small J_3 exchange is the only source of frustration. The site-percolation threshold for the $Fe_xZn_{1-x}F_2$ system with nnn interactions is $x_p \approx 0.24$. So below this concentration we would expect our old nemesis, namely, independent antiferromagnetic domains gradually becoming blocked at low temperatures in analogue to super-paramagnetism. But what about concentrations slightly above x_p, e.g. ≈ 0.3? At present there is much interest in this query. Can a spin glass be formed without frustration? Some experimental indications exist for spin-glass behaviour, but others seem to suggest a non-co-operative freezing more like the above blocking of discrete domains. Physically, if no other exchange were present except J_2, then we would anticipate a long-range, percolative antiferromagnetic transition at some low temperature. And neutron scattering with the observable Bragg peaks has shown this long-range order at $x = 0.31$. Notice that this situation is somewhat similar to that discussed previously (section 4.4) with our all-antiferromagnetic dilute magnetic semiconductors. However, for $Fe_xZn_{1-x}F_2$ there is only *one* predominant superexchange coupling, not two. Again, with better samples, time and experiment will tell us the correct answer as to the low-temperature phase.

We could continue further with the many additional combinations of mixed or random compounds which are insulating and have well-defined superexchange. Such will certainly increase the generality of our spin-glass phenomenon, but with our chemists able to supply 'custom made' compounds, we should restrict ourselves to truly ideal or model spin glasses. This topic will be dealt with in Chapter 6 once we have explained the various theoretical models for spin glasses. Only then can we judge what real materials come closest to the theories. And this will afford us the luxury of making a direct and meaningful experimental-theoretical comparison without the numerous complications and approximation of the many, many non-ideal spin glasses.

In concluding this section we must mention the sundry magnetic insulators which exhibit little or no exchange or dipolar coupling. Most of these superparamagnets are distinguished by their peculiar chemical structure. For instance, rock magnets are tiny magnetic particles diffusely dispersed within a non-magnetic 'rock'. An arrangement of CoO particles sufficiently spaced is another example; ferrofluids, aerosol particles, nanocrystals, etc. are more possibilities. In what follows we touch upon two specific exemplifications.

An insulating helium-borate glass has been intensively studied via ac-susceptibility and its frequency and temperature dependences (see Table 3.1). The resulting behaviour was fully describable within the Néel

4.6 What is a good spin glass?

superparamagnetic model using a gaussian distribution of relaxation times. On the other hand, when similar measurements were performed on cobalt-aluminosilicate glasses definite indications were found for interaction effects. For example, as the temperature is lowered there was a more rapid shift of the (average) relaxation times than simple Arrhenius activation. Yet, when these relaxation-time shifts were compared to CuMn et al. as in Table 3.1, they were not nearly as swift. So we conclude that the helium-borate glass is essentially non-interacting, while the cobalt-aluminosilicate glass is a weakly coupled system. We remove these two glasses and similar, non or flimsily interacting, spin systems from further consideration in this treatise.

4.6 WHAT IS A GOOD SPIN GLASS?

We revisit this question and try to pose an answer with respect to the different amounts of magnetic coupling between the randomly distributed magnetic entities. Table 4.6 collects, in summary version, the results of the

Table 4.6 Examples of differently coupled random magnetic systems

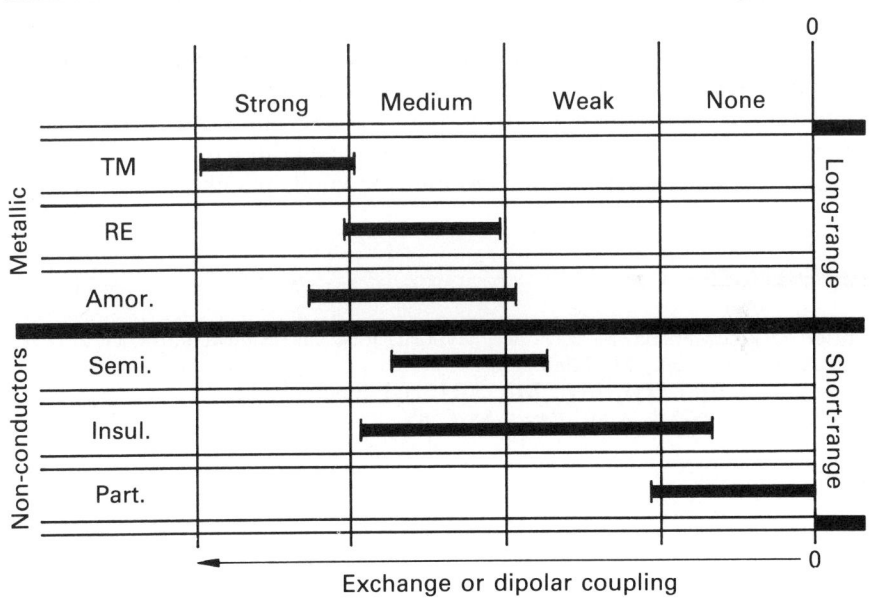

Exchange or dipolar coupling

TM = CuMn, AuFe, PtMn, etc.
RE = LaGd, $(Y$Gd$)$Al$_2$, (CeGd)Ru$_2$, etc.
Amor. = a-(FePd)$_{80}$ P$_{20}$, a-FeSn etc.
Semi. = (EuSr)S, CdMnTe etc.
Insul. = (MgFe)Cl$_2$, ZnFeF$_2$, a-CoO·Al$_2$O$_3$SiO$_2$ etc.
Part. = CoO, a-(Ho$_2$O$_3$)(B$_2$O$_3$), etc.

previous five subsections. A rough estimation is given for the range of the exchange coupling in the different classes of materials which we have encountered. The interactions are basically RKKY versus superexchange. We list the material entries according to the two types of magnetic elements, 3d and 4f, that go into the RKKY random alloys. Here the 3d-impurities are noticeably stronger than the 4f ones. The amorphous magnetic metals follow with their similar expanse of interaction strengths. Then come the not-so-dilute magnetic semiconductors with a more narrow range comparable to the rare earths. The insulators appear next, occasionally the superexchange can be rather large, but it is too often only antiferromagnetic. Finally, we have the particles or glasses, which possess little or no coupling energy and thus can be ignored with respect to their spin-glass properties.

So once again a spin glass is a random, mixed-interacting, with sufficient strength, magnetic system characterized by a random, yet co-operative, freezing at a well-defined T_f below which a highly irreversible, metastable frozen state occurs without the usual long-range magnetic order. As this chapter has demonstrated, there are myriad systems which satisfy this definition. And we have been calling such systems the good spin glasses where the special case of *canonical* denotes the infinite-range RKKY noble-metal alloys upon which so much experimentation has been performed. The word 'ideal' we reserve for those materials which closely conform to a specific theoretical model as we shall enumerate in Chapters 5 and 6. Nevertheless, there are other materials, discussed previously, which lie in the grey area, and when further progress is gained through more experimentation they may possibly be shifted into the good spin-glass category.

REFERENCES

Hüser, D., Rewiersma, M. J. F. M., Mydosh, J. A. and Nieuwenhuys, G. J. (1983) *Phys. Rev. Lett.*, **51**, 1290.
Larsen, U. (1978) *Phys. Rev.*, **B18**, 5014; and unpublished.
Maletta, H. (1982) *J. Appl. Phys.*, **53**, 2185.

5
Models and theories

In this chapter we wish to introduce the flavour of spin-glass theory and the associated models. It is not our purpose to review in detail the entire development of the various theories but to offer the salient features. There are simply too many contributions mainly spanning the decade 1975–1985 to present a complete account in one chapter. Besides several monographs and review articles (see Further Reading and References at the end of this chapter) have focused upon an in-depth description of the theory including the necessary mathematical manipulations. Instead, we shall survey the important theoretical concepts without mathematical rigour, yet with particular emphasis on the physical content. This means the mainstream models and concepts will be explained simply by words and pictures employing a limited number of equations. Derivations will not be carried out, but hopefully the basic physics of the spin glasses and its unique nature can be gleaned from our treatment.

As we shall see present-day theory is still developing and has a long journey to make before the 'solution' of the spin-glass problem is reached. Also many of the new ideas and calculations are as yet preliminary and certainly there has been no complete contact with experiment to establish the final validity of a particular model. The spin-glass problem is a tough one and requires some radical new concepts and sophisticated insights. Essentially, a new form of statistical mechanics needs to be constructed for even the first-order (mean-field) approximation of the freezing transition. Well with all this fanfare as a preamble, let's sketch how we shall proceed with the various sections of this chapter.

We begin by listing four historical developments which preceded the name 'spin glass' or the possibility of a phase transition. These then set the stage for the Edwards–Anderson (EA) model and order parameter, to be immediately followed by the Sherrington–Kirkpatrick (SK) mean-field model and solution. These theories directly confront the freezing process as a phase transition with a special order parameter. Proceeding chronologically we discuss the instability of the SK solution and the de Almeida–Thouless line. A different approach is the Thouless–Anderson–Palmer (TAP) solvable model of a spin glass. Here the solution gives a profusion of equivalent ground states. The cause of the instability of the SK model lies in the replica-symmetric order parameter and a replica-symmetry-breaking (RSB) scheme must be incorporated into the theory. At this point, an unconventional, yet mean field, order parameter arises with fundamentally new and subtle physics. We try to dwell on the pictorial representation of RSB and its physical meaning. Afterwards, we move onto the dynamics of the mean-field model and the time-dependent interpretation of the order parameter. We hope these seven sections will present an introductory overview of the mean-field theory (only a first approximation) for the spin-glass transition. Remember the theory took more than five years to evolve and at least a thousand publications to elucidate.

The remaining sections of the chapter summarize two more recent treatments of spin glasses, namely the droplet model which is diametrically opposed to the RSB standpoint, and the fractal-cluster model which closely resembles our experimentalist's interpretation, yet can quantitatively derive the results. Then, since all of the above models and calculations have usually relied on the Ising simplicity, we briefly consider non-Ising spin glasses and what kinds of new effects can be expected. Finally, the all-important computer simulations of spin glasses are described and the various results collected for comparison (in Chapter 6) with experiment.

5.1 SOME HISTORICAL PERSPECTIVES

In the old days (1960s) before the name 'spin glass' was coined, a number of theoretical ideas were afloat helping to explain random magnetic systems. Four of these early approaches are now recapitulated, particularly because we have already utilized them in the previous chapters. So we begin this theory chapter with a bit of history by collecting the four notions and briefly outlining their physical content.

The concept of concentration and range scaling for the RKKY interaction was exploited by Blandin (1961) to derive some universal properties of the noble-metal alloys. All thermodynamic quantities follow from the partition function

$$Z = Tr \exp(-\mathcal{H}/k_B T) \tag{5.1}$$

where for an RKKY system, recall (1.2), the Hamiltonian can be approximated with leading term and external field H

$$\mathcal{H} = \sum_{ij} \frac{J_o \cos(2k_F r_{ij})}{(2k_F r_{ij})^3} \mathbf{S}_i \cdot \mathbf{S}_j - g\mu_B H \Sigma S_i \tag{5.2}$$

If we simultaneously multiply H and T by the same coefficient $(1/x)$, the inverse concentration, the partition function remains unchanged. This follows from the introduction of a length scale R_c, which is that of the impurities

5.1 Some historical perspectives

– not the lattice, so defined that its volume always contains a constant number of impurity spins, xR_c^3. If we divide this invariant by the constant volume V_o determined by the number of conduction electrons N, we can write

$$\frac{xR_c^3}{V_o} = \frac{x(k_F R_c)^3}{3\pi^2 N} = \text{const} \quad \text{or} \quad (k_F R_c)^3 \propto \frac{1}{x} \tag{5.3}$$

using

$$\frac{N}{V_o} = \frac{k_F^3}{3\pi^2} \tag{5.4}$$

Therefore, the first (RKKY) term in our Hamiltonian (5.2) also scales with the inverse concentration, and after dividing by x it becomes an invarient with respect to (r_{ij}/R_c). Since all terms scale with x^{-1}, the partition function may be expressed in scaling form as

$$Z = \left[z\left(\frac{T}{x}, \frac{H}{x}\right) \right]^x \tag{5.5}$$

Hence the free energy obeys the scaling relation

$$F = -k_B T \ln Z = -k_B T x \ln\left[z\left(\frac{T}{x}, \frac{H}{x}\right) \right] = -k_B x^2 f\left(\frac{T}{x}, \frac{H}{x}\right) \tag{5.6}$$

By taking the appropriate deviatives with respect to H or T, the thermodynamic properties for all concentrations can be calculated, if they are known for a single x. And experimentally we used these scaling equations already in sections 2.4 and 2.7.

The second of our historical ideas is related to the first and was proposed by Larkin and Khmel'nitskii (1970). It utilized a virial expansion of the same Hamiltonian in powers of x/T or H/T. The first term (linear in x) gives the usual paramagnetic contribution to the free energy $F^{(1)}$, and higher-order terms $F^{(m)}$ reflect the effects of impurity–impurity interactions. Summing over all the terms establishes the total free energy

$$F = \Sigma F^{(m)} = -Nk_B T \Phi\left(\frac{xJ_o}{T}, \frac{\mu_B H}{T}\right) \tag{5.7}$$

where Φ is the scaling function of the virial expansion. $F^{(2)}$ and $F^{(3)}$ may be calculated in certain limits and, accordingly, various prognostications are acquired for the specific heat, magnetization and susceptibility. These are only valid at very low concentrations and rather high temperatures for diverse values of the applied field. By studying sufficiently dilute magnetic alloys, a number of these predictions have been verified. For example, the magnetization should approach saturation in a large field as

$$M = M_{\text{SAT}} \left(1 - \frac{aT + B}{H}\right) \tag{5.8}$$

and the specific heat at high temperature is $C_m \propto 1/T$. Nevertheless, it becomes very difficult to calculate the $F^{(m)}$ virial coefficients beyond the third ($m = 3$) term and this limits the x, T and H ranges of application. Furthermore, there is no occurrence in this method of the phase-transition-like freezing phenomenon of our canonical spin glasses.

Proceeding to the third approach, we return to the early 1960s when Marshall (1960), and Klein and Brout (1963) attempted to calculate the free energy of what is now known as a spin glass by using a distribution $P(H)$ of local, random magnetic fields. The non-trivial problem is how to determine the $P(H)$ prescribed by the random impurities and RKKY interactions. At $T = 0$ a static molecular field at the spin-site i can be written as

$$H_i = \sum_j \frac{J_o \cos(2k_F r_{ij})}{(2k_F r_{ij})^3} S_j \qquad (5.9)$$

For these competing interactions in an Ising (up/down) model, the modulus of the field and its energy may be related to the random-walk problem. So it is not surprising that a gaussian $P(H)$ results. However, we should go a step further and include the effects of correlations, i.e., the influence of the original orientation of S_i on its neighbouring spins, usually called the Onsager reaction field. Thus, we must split up the RKKY into near (reaction field) and far (cavity field) parts:

$$\frac{H_i}{J_o S_i} = \sum_{r_{ij} < R_c} \frac{|\cos(2k_F r_{ij})|}{(2k_F r_{ij})^3} + \sum_{r_{ij} > R_c} \frac{\cos(2k_F r_{ij})}{(2k_F r_{ij})^3} \qquad (5.10)$$

where R_c is used to limit the range of correlated spins, and, due to the previous scaling arguments, varies like $(x)^{-\frac{1}{3}}$; therefore, contains a constant number of impurities. The final distribution $P(H)$ is a convolution of reaction-field distribution with the cavity-field distribution and results in a broadened, more lorentzian-like shape with a significant value at $H = 0$. For the Heisenberg case all 3D orientations are possible and we must use a 3D random walk. But now the probability density $P(\mathbf{H})$ (note the vector $\mathbf{H}$), which is constant near $\mathbf{H} = \mathbf{0}$, should be integrated over the appropriate volume element $4\pi H^2\, dH$ to obtain the scalar molecular-field distribution $P(H)$. This gives $P(H) \to 0$, as $H \to 0$, and the full distribution is shown in Fig. 5.1 a. Here $P(H)$ begins as H^2 and reaches symmetric maxima at $\pm H_o$. Thus, most of the moments are aligned in finite field $|H_o|$. Nevertheless, for the Heisenberg case there will be a transverse response to a field H and this creates elementary excitations (localized magnons) of the random system. Employing numerical simulations of randomly distributed spins coupled via the RKKY interaction, the density of these excitations at various energies may be estimated. This $N(\omega)$ is shown in Fig. 5.1 b from the computer results. Note that now a finite density of excitations is found as $\omega \to 0$. Such an $N(0) \neq 0$ requirement is essential for the agreement of

5.2 The Edwards–Anderson model

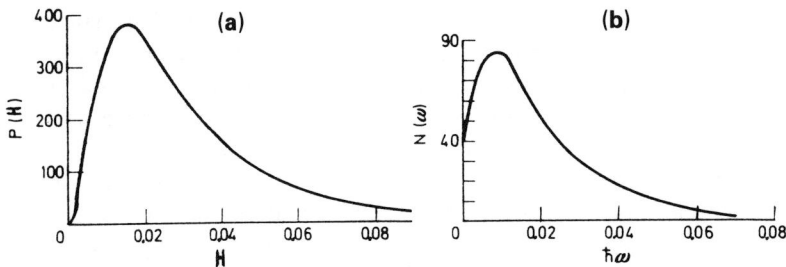

Fig. 5.1 (a) Expected shape for the internal (molecular) field distribution $P(H)$ for a Heisenberg model in $d = 3$; note the initial H^2 dependence. (b) Density of elementary excitations from the computer simulation for an RKKY interacting $x = 0.3$ at. % fcc lattice. From Walker and Walstedt (1977).

experimental data, e.g. susceptibility and specific heat. And it further justifies the *ad hoc* postulate of a finite (constant) density of excitations as $\omega \to 0$ used in the two-level model. See below and section 3.3.2 with Fig. 3.37.

The fourth and final ideal has already been considered. It is the above two-level system of Anderson, Halperin and Varma (1972), and Phillips (1972) which sought to explain the linear dependence of the specific heat at low temperatures for all disordered materials, glasses as well as spin glasses. The two asymmetric wells model of Fig. 3.37 with a constant density of low lying excitations and quantum mechanical tunnelling provides the necessary ingredients to generate this special linear term. Excess contributions to the specific heat (and other experimental quantities) have been ascertained for a variety of real glasses and spin glasses and they served as a bridge to connect spatial (lattice) disorder with spin disorder.

The above-discussed, four theoretical approaches (scaling, virial expansion, random molecular field and two-level systems) all came into being before the acceptance by the theorists of a possible phase transition demonstrated by the ac-susceptibility cusp. Up until the early 1970s there were no indications of sharp or dramatic behaviour in any of the spin-glass properties. Most measurements of the susceptibility were smeared out with the use of too large an external field and the specific heat does not exhibit any acute effects. Consequently, it was only in the mid-1970s that a flurry of theoretical activity began to treat critical phenomenon in a disordered spin system. But we have said all this before so let's move on to the theories which confront this possibility.

5.2 THE EDWARDS–ANDERSON MODEL

How can we describe the sudden random freezing of a spin glass? In 1975 Edwards and Anderson (EA) proposed the following picture. Each spin $\mathbf{S}_i$

becomes locked into a preferred direction whose orientation is random over the distribution of sites i. This transition from a paramagnet to the frozen ground state requires an order parameter. Since there is no long-range order, conventional order parameters reflecting spatial correlations are unusable. So EA focused on the time order. If a spin at site i is frozen into its ground state, its orientation will remain the same tomorrow as it was today. Thus, we can try a time autocorrelation function:

$$q = \lim_{t \to \infty} \langle \langle \mathbf{S}_i(0) \cdot \mathbf{S}_i(t) \rangle_T \rangle_C \tag{5.11}$$

where the inner angular brackets represent a thermal averaging (T) and the outer a configurational average (C) over all spins. Naturally, $q = 1$ at $T = 0$ and $q \to 0$ as $T \to T_f$. A way of visualizing this state is to compare snapshots taken at different time intervals of our spin glass. If the various $\mathbf{S}_i$ point in the same direction in the different photos, then the spin glass is frozen and $q \neq 0$. If there are changes of orientation between the photos, then $q = 0$ and we have a paramagnet. For ergodic systems the local time correlation is identical to

$$q = \langle \langle \mathbf{S}_i \rangle_T^2 \rangle_C \tag{5.12}$$

where we have squared the thermal (ensemble) average before taking the configurational one. Recall section 2.4 for our spin glass: $\langle \langle \mathbf{S}_i \rangle_T \rangle_C = 0$, so our system is neither ferro- nor antiferromagnetic on any scale. q, then, represents the Edwards–Anderson order parameter which will characterize our spin-glass phase. Now we need a model which will allow the free energy to be calculated in terms of q and EA have also provided us with such a model and its molecular-field approximation.

The EA model innocuously starts by writing the standard hamiltonian

$$\mathcal{H} = -\sum_{ij} J_{ij} \mathbf{S}_i \cdot \mathbf{S}_j - \sum_i \mathbf{H}_i \cdot \mathbf{S}_i \tag{5.13}$$

for a random-bond, 3D, square lattice. The classical spins on site i and j interact via the exchange coupling J_{ij}. These are randomly chosen according to a gaussian distribution

$$P(J_{ij}) = \frac{1}{(2\pi\Delta^2)^{\frac{1}{2}}} \exp\left(-\frac{J_{ij}^2}{2\Delta^2}\right) \tag{5.14}$$

where Δ is the variance. We now need to determine the free energy which is given in terms of the partition function Z

$$F = -k_B T \ln Z = -k_B T \, Tr\left(\exp \frac{-\mathcal{H}}{k_B T}\right) \tag{5.15}$$

The model system has *quenched disorder*, i.e., the impurity degrees of freedom are rigidly frozen, meaning there is no change in the randomness

5.2 The Edwards–Anderson model

of the spin sites (sample structural disorder is frozen in), only the spin orientations can vary. For such a system one must average ln Z over the distribution $P(J_{ij})$ which is a difficult task. However, it is relatively easier to average the partition function raised to some power Z^n where n is an integer. Here the so-called replica trick can be employed

$$\ln Z = \lim_{n \to 0} \left[\frac{1}{n}(Z^n - 1) \right] \quad (5.16)$$

For positive n, we can express $Z^n\{x\}$ ($\{x\}$ represents the set of bonds describing the disorder) in terms of n identical replicas of the system

$$Z^n\{x\} = \prod_{\alpha=1}^{n} Z_\alpha(x) = \prod_{\alpha=1}^{n} \exp\left[-\mathcal{H}\{x, S_i^\alpha\}/k_B T\right]$$

$$= \exp\left[-\sum_{\alpha=1}^{n} \mathcal{H}\{x, S_i^\alpha\}/k_B T\right] \quad (5.17)$$

where Z_α is the partition function of the αth replica.

The configurational average of $Z^n\{x\} \equiv \langle Z^n \rangle_C$ over the disorder is computed by

$$\langle Z^n \rangle_C = \mathop{\rm Tr}_{\{S_i^\alpha\}} \exp\left[-\sum_{\alpha=1}^{n} \mathcal{H}\{x, S_i^\alpha\}/k_B T\right] \quad (5.18)$$

Substitution of our starting hamiltonian (with $H_i = 0$ for simplicity) leads to

$$\langle Z^n \rangle_C = \mathop{\rm Tr}_{\{S_i^\alpha\}} \prod_{\langle i,j \rangle} dJ_{ij} P(J_{ij}) \exp\left[\frac{J_{ij}}{k_B T} \sum_{\alpha=1}^{n} S_i^\alpha S_j^\alpha\right] \quad (5.19)$$

$$\langle Z^n \rangle_C = \sum_{\{S_i^\alpha\}} \int_{-\infty}^{+\infty} \left[\prod_{\langle i,j \rangle} dJ_{ij} P(J_{ij}) \exp\left[\frac{1}{k_B T} \sum_{\langle i,j \rangle} J_{ij} \sum_{\alpha=1}^{n} S_i^\alpha S_j^\alpha\right]\right] \quad (5.20)$$

The integral over J_{ij} can be performed for the gaussian $P(J_{ij})$ by completing the square

$$\langle Z^n \rangle_C = \sum_{\{S_i^\alpha\}} \exp\left[\sum_{\langle i,j \rangle} \frac{1}{2}\left(\frac{\Delta}{k_B T}\right)^2 \sum_{\alpha,\beta} S_i^\alpha S_j^\alpha S_i^\beta S_j^\beta\right] \quad (5.21)$$

This four-spin product in the exponential arises from the addition and subtraction of a term

$$\left(\sum_{\alpha=1}^{n} S_i^\alpha S_j^\alpha\right)^2 \quad (5.22)$$

to complete the square. Now the free energy may be obtained in the mean-field approximation, assuming $\langle\langle S_i^\alpha S_j^\beta\rangle_T\rangle_C = 0$, and setting $q = \langle\langle S_i\rangle_T^2\rangle_C$. An expression $F(q)$ results from which $q(T)$ can be determined via the condition $\partial F/\partial q = 0$. These original EA equations are not simple and are only soluble in the limits $T \to 0$ and $T \to T_f$. The results are

$$q(T \to 0) = 1 - \left(\frac{2}{3\pi}\right)^{\frac{1}{2}} \frac{T}{T_f} \tag{5.23}$$

and

$$q(T \to T_f) = -\frac{1}{2}\left[1 - \left(\frac{T_f}{T}\right)^2\right] \tag{5.24}$$

We can write the susceptibility via the fluctuation-dissipation theorem as

$$\chi(T, H = 0) = \frac{(g\mu_B)^2}{3k_B T} \sum_{i,j} \left[\langle\langle S_i^2\rangle_T\rangle_C - \langle\langle S_i\rangle_T^2\rangle_C\right] \tag{5.25}$$

where all $i \neq j$ terms have to be set equal to zero (no long- or short-range correlations). Since $\langle\langle S_i^2\rangle_T\rangle_C = 1$ and $\langle\langle S_i\rangle_T^2\rangle_C = q$ we obtain

$$\chi(T, H = 0) = \frac{(g\mu_B)^2}{3k_B T}(1 - q(T)) \approx \chi_{ac}(T) \tag{5.26}$$

Substituting the limiting results for $q(T)$ and since $H = 0$, we write for the ac-susceptibility

$$\chi_{ac}(T \lesssim T_f) = \frac{(g\mu_B)^2}{3k_B T_f} - O(T_f - T)^2 \tag{5.27}$$

and

$$\chi_{ac}(T \to 0) = \frac{(g\mu_B)^2}{3k_B T}\left(\frac{2}{3\pi}\right)^{\frac{1}{2}} \frac{T}{T_f} = \text{const} \tag{5.28}$$

According to these equations an asymmetric cusp is formed in the susceptibility at T_f, while χ should approach a constant value at very low temperatures.

Once we have the free energy $F(q)$, the internal energy U can easily be calculated. Since the specific heat C_m is just $\partial U/\partial T$, $C_m(T)$ follows from differentiating and knowing $q(T)$. In the original EA estimates a cusp was implied at T_f for the specific heat. The calculations of the EA model were extended by Fischer (1975) who employed a different technique within the mean-field approximation and used quantum spins ($S = \frac{1}{2}$) instead of classical ones ($S = \infty$). In Fig. 5.2 we show the results of the susceptibility and specific heat for these two spin values. Note the quantum theory gives sharp cusps in both these quantities at T_f. The $\chi(T)$ for $S = \frac{1}{2}$ nicely resembles the generic measured behaviour of χ_{ac} even down to the low-temperature constant. In contradistinction, the theoretical specific heat for

5.2 The Edwards–Anderson model

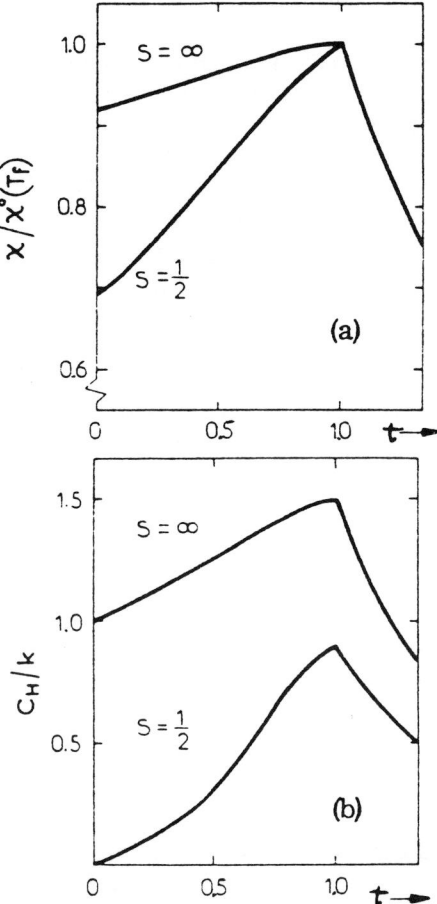

Fig. 5.2 (a) susceptibility and (b) specific heat versus reduced temperature calculated from the EA model for two spin values; from Fischer (1975).

$S = \frac{1}{2}$ completely disagrees with experiment except for the low-temperature linear dependence. So we clearly have a problem with the EA model and its mean-field approximation. Nevertheless, this was the first provocative attempt at coming to grips with the 'cusp' and a possible phase transition. The model is really very simple, yet elegant: replace the site disorder and RKKY interaction by a random set of bonds which satisfy a gaussian distribution. The clever definition of an order parameter related to a time-correlation function makes the freezing transition tractable with a statistical mechanics treatment. What still remains is to establish the true mean-field theory for this model.

5.3 SHERRINGTON–KIRKPATRICK MODEL

A mean-field theory (MFT) is usually a first-order approximation for describing a second-order phase transition. For ferromagnetism the MFT becomes exact in the limit of infinite-range interactions. Another applicable example is superconductivity where the interaction between the electrons forming the Cooper pair is long range. Once the MFT has been established, the necessary additions or corrections, e.g. fluctuations, short-range interactions, etc., can be included to treat real systems. Sherrington and Kirkpatrick (SK) in 1975 proposed that the proper MFT of spin glasses should be the exact solution of an infinite-range EA model where every spin (N is their number) couples equally with every other spin. This means the probability distribution $P(J_{ij})$ is assumed (unphysically) to be the same for all i–j pairs of spins independent of how far they are apart.

SK considered an Ising spin glass with a gaussian distribution

$$P(J_{ij}) = \frac{1}{\sqrt{2\pi\Delta'^2}} \exp\left[-(J_{ij} - J'_o)^2/2\Delta'^2\right] \quad (5.29)$$

where a mean, J'_o, has been included for the possibility of ferromagnetism in the gaussian function. Infinite-range interactions require the scaling of the variance and the mean according to $\Delta' = \Delta/N^{\frac{1}{2}}$ and $J'_o = J_o/N$, so that both our new Δ and J_o are intensive quantities. Thus

$$P(J_{ij}) = \left(\frac{N}{2\pi}\right)^{\frac{1}{2}} \frac{1}{\Delta} \exp\left[-N(J_{ij} - J_o/N)^2/2\Delta^2\right] \quad (5.30)$$

Repeating the 'replica trick', i.e., calculate $\langle Z^n \rangle_C$ instead of $\langle \ln Z \rangle_C$, as with the EA calculations and after a lot of mathematics, which nobody likes to show, SK arrived at a rather complicated expression for $\langle Z^n \rangle_C$, the configurational-averaged, replica partition function. Substituting this expression into the replica equation, we obtain

$$F = -k_B T \langle \ln Z \rangle_C = -k_B T \lim_{n \to 0} \frac{1}{n} (\langle Z^n \rangle_C - 1) \quad (5.31)$$

$$-F/k_B T = \lim_{n \to 0} \frac{1}{n} \left\{ \exp\left[\frac{\Delta^2 N n}{4(k_B T)^2}\right] \int\int_{-\infty}^{+\infty} \prod_{(\alpha\beta)} \left(\frac{N}{2\pi}\right)^{\frac{1}{2}} \right.$$

$$\frac{\Delta}{k_B T} dy^{(\alpha\beta)} \prod_{(\alpha)} \left(\frac{NJ_o}{2\pi k_B T}\right)^{\frac{1}{2}} dx^{(\alpha)}$$

$$\times \exp\left[-\frac{N\Delta^2}{(k_B T)^2} \sum_{(\alpha\beta)} \frac{1}{2}(y^{(\alpha\beta)})^2 - \frac{NJ_o}{k_B T} \sum_{(\alpha)} \frac{1}{2}(x^{(\alpha)})^2\right]$$

$$\times \exp\left[N \ln Tr_S \exp\left[\left(\frac{\Delta}{k_B T}\right)^2 \sum_{(\alpha\beta)} y^{(\alpha\beta)} S^{(\alpha)} S^{(\beta)}\right.\right.$$

5.3 Sherrington–Kirkpatrick model

$$+ \frac{J_o}{k_BT} \sum_{(\alpha)} x^{(\alpha)} S^{(\alpha)} \Bigg]\Bigg] - 1 \Bigg\} \tag{5.32}$$

Here $y^{(\alpha\beta)}$ and $x^{(\alpha)}$ are the dummy variables of integration where $(\alpha\beta)$ label distinct pairs of replicas α and β which take on values from 1 to n. The trace is over 2^n values of $S^\alpha = \pm 1$ at a single site. SK first took the thermodynamic limit $N \to \infty$, then the replica limit $n \to 0$ in order to perform more easily the integration (method of steepest descent). They considered the various replicas to be indistinguishable. This is called the *replica-symmetric* solution with

$$q = q_{\alpha\beta} = \langle\langle S^\alpha S^\beta\rangle_T\rangle_C \tag{5.33}$$

for the spin-glass order parameter and

$$m = m_\alpha = \langle\langle S^\alpha \rangle_T\rangle_C \tag{5.34}$$

for the ferromagnetic one.

The problem is now to calculate $F(q, m)$ in the limit $n \to 0$ and by differentiating with respect to q and m determine the self-consistent simultaneous equations for q and m. The SK results are

$$q = \frac{1}{\sqrt{2\pi}} \int \exp\left(\frac{-z^2}{2}\right) \tanh^2\left[\frac{\Delta q^{\frac{1}{2}}}{k_BT}z + \frac{J_o m}{k_BT}\right] dz \tag{5.35}$$

$$m = \frac{1}{\sqrt{2\pi}} \int \exp\left(\frac{-z^2}{2}\right) \tanh\left[\frac{\Delta q^{\frac{1}{2}}}{k_BT}z + \frac{J_o m}{k_BT}\right] dz \tag{5.36}$$

Hence for given ratios of J_o/Δ, $q(T)$ and $m(T)$ may be calculated and a magnetic phase diagram is thereby established. Figure 5.3 shows this T

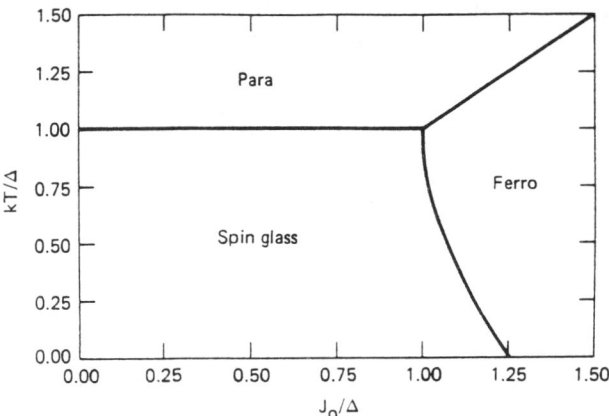

Fig. 5.3 Magnetic phase diagram for Ising spins interacting via an infinite-ranged gaussian distribution of exchange forces with variance Δ and mean J_o; from Sherrington and Kirkpatrick (1975).

versus J_o/Δ plot for Ising spins interacting via an infinite-ranged gaussian distribution of exchange forces centred at J_o with width Δ. Note the three possibilities for phase transitions which are predicted from the SK model: (i) paramagnetic → spin glass; (ii) paramagnetic → ferromagnetic; and (iii) double (or re-entry) transitions paramagnetic → ferromagnetic → spin glass.

The differential (or ac-) susceptibility may be computed from the $q(T)$ function using

$$\chi(T) = \frac{[1 - q(T)]}{k_B T - J_o[1 - q(T)]} \qquad (5.37)$$

By including the applied field H from our original hamiltonian in all the SK calculations (something we omitted for simplicity), we can also obtain the field dependence of $\chi(T)$. Figure 5.4 exhibits the susceptibility behaviour for $J_o/\Delta = 0$ and 0.5 with and without a field $H = 0.1 \Delta$. Once again, conforming to experiment, we have a cusp in the ac-χ which becomes rounded and shifted downward in a dc field. But, when the specific heat is calculated according to the same energy-derivative procedure used in the EA model, there is also a cusp in the predicted $C_m(T)$ at T_f. For $T > T_f$, $C_m = Nk_B\Delta^2/2(k_B T)^2$, hence a tail in $C_m \propto 1/T^2$ persists to higher temperatures. (Remember from the virial expansion treatment of the specific heat at high temperatures (section 5.1) $C_m \propto 1/T$.) The SK result is in stark contrast to the usual mean-field-theory conclusion for a pure system where $C_m = 0$ for $T > T_f$. The leading term for the spin-glass specific heat at low temperatures is proportional to T. However, the entropy S, when determined

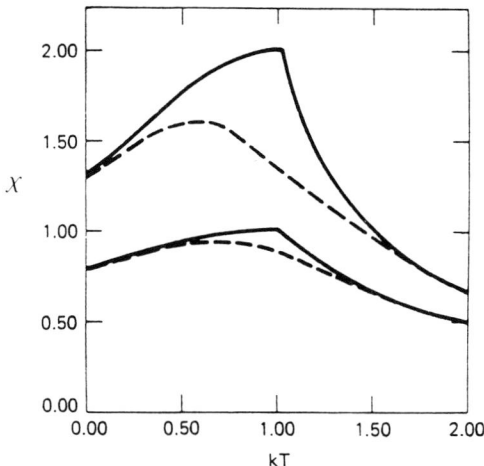

Fig. 5.4 Differential susceptibility as a function of temperature calculated from the SK model: solid curves $H = 0$; dashed curves $H = 0.1 J_o$; lower curves $J_o/\Delta = 0$; upper curves $J_o/\Delta = 0.5$. From Sherrington and Kirkpatrick (1975).

5.4 Instability of the SK solution

from the SK model, goes to a negative limit: $-Nk_B/2\pi$ at $T = 0$. This is a most unphysical and disturbing feature of the model.

One of the impressive experimental realizations of the SK model is the ternary $Pd_{1-y-x}Fe_yMn_x$ system. Here Fe_y controls the amount of ferromagnetism present, viz. the J_o, and essentially Mn_x creates a mixed interaction which governs Δ. With a suitable choice of y and x, a phase diagram closely resembling (mirror image) that of Fig. 5.3 is generated and experiment clearly shows the three distinct types of 'phase-transitions' listed above. Furthermore, there is also a sound qualitative agreement between the SK-model results and the measured susceptibility and its field dependences for the various concentration regions.

5.4 INSTABILITY OF THE SK SOLUTION

Theoretically the first warning that something was wrong with the SK solution came from the negative, low-temperature entropy. Other difficulties became apparent with the free energy which turns out to be a maximum with respect to q for the solutions $q = 0$, $T > T_f$ and $q \neq 0$, $T < T_f$. Moreover, the $q = 0$ solution, if analytically continued below T_f has the lower free energy than the spin-glass state ($q \neq 0$). These results are opposite to what is expected for a conventional second-order phase transition.

Finally, after a few years, de Almeida and Thouless (1978) performed a detailed analysis of the SK solution and showed it to be *unstable* at low temperatures both in the spin-glass and ferromagnetic phases. According to the SK solution the spin-glass susceptibility is really negative. This is not only in contradiction with experiment, but also with the correlation-function susceptibility which is positive definite. In the presence of an applied field ($H \neq 0$) the instability line of the SK-solution for the spin-glass phase extends all the way from T_f down to 0. Such behaviour is shown in Fig. 5.5 where we plot the H-T line (also called the AT line) that gives the stability limits of the SK solution. The functional form is

$$\frac{T_f - T_{AT}(H)}{\Delta} = \left(\frac{3}{4}\right)^{\frac{1}{3}} \left(\frac{H}{\Delta}\right)^{\frac{2}{3}} \tag{5.38}$$

for $T_f \approx T_{AT}(H)$ and an Ising spin glass. This becomes exponential in $-H^2$ as $T \to 0$. An experimental consequence is that below the stability line irreversibilities in the magnetic properties should appear. As we know from Chapter 3 these irreversibilities are indeed a decisive characteristic of the frozen spin-glass state.

The reason for the instability has to do with treating all the replicas as indistinguishable. Remember SK set $q = q_{\alpha\beta}$ in (5.33) – a replica symmetric solution, and such an assumption leads to an *invalid* solution of the mean-field EA model. Despite these theoretical difficulties, the SK model offers

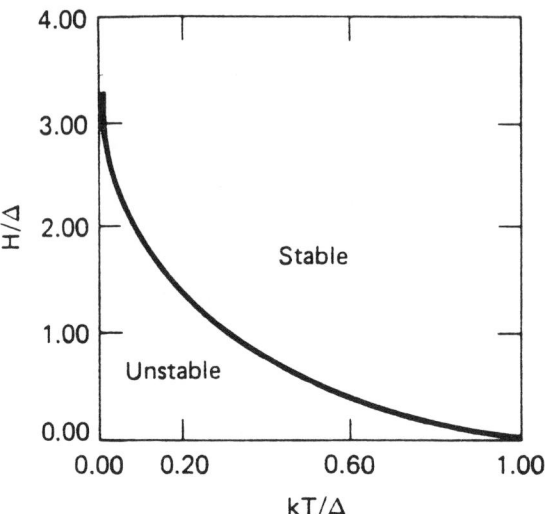

Fig. 5.5 H–T phase diagram (or AT line) illustrating the stability limits of the SK solution for the case of $J_o = 0$; from de Almeida and Thouless (1978).

a reasonable first basis for comparison with experiment. As we mentioned above (section 5.3), the predicted phase diagram can be nicely mimicked by real spin-glass materials, and the calculated susceptibility is in qualitative agreement with measurement. This accord seems to be due to the SK-solution being correct above $T_{AT}(H)$. And for certain experimental quantities (e.g. the ac-susceptibility), which are not strongly dependent on the irreversibilities or energy of the frozen state, the theory is valid. Yet for the majority of other measured parameters (specific heat, magnetization, etc.) the above instability causes the SK-model to break down and more is needed to describe the subtleties of the frozen spin glass.

Theoretically what do we do now? After using so much mathematics to derive a simple mean-field theory that looked attractive from an experimental point of view, we find the solution to be unstable. One answer is to find a scheme and solution which breaks the replica symmetry. Nonetheless before proceeding with devising an RSB scheme, we consider a different approach that does not rely on the replica trick.

5.5 THE TAP APPROACH

Let us recall what the usual mean-field equations for a magnet (ferro or antiferro) are

$$\langle S_i \rangle_T = \tanh\left[\frac{1}{k_B T}\left(\sum_j J_{ij}\langle S_j \rangle_T + g\mu_B H_i\right)\right] \tag{5.39}$$

5.5 The TAP approach

where the thermal averaging of the hyperbolic tangent has been replaced by the thermal average of S_j, namely, $\langle f(S_j) \rangle_T \approx f(\langle S_j \rangle_T)$. And such is the standard mean-field approximation. However, for a spin glass we must subtract from the 'effective field' given above a *reaction term*. This reaction field represents the influence of the ith spin on the polarization of the neighbouring spins at site j. A spin $\langle S_i \rangle_T$ produces a 'field' $J_{ij}\langle S_i \rangle_T$ at site j and induces a moment $J_{ij}\langle S_i \rangle_T \chi_{jj}$ at that site (where χ_{jj} is the local susceptibility). The induced moment in turn produces a reaction field back at site i:

$$J_{ij} \cdot J_{ij}\langle S_i \rangle_T \chi_{jj} = J_{ij}^2 \langle S_i \rangle_T \frac{1}{k_B T}[1 - \langle S_j \rangle_T^2] \quad (5.40)$$

using the linear response form for the susceptibility

$$\chi_{jj} = \frac{1}{k_B T}[1 - \langle S_j \rangle_T^2] \quad (5.41)$$

By including this correction term in our original mean-field equation, we now have

$$\langle S_i \rangle = \tanh\left\{\frac{1}{k_B T}\sum_j J_{ij}\langle S_j \rangle_T - \left(\frac{1}{k_B T}\right)^2 \sum_j J_{ij}^2[1 - \langle S_j \rangle_T^2]\langle S_i \rangle_T\right\} \quad (5.42)$$

where $i = 1, 2, 3, \ldots, N$ and we have neglected the local applied field H_i. This set of N coupled equations has been proposed and studied by Thouless, Anderson and Palmer (1977) as a way of avoiding the replica method. The resulting free energy is

$$F_{TAP} = -\sum_{ij} J_{ij}\langle S_i \rangle_T \langle S_j \rangle_T - \frac{1}{2}\left(\frac{1}{k_B T}\right)\sum_{ij} J_{ij}^2 [1 - \langle S_i \rangle_T^2][1 - \langle S_j \rangle_T^2]$$

$$+ \frac{1}{2}\frac{1}{k_B T}\sum_i \left[(1 + \langle S_i \rangle_T) \ln^{\frac{1}{2}}(1 + \langle S_i \rangle_T)\right.$$

$$\left. + (1 - \langle S_i \rangle_T) \ln^{\frac{1}{2}}(1 - \langle S_i \rangle_T)\right] \quad (5.43)$$

which may be verified by taking $\partial F_{TAP}/\partial \langle S_i \rangle_T = 0$ and arriving back at the $\langle S_i \rangle_T$ equation. Also the TAP approach is only valid near T_f around the region $1 - q \leq T/T_f$ and in the low-temperature limit ($T \to 0$). Here q is the EA order parameter defined now [compare with (5.12)] as

$$q = \frac{1}{N}\sum_i \langle S_i \rangle_T^2 \quad (5.44)$$

The physical meaning of F_{TAP} in (5.43) is the following. The first term represents the energy of the frozen spin glass, the second is the correlation energy of the fluctuations which are reduced from the paramagnetic result by $[1 - \langle S_i \rangle_T^2]$ for each spin; and the last term gives the entropy of N Ising spins possessing the mean values $\langle S_i \rangle_T$.

At $T \ll T_f$, TAP find that the order parameter behaves like

$$q(T) = 1 - \alpha_1 \left(\frac{T}{\Delta}\right)^2 \tag{5.45}$$

which is different from the linear SK dependence $q = 1 - \alpha_2 T/\Delta$ derived from (5.35). Near T_f there is a saddle-point behaviour defining the accessible region $q_0 \geq q = 1 - (T/T_f)$, below which divergences occur in the F_{TAP}. Thus, a linear temperature dependence of q occurs just below T_f. According to these $q(T)$ variations we can calculate the low-T susceptibility as $\langle \chi_i \rangle_T = \alpha_1 T/\Delta$ which goes to zero as $T \to 0$ (recalling χ_{SK} remains finite as $T \to 0$ in agreement with experiment). The entropy S may also be determined: $S_{TAP} = -\partial F_{TAP}/\partial T$, and in the low-$T$ limit

$$S_{TAP}(T \to 0) = -\frac{1}{4} N k_B \Delta^2 \lim_{T \to 0} \left(\frac{1}{k_B T}\right)^2 [1 - q(T)]^2 \tag{5.46}$$

Substituting the low-T quadratic form for $q(T)$ in (5.45), $S_{TAP} \to 0$ as $T \to 0$, in stark contrast to the negative constant entropy at $T = 0$ for the SK result. Therefore, the TAP model has a physically meaningful entropy and its configurational-averaged free energy also tends to zero as $T \to 0$, i.e., the TAP ground-state energy vanishes at $T = 0$. These distinctions between the SK and TAP models lie in the two dissimilar temperature dependences of $q(T)$ and in the breakdown of linear-response theory for spin glasses. Even so both models incorrectly predict a cusp in the specific heat at T_f.

While the TAP approach does not have the fundamental theoretical drawbacks of the SK solution, it offers *many* physical and stable solutions below the AT-line with different free energies. We can interpret each of them as a locally stable thermodynamic state whose free energy is not necessarily a global minimum. But what are the exact meaning and relevance to experiment of all these different thermodynamic states for real spin glasses? Our attempt at an answer will emerge below. Nevertheless, above the irreversibility or AT line (shown in Fig. 5.5), similar results for the 'paramagnetic' phase are obtained from both SK and TAP models. The hint here is clear, the frozen spin-glass state is a unique phase that requires new physics, e.g. an *unconventional* order parameter and statistical treatment. Consequently, a new approach to the theory and interpretation is needed.

5.6 HOW TO BREAK THE REPLICA SYMMETRY

Since we know where the trouble lies let's try to overcome the replica symmetry by introducing a scheme that breaks it. After a number of

5.6 How to break the replica symmetry

attempts, the correct one was finally proposed by Parisi in 1979. His procedure for breaking the symmetry of the order-parameter matrix was as follows. Consider an $n \times n$ replica symmetric matrix q_o (for illustrative purposes we take $n = 8$). We then divide the $(n \times n)$ matrix of elements q_o into constant blocks $[(n/m_1) \times (n/m_1)]$ of size $m_1 \times m_1$. Along the diagonal we introduce sub-matrix q_1. The off-diagonal blocks are left unchanged with elements q_o. This process of dividing 8×8 matrices into 4 blocks of 4×4 matrices is portrayed in Fig. 5.6 and is only the first of many such steps. The division is again repeated along the diagonal: $m_1/m_2 \times m_1/m_2$ (or 4) new blocks of $m_2 \times m_2$ (2×2) sub-matrices are created with order parameter q_2 for the new diagonal ones. And so on (see Fig. 5.6) the procedure is reiterated with smaller and diagonal blocks each with its own order parameter q_k. (In our 8×8 matrix shown in Fig. 5.6, the next iteration results in all zeros along the diagonals.) Finally, for a much larger matrix the process is terminated after R such iterations with the smallest diagonal block of size $m_R \times m_R$. Throughout this construction, the successive sizes of the blocks are

$$n \geq m_1 \geq m_2 \cdots \geq m_R \geq 1 \tag{5.47}$$

which in the limit as $n \to 0$ becomes 'turned around'

$$0 \leq m_1 \leq m_2 \cdots \leq m_R \leq 1 \tag{5.48}$$

By taking R, the number of iterations, to be very large, we obtain a *continuous* variation and so

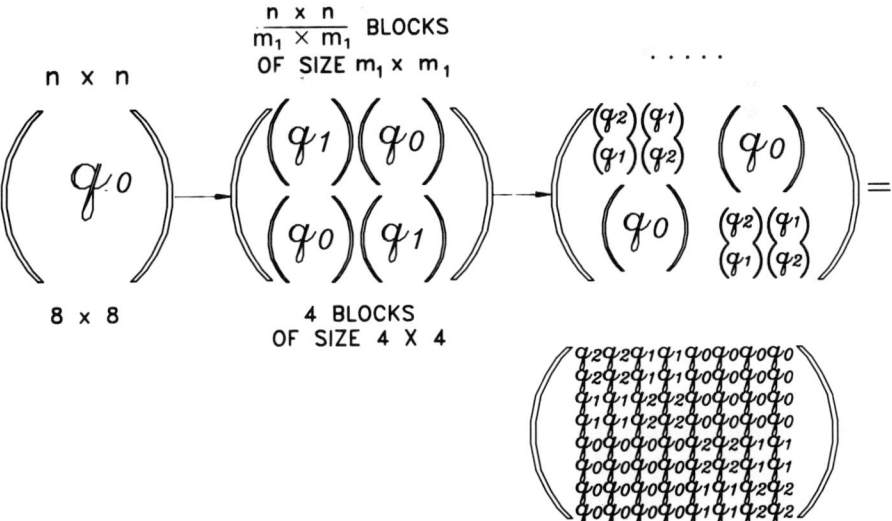

Fig. 5.6 Replica symmetry breaking (RSB) scheme for $q_{\alpha\beta}$ with two levels of breaking ($n = 8$, $m_1 = 4$; $m_2 = 2$).

$$m_k/m_{k+1} \to 1 - dx/x \quad \text{and} \quad q_k \to q(x) \tag{5.49}$$

where x is defined over the unit interval $0 \le x \le 1$. In other words, we now have an infinite number of order parameters. Note that the SK solution corresponds to the $k = 0$ step or original q_o matrix which means that $q(x)$ is independent of x.

There is an alternative way of visualizing the Parisi replica-symmetry-breaking scheme. The order-parameter matrix $q_{\alpha\beta}$ can be represented by a tree with emanating branches. Again we utilize the (8 × 8) or 8 replica matrix of Fig. 5.6 and sketch our tree construction in Fig. 5.7. At lowest order q_o all the elements are equal and contained in one big branch. The sub-division process breaks up the branch into two smaller ones which in turn may be further broken up into two yet smaller branches, etc. The integers represent the $n = 8$ replicas and as they are divided q increases towards 1. In order to find the value of $q_{\alpha\beta}$ for any two of our replicas (in Fig. 5.7, α is replica 1 and β is 4), we trace these back until they join at q_1. The overlaps of the different replicas are all equal if the point of encounter lies along the same horizontal line. For example in Fig. 5.7, q_{12}, q_{34}, q_{56}, q_{78} all have equal overlap with value q_2, however, q_{18}, i.e., replicas 1 and 8, has a different overlap since it finally joins branches at q_o.

The concept of replica overlap has already been treated in pure mathematics and is known as an ultrametric space. It is defined by the so-called ultrametric inequality

$$d(A, C) \le \text{Max}\{d(A, B), d(B, C)\} \tag{5.50}$$

where the ds represent the distances AC, AB and BC. This is a much stronger inequality than the usual triangular one with the same three distances

$$d(A, C) \le d(A, B) + d(B, C). \tag{5.51}$$

For a spin glass, which was shown to satisfy the ultrametric inequality, we mean the 'overlap' (equivalent to the above distance) between two spin configurations or replicas

$$q^{\alpha\beta} = \frac{1}{N} \sum_{i=1}^{N} S_i^\alpha S_i^\beta \tag{5.52}$$

(Ising spins). The ultrametric inequality now becomes

$$q^{\alpha\beta} \ge \text{Min}\{q^{\alpha\beta}, q^{\beta\gamma}\}. \tag{5.53}$$

This inequality implies that the space of states can be divided into clusters of states, each cluster of state may be subdivided into subclusters, and so on. Figure 5.8 depicts such a space of clusters with separation into smaller and smaller clusters. Note there are no overlappings among the different clusters of similar order, i.e., each point lies in just one cluster. The sketch in Fig. 5.8 is nothing but another way of looking at our tree diagram of

5.7 Physical meaning of RSB

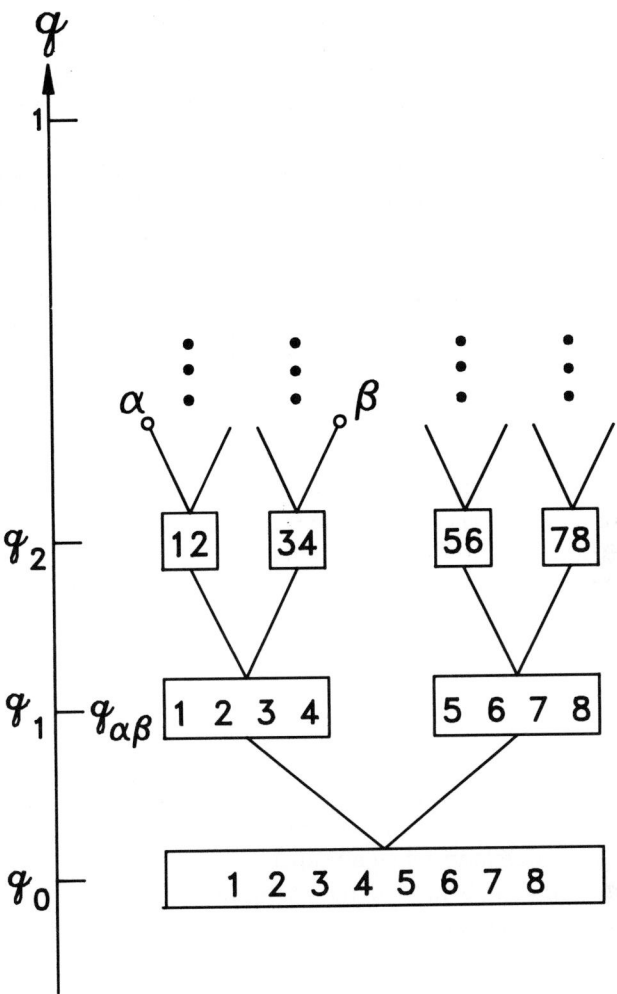

Fig. 5.7 Tree representation of Parisi's RSB scheme. To find $q_{\alpha\beta}$ trace back along the branches of the tree from α and from β until they join: $q_{\alpha\beta} = q_1$ is the value of q at this point.

Fig. 5.7, both of which are a direct result of ultrametricity. And this new 'space', called ultrametric, is a natural consequence of the replica-symmetry breaking required to remove the defects of the SK solution

5.7 PHYSICAL MEANING OF RSB

Returning briefly to experiment, let us try and see how this highly sophisticated mathematical theory makes contact with what we have

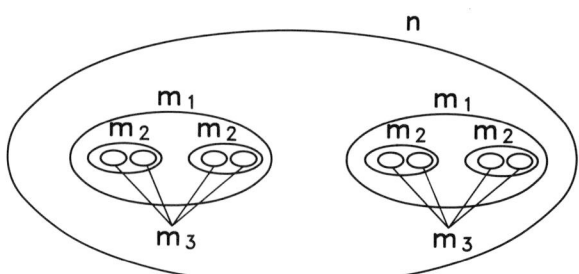

Fig. 5.8 Sketch of the cluster structure of ultrametric space. n = 8 {1,2,3,4,5,6,7, 8}; m_1 = 4 {1,2,3,4} and {5,6,7,8}; m_2 = 2 {1,2,}, {3,4}, {5,6} and {7,8}; m_3 = 1 {1}, {2}, {3}, {4}, {5}, {6}, {7} and {8}. Note how closely this subdivision follows the tree representation and the replica symmetry breaking scheme of our 8 × 8 matrix.

measured on real spin glasses. The key information after the RSB has been accomplished is contained in our now-continuous Parisi order parameter $q(x)$. This is obtained in the interval 0 to 1 via the matrix construction and limit taking procedures

$$-\lim_{n \to 0} \frac{1}{n} \sum_{\alpha\beta} q_{\alpha\beta} = \int_0^1 q(x)\, dx \tag{5.54}$$

A number of calculations to determine $q(x)$ have been performed at different temperatures and fields. In Fig. 5.9 we illustrate the various behaviours of $q(x)$ and $P(q) \equiv dx/dq$ (see below) for several interesting cases spanning a range of T and H values. A careful perusal of this figure is necessary to understand its significance.

According to the RSB model, the susceptibility and the internal energy are, respectively, given by

$$\chi = \frac{1}{k_B T} \int_0^1 [1 - q(x)]\, dx \tag{5.55}$$

and

$$U = -\frac{1}{2k_B T} \int_0^1 [1 - q^2(x)]\, dx \tag{5.56}$$

We call attention to the breakdown of linear-response theory, and its corresponding susceptibility $\chi = (C/T)(1 - q)$, in the RSB scheme. By substituting the results for $q(x)$ at several different temperatures, we can calculate $\chi(T)$ by evaluating the integral. The Parisi and SK results are compared in Fig. 5.10 for the low-field susceptibility. Notice the dissimilarity with respect to the SK form, for, χ(RSB) is constant for all $T < T_f$, and this T-independent susceptibility certainly resembles our field-cooled one,

5.7 Physical meaning of RSB

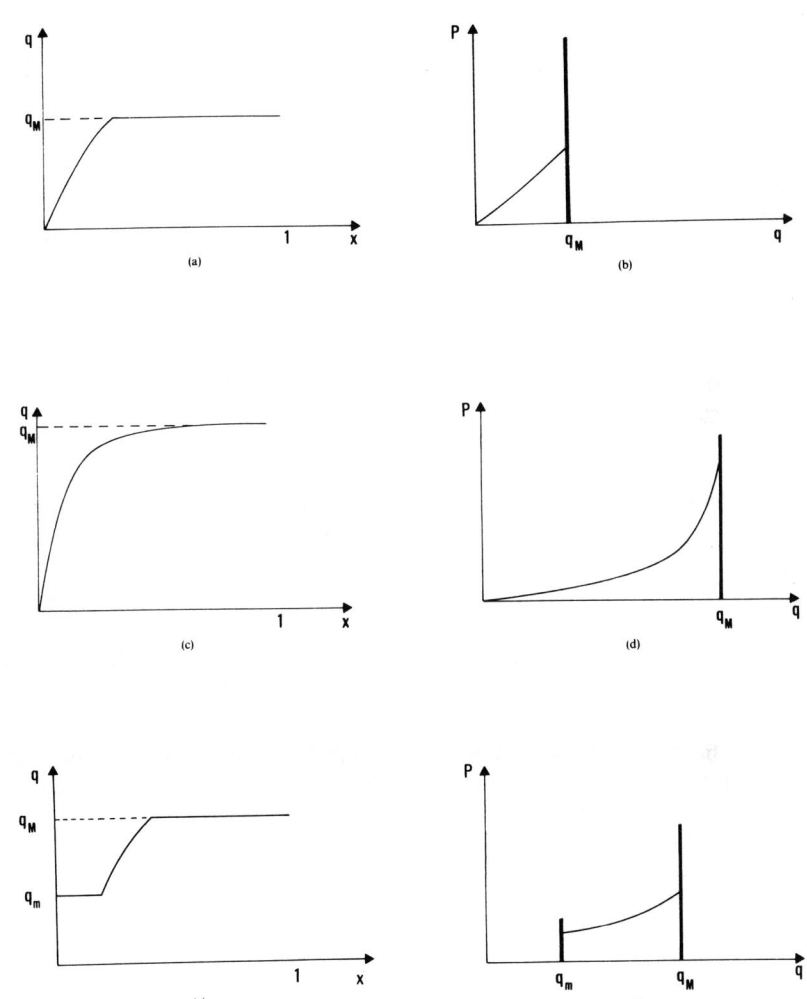

Fig. 5.9

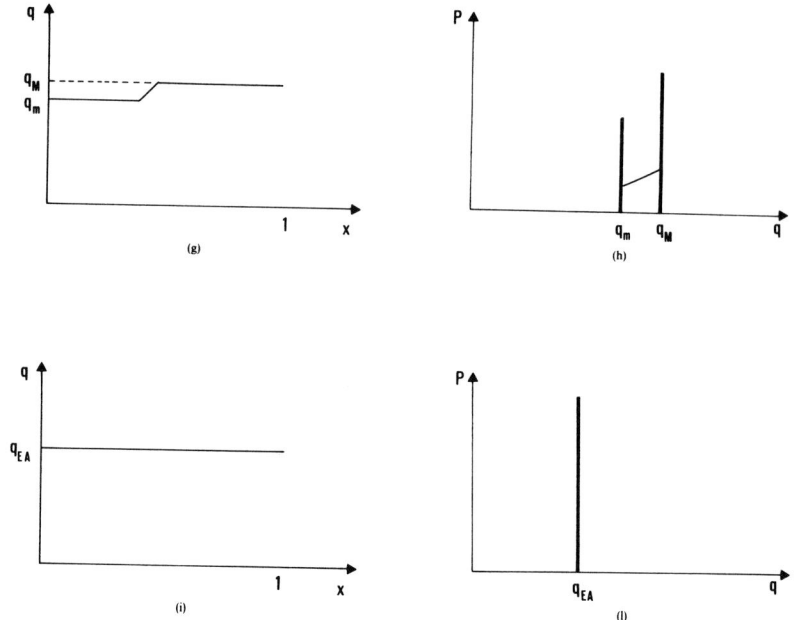

Fig. 5.9 Continued Behaviour of the order-parameter function $q(x)$ and the probability distribution of overlaps $P(q)$ for various cases. (*a*) and (*b*) just below T_c in zero field: $q_M \propto T_c - T$ and the initial slope of $q(x)$ is roughly independent of temperature. (*c*) and (*d*) at low-temperature $T \ll T_c$ and in zero field: $q_M \approx 1$. (*e*) and (*f*) for small magnetic fields: $q_m \propto H^{\frac{4}{3}}$. (*g*) and (*h*) at larger magnetic fields near the AT line: $(q_M - q_m)$ is proportional to the distance from this line. And (*i*) and (*j*) in the replica-symmetry region $T > T_f$ at non-zero magnetic field. Note the value of q_{EA} vanishes when $H \to 0$. From Mezard et al. (1987).

χ_{FC}, *not* the ZFC or ac-χ. Further (computer) computations show that $U(T = 0)$ reaches a negative value while $S(T = 0) = 0$, so we have removed via RSB the unphysical negative entropy. Finally, the stability of the Parisi solution has been carefully analysed and has been found to be at least marginally stable.

In order to avoid the subtleties of the RSB approach we digress for a paragraph and use two results of RSB in combination with the stable SK solution (above the instability line) which we project into the spin-glass phase. This is called the 'projection hypothesis' which begins with valid SK mean-field theory. The two features of RSB are that $\chi(T)$ = constant in the spin-glass phase and that linear response is destroyed at low T. Accordingly, we can assume for $T < T_f$:

$$S(T, H) = S(T)$$
$$M(T, H) = M(H)$$
$$q(T, H) = q(T) \qquad (5.57)$$

5.7 Physical meaning of RSB

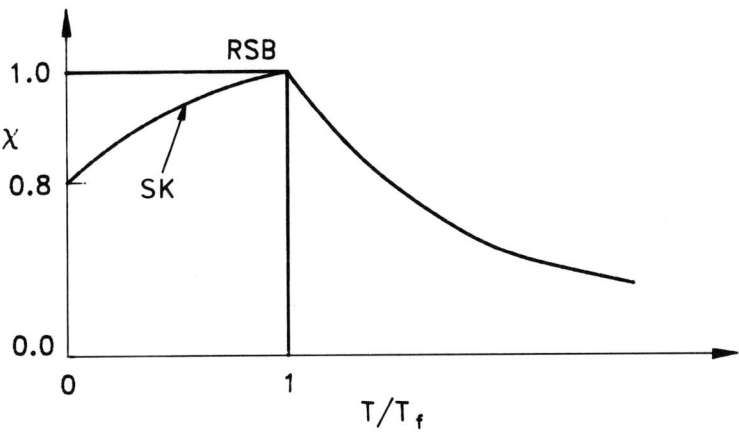

Fig. 5.10 Difference in the temperature dependence of the susceptibility as calculated from the RSB-solution of the SK model and the SK-solution without RSB.

where the latter represents the field independence of the EA order parameter. A Maxwell relation unites the first two assumptions via $(\partial S/\partial H)_T = (\partial M/\partial T)_H$. Starting from the instability line, where the replica symmetric (SK) solution is valid, and using the above projections, expansions can be developed for the magnetization $M(T, H)$ around T_f.

$$M = \frac{H}{T}\left[1 - \frac{H^2}{3T^2}\frac{T^2 + 2T_f^2}{T^2 - T_f^2} + \cdots\right] \quad \text{for } T > T_f \quad (5.58)$$

$$M = \frac{H}{T_f}\left[1 - \frac{|H|}{\sqrt{2T_f}} + \cdots\right] \quad \text{for } T = T_f \quad (5.59)$$

$$M = \frac{H}{T_f}\left[1 - \left(\frac{3}{4}\right)^{\frac{2}{3}}H^{\frac{4}{3}} + \cdots\right] \quad \text{for } T < T_f \quad (5.60)$$

This last equation, describing a non-analytic regime and being independent of T, suggests that a line of singularities extends over the entire interval from T_f down to 0. Equation (5.58) for $T > T_f$ shows that, while there is a cusp (or really a plateau) in $\chi = M/H$; the non-linear susceptibility

$$\chi_{n1} \equiv \frac{\partial^3 M}{\partial H^3} = \frac{2}{T^3}\frac{T^2 + 2T_f^2}{(T + T_f)(T - T_f)} \propto (T - T_f)^{-1} \quad (5.61)$$

diverges according to the mean-field critical exponent $\gamma = 1$. In addition the above equations exhibit the field rounding of χ (and also for C_m) around T_f. So here we have established a few preliminary points of contact with experiment, some in agreement as with the FC susceptibility and field smearing, and others clashing as with the specific heat. But much more is needed in the interpretation of RSB.

Is there a simple physical picture for the replica symmetry breaking (RSB) and the resulting ultrametricity? Well let's attempt to sketch one. During the freezing of a spin glass, the various spins can align their directions in many different ways such that all these arrangements have similar values of the free energy. Hence, a frozen spin glass possesses a multi-degenerate ground state. The configurations which have lower free energies correspond to *pure* equilibrium states. Other configurations with higher free energies correspond to the metastable states. How do we then characterize the ensemble of equilibrium state? Answer: use a weighting probability w_α for the state α to appear in the ensemble

$$w_\alpha \propto \exp\left(\frac{1}{k_B T} F_\alpha\right) \tag{5.62}$$

where F_α is the free energy of equilibrium state α. Another question is how does one equilibrium state α differ from a second one β. Answer: define a 'distance' $d_{\alpha\beta}^2$ between these states as the difference in their average spin values square

$$d_{\alpha\beta}^2 = \frac{1}{N}\sum_i (\langle S_i\rangle_\alpha - \langle S_i\rangle_\beta)^2 \tag{5.63}$$

In a similar manner we can construct an 'overlap' between two states α and β

$$q_{\alpha\beta} = \frac{1}{N}\sum_i \langle S_i\rangle_\alpha \langle S_i\rangle_\beta \tag{5.64}$$

Here the EA order parameter q_{EA} represents states which have the same overlap with themselves, i.e., $q_{\alpha\alpha} = q_{\beta\beta} = q_{EA}$. Thus

$$d_{\alpha\beta}^2 = 2(q_{EA} - q_{\alpha\beta}) \tag{5.65}$$

So we have arrived at a simple relation between distance and overlap via (5.65).

Now we must consider the probability distribution $P(q)$ of the various overlaps among the pure equilibrium states

$$P(q) = \overline{P_J(q)} = \overline{\sum_{\alpha\beta} w_\alpha w_\beta \,\delta(q_{\alpha\beta} - q)} \tag{5.66}$$

where the *bar* averaging is over different values of the couplings J. In other words, $P(q)$ is the probability of finding two states with overlap q after weighting each state with their ensemble probabilities $w_{\alpha,\beta}$. The physical interpretation of RSB is grounded in the non-trivial result that $P(q)$, defined above, and $P(q) = dx/dq$ (introduced at the beginning of this section and shown in Fig. 5.9) are identical.

5.7 Physical meaning of RSB

Let us examine a most simple example via this formalism, namely, the ferromagnetic Ising model. In zero magnetic field at $T < T_c$, $P(q)$ contains two delta functions at $q = \pm m^2$ where m is the spontaneous magnetization. If an up (or +) external field is applied, then only one delta function at $q = +m^2$ remains. For the former case there are two equilibrium states; for the latter only one. Now we return to the spin glasses. At $T > T_f$ one delta function at $q = 0$ exists (the paramagnetic state). However, as T is lowered below T_f in zero field, a delta function survives but along with it is a smooth region of probabilities in between 0 and q_M – see Fig. 5.9. Here, when $P(q)$ is *not* an array of discrete delta functions, we say that the replica symmetry is broken. This usually occurs in a very smooth way as the temperature is reduced through T_f. Smeared out transitions? Consequently, there are a significant number (not just one or two) of pure equilibrium states which compose the spin-glass ground state. And the weight and overlap of these states generate the continuous function $P(q)$.

Figure 5.11 gives a schematic, 1D picture of the free-energy landscape that exists in multi-dimensional configurational space. The lowest-lying minima represent the pure equilibrium states. At higher energies are the many metastable states. Upon cooling, a spin glass may become stuck in one of these states. If the system cannot explore all of phase space, then we call it non-ergodic. RSB leads naturally to ultrametricity which in turn is pictured by the multi-valley free-energy landscape. And when the barriers between the valleys become infinite, ergodicity is lost. Note that the time it takes to go from one valley to another is the exponential of the height of the lowest saddle point between these two valleys. So it is not surprising

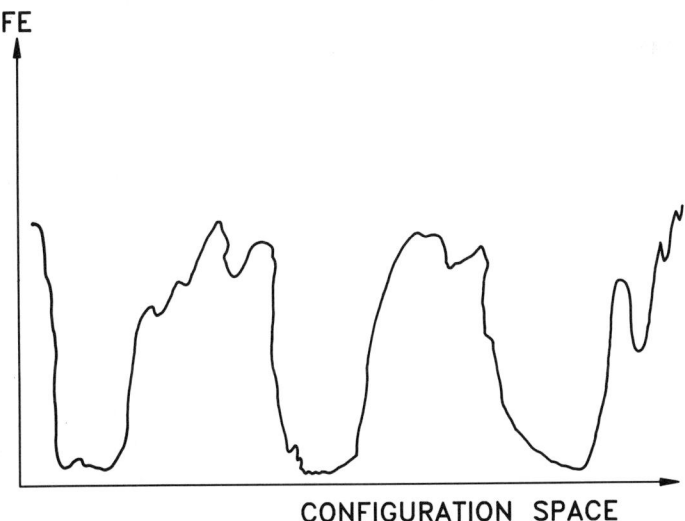

Fig. 5.11 Multi-valley 'landscape' for the free energy according to the RSB scheme.

that there are many different relaxation times present, and we should now move on to examine the dynamics of spin glasses within the SK and RSB models. However, before we do, we must mention a final consequence of the RSB. This is the lack of self-averaging.

Usually in a random system we can average over the extensive quantities by dividing it into a large number of macroscopic subsystems, each containing a different set of random variables $\{x\}$. If we then average over the subsystems, we expect to obtain a result very similar to performing a complete configurational average over all choices of $\{x\}$. Taking the magnetization per spin as our extensive quantity, self-averaging may be written as

$$M\{x\} - \langle M \rangle_C \to 0 \quad \text{for } [N\text{-spins}] \to \infty \tag{5.67}$$

where the chosen set $\{x\}$ occurs with reasonable probability. In other words, one large random system gives the same result for the magnetization as a configurational average over all choices of $\{x\}$. Clearly, when we measure the magnetization on various samples of CuMn having the same concentration, there are no differences in the experimental values, even though each sample has a different set $\{x\}$. The configurational-averaged magnetization is expressed as

$$\langle M \rangle_C = \frac{1}{N} \sum_i \langle \langle S_i \rangle_T \rangle_C \tag{5.68}$$

But after performing the configurational average on the right-hand side, translational invariance is restored to the spin system. So each term in the summation is equal. Thus $\langle M \rangle_C = \langle \langle S_i \rangle_T \rangle_C$. And using our definition in (5.67), of self-averaging $M\{x\} = \langle M_C \rangle$ for $N \to \infty$, we have

$$\frac{1}{N} \sum_i \langle S_i \rangle_T = \langle \langle S_i \rangle_T \rangle_C \tag{5.69}$$

where the summation runs over all the spin sites for a single large sample.

Although self-averaging is intuitive and seems almost obvious, it does not work when a phase transition occurs which makes the system depend crucially on the boundary conditions. For our spin glass the RSB solution creates the existence of many degenerate ground states and this causes the lack of self-averaging. Such quantities as $\langle q_{ij}^2 \rangle_C - q^2$ (overlaps) and $\langle P_J(q_1) P_J(q_2) \rangle_C - P(q_1) P(q_2)$ (weights) do not equal zero as RSB calculations have shown. Here the weights of the thermodynamic states are the important factor. Extensive quantities depending on these weights will not be self-averaging. This means different bond configurations will give different results even for $N \to \infty$. It turns out in RSB theory that the magnetization is self-averaging, but the susceptibility is not:

$$\chi_J\{J\} = \frac{1}{NT} \sum_{ij} [\langle S_i S_j \rangle_T - \langle S_i \rangle_T \langle S_j \rangle_T] \neq \langle \chi_J \rangle_J \tag{5.70}$$

5.8 Dynamics of the mean-field model

And this is because the thermodynamics weights change rapidly as the field is varied. It is too bad experiment cannot distinguish these two susceptibilities.

5.8 DYNAMICS OF THE MEAN-FIELD MODEL

Since the SK-model considers Ising spins that have no natural intrinsic dynamics, we must introduce time dependences artificially via Glauber dynamics or the soft-spin gaussian-noise term. Near a second-order phase transition time-dependent correlations take the form predicted by dynamical-scaling theory. The autocorrelation function is defined by

$$q(t) = \langle\langle S_i(0)S_i(t)\rangle_T\rangle_C \tag{5.71}$$

where the thermal average $\langle \ \rangle_T$ can be replaced by an average over sufficiently long observation times. Dynamical scaling predicts that the decay of $q(t)$ is governed by a characteristic time τ which (power-law) diverges at T_f. Accordingly, we can write

$$q(t) \propto t^{-\lambda} Q_{\pm}(t/\tau) \tag{5.72}$$

where $Q_{\pm}(x)$ are universal scaling functions and λ is a critical exponent. We now relate the characteristic time to a characteristic length ξ_{SG}, the correlation length for our spin glass,

$$\tau \propto \xi_{SG}^z \propto \left(\frac{T-T_f}{T_f}\right)^{-z\nu} \quad \text{since} \quad \xi = \left(\frac{T-T_f}{T_f}\right)^{-\nu} \tag{5.73}$$

Here ν is the correlation-length critical exponent and z is the dynamical one. In mean-field (SK) theory $\nu = \frac{1}{2}$. Taking the limit of $t \to \infty$ for the autocorrelation function $q(t)$ we have

$$\lim_{t\to\infty} q(t) = q_{EA} \propto \left(\frac{T-T_f}{T_f}\right)^{\beta} \tag{5.74}$$

where β is the order-parameter exponent. In order for the scaling function $Q(x)$ to cancel the $t^{-\lambda}$ dependence in the limit $t \to \infty$, we obtain a relation between the exponents $\lambda = \beta/(z\nu)$.

Above the AT-line where SK is valid, various calculations have all shown for $H = 0$

$$q(t) \propto t^{-\frac{1}{2}} Q_+\left[t\left(\frac{T-T_f}{T_f}\right)^2\right] \tag{5.75}$$

with the proper scaling function $Q_+(x) \propto \exp(-x)$. Thus, for $T > T_f$ and as $t \to \infty$

$$q(t) \propto \frac{1}{\left(\frac{T-T_f}{T_f}\right)^2 t^{\frac{3}{2}}} \exp\left[-t\left(\frac{T-T_f}{T_f}\right)^2\right] \quad (5.76)$$

The scaling function reduces to $q(t) \propto t^{-\frac{1}{2}}$ at T_f. For $H \neq 0$ one can begin with the dynamical-scaling form and define a critical temperature according to $[T - T_{AT}(H)]/T_{AT}(H)$. However, both the scaling functions and the exponents are non-universal and vary along the AT-line. This suggests a marginality of the transition in a field. By examining the dynamical susceptibility near the AT line, theory predicts the frequency dependence of $\chi''(\omega) \propto \omega$ for $T \gg T_f$ crossing over to $\chi''(\omega) \propto \omega^{\frac{1}{2}}$ at T_f.

Below the AT-line, instabilities also occur in the dynamics of the SK model, so a new and different approach is required here. And Sompolinsky (1981) has provided the method with a stable solution entirely within a dynamical framework. We now write our continuous order parameter as

$$q(x) = \langle\langle S_i(0)S_i(\tau_x)\rangle_T\rangle_C \quad 0 \leq x \leq 1 \quad (5.77)$$

where $\langle\cdot\rangle_T$ and $\langle\cdot\rangle_C$ represent thermal and configurational averages, respectively, and τ_x represent a time in the broad continuous spectrum of relaxation times. Note that $\tau_{x_1} \gg \tau_{x_2}$, if $x_2 > x_1$. So as $x \to 0$ we obtain the very long relaxation times necessary to equilibrate the system. This gives the statistical-mechanics order parameter

$$q = q(0) = \lim_{N\to\infty} \lim_{t\to\infty} q(t) \quad (5.78)$$

which Sompolinsky assumed goes to zero in zero field, i.e., all correlations have decayed away. On the other hand, for $x \to 1$ we have the short times scales characteristic of the EA (or confined to one valley of Fig. 5.11) order parameter

$$q(1) = q_{EA} = \lim_{t\to\infty} \lim_{N\to\infty} q(t) \quad (5.79)$$

It is which *limit* comes first that distinguishes these order parameters.

The dynamical susceptibility (measured at frequency $\omega_x = 1/\tau_x$) may be calculated with the help of another function $\Delta(x)$

$$k_B T \int_0^{\tau_x} \chi_{ii}(t)\, dt = 1 - q_{EA} + \Delta(x) \quad (5.80)$$

We assume that linear-response theory (or equivalently, the fluctuation-dissipation theorem) is valid at finite (short) $x \to 1$ times, but not at the very long times $x \to 0$ needed for equilibrium. Hence, at $x = 1$, $\Delta(1) = 0$ and (5.80) reduces to its EA-form. By establishing self-consistent equations for $q(x)$ and $\Delta(x)$, Sompolinsky has determined the following stability condition for his solution, using $M = \tanh(\tilde{H}/k_B T)$ where $\tilde{H}$ is a local field made up of components which are frozen on a time scale τ_x:

5.8 Dynamics of the mean-field model

$$\frac{J^2}{(k_B T)^2}[1 - 2q(1) + \langle M^4 \rangle] = 1 \tag{5.81}$$

The limit is $\Delta(x) \to 0$ (or $x \to 1$) and J is now the width or variance of our gaussian-distribution of exchange interactions. The above equation corresponds to the TAP stability condition which is marginally stable. For $\Delta(x) = 0$ the SK solution is revived since $q(1) = q_{EA}$.

As the external field H goes to zero, the only dynamically stable solution is given by the equilibrium (or long-time) susceptibility $\chi = \chi_{ii}$

$$k_B T_\chi = 1 - q(1) + \Delta(0) \tag{5.82}$$

where in the limit of $x \to 0$, $\Delta(x) = q(1) - 1 + (k_B T)/J$. Hence, $\chi = 1/J$ (a constant below T_f) and we have recovered a dynamical form of Parisi's susceptibility. The interpretation here is that in the short-time limit (e.g. ac-susceptibility) we do obtain a *quasi*-equilibrium (linear-response) χ with its cusp

$$\chi(\text{quasi-equil}) = \frac{1}{k_B T}(1 - q_{EA}) \tag{5.83}$$

However, for the *true* equilibrium susceptibility

$$\chi(\text{equil}) = \frac{1}{k_B T}[1 - q_{EA} + \Delta(0)] \tag{5.84}$$

The time scales necessary for (5.84) to be valid may be far beyond our experimental reach of maximum 10^6 s. Nevertheless, we can measure a field-cooled susceptibility which although not the true equilibrium one, does indeed mimic it in a field. The FC-field seems to drive the spin glass towards its long-time equilibrium, and thus, $\chi_{FC} \doteq$ constant at $T \leq T_f$. For the intermediate cases we must return to the local dynamical susceptibility measured at frequency $\omega_x = 1/t_x$ according to (5.80) and explicitly calculate $\Delta(x)$.

For our practical purposes the Parisi and Sompolinsky results may be placed within the model of RSB and benefit from its consequences. Therefore, we can return to our rugged landscape of Fig. 5.11 and use it to describe the spin-glass dynamics. Two main dynamical processes can be distinguished in the frozen state. Firstly, confined to a single deep valley are the non-exponential-relaxation processes within one phase, i.e., the transversing of small bumps composing the floor of the 'equilibrium' valley. In contradistinction, there are the hopping processes between the equilibrium phases or from one deep valley to another. These appear on much longer time scales and they are estimated to diverge with the size of the system as $\log \tau \sim N^{\frac{1}{4}}$. Now according to RSB, hierarchical distributions of time scales manifest themselves and the hopping has taken place in an ultrametric space.

5.9 DROPLET MODEL

During the mid-1980s a series of numerical studies of domain walls and their scaling properties by McMillan (1980, 1984), and Bray and Moore (1984) stimulated a completely different approach to the theory of spin glasses. Fisher and Huse (1986) advanced these methods and presented the *droplet model*, as a phenomenological scaling theory of droplet excitations in short-range, Ising spin glasses. Central to this theory is the belief that an understanding of the spin-glass phase (which should exist at $T = 0$) can be obtained from its ground-state properties. The basic idea is to define a 'droplet' in the ground state as the lowest-energy excitation of length scale L around a particular point x_j. Figure 5.12 illustrates the 'droplet' which consists of having all its spins placed in the opposite direction with respect to those in the ground state Γ. So $\bar{\Gamma}$ is a global reversal of spins (or spin flips) inside the length scale L. Note that the model assumes a phase transition below T_f where the global spin-reversal symmetry is broken and there are only two pure-equilibrium states related by this global symmetry. The above droplets are the dominant low-lying excitations in each state.

Now we study these low-lying droplet excitations with this scale L. They have a broad distribution of free energies $F_L \propto Y(T) L^\theta$ where Y is the stiffness constant and θ a new exponent. If $\theta < 0$, as is expected for $d = 2$, then $F \propto L^{-|\theta|}$ and the low-lying excitations can be created on longer and longer length scales, thereby destroying the frozen spin-glass phase. Computer simulations indicate that θ becomes positive just below $d = 3$, so a marginal phase transition is anticipated for a 3D spin glass. A distribution of droplet free energies $\rho(F_L)$ can be determined from the above scaling ansatz, and then correlations between the droplet free energies

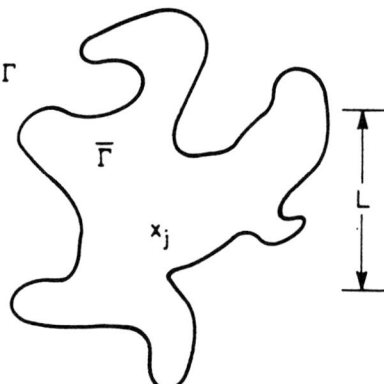

Fig. 5.12 Schematic picture of the droplet model. A droplet of length scale L (containing site j) has all its spins reversed (global spin flip) creating ground state $\bar{\Gamma}$. Outside the droplet the spins are aligned as in ground state Γ. The surface of the droplet is fractal. From Fisher and Huse (1988).

5.9 Droplet model

are calculated. This leads to a non-linear susceptibility χ_{nl} which is infinite when $d > (1 + \phi)\theta$ (ϕ is an exponent governing the dependence of ρ on F_L). Hence, we would expect a true phase transition if this condition is satisfied.

The effect of a uniform magnetic field on the Ising spin-glass state is to destroy it by breaking the global spin-flip symmetry. So argue Fisher and Huse that the two pure-equilibrium state becomes unstable, χ_{nl} is finite, and the system falls out of equilibrium even in the smallest fields. Consequently, in their interpretations there is no true AT line, only a dynamical line which may be accessible to experiment depending on the available measurement time where non-equilibrium behaviour sets in. Although T_f scales with H in the same way as the AT line, the position of this freezing line is weakly dependent on the time allowed for equilibration.

Annihilation and creation of droplet excitations determine the equilibrium low-frequency dynamics of the ordered phase. Here thermal activation over energy barriers is assumed for those droplets which are thermally active, i.e., their free energy is anomalously low: $F_L \approx k_B T \ll Y L^\theta$. The typical free-energy barriers according to the length-scaling ansatz are $B \sim L^\psi$ where ψ is a new independent exponent $\theta \leq \psi \leq d-1$. The characteristic time for the droplet to form or grow to scale L is that necessary to surmount this energy barrier. Thus, a droplet of scale L will last for

$$\tau = \tau_o \exp\left(\frac{B}{k_B T}\right) \quad \text{or} \quad \ln\left(\frac{\tau}{\tau_o}\right) \sim \frac{L^\psi}{k_B T} \tag{5.85}$$

When the autocorrelation function $C(t)$ is derived, the above thinking leads to an extremely slow logarithmic decay of temporal correlations

$$C(t) \sim (\ln t)^{-\theta/\psi} \tag{5.86}$$

Further calculations within the droplet model give a '1/f-noise' spectrum proportional to $1/\omega|\ln \omega|^{1+(\theta/\psi)}$ and real and imaginary parts of the susceptibility

$$\chi'(\omega) \sim |\ln \omega|^{-\theta/\psi} \quad \text{and} \quad \chi''(\omega) \sim |\ln \omega|^{-1+(\theta/\psi)} \tag{5.87}$$

Finally, the ac non-linear susceptibility scales as

$$\chi_{nl}(3\omega; \omega) \sim |\ln \omega|^{[d-(1+\phi)\theta]/\psi} \tag{5.88}$$

and this diverges for $\omega \to 0$ if $d > (1 + \phi)\theta$. For the 3D spin glasses the droplet model has assumed that conventional dynamical scaling is operative. As we shall see in the next chapter there exists a non-conventional dynamical scaling for 2D spin glasses.

Since the theory is so recent there have been very few meaningful experimental tests of the above predictions. One problem is the lack of an 'ideal' 3D short-range Ising spin glass (see Chapter 6). Another is the popularity of the Parisi RSB solution to the mean-field-SK model on which most measurements have concentrated. So at this stage it is impossible to say which of the two gives a better description of the spin-glass phase, although the 'noise' people are strictly for RSB.

We proceed now to consider the non-equilibrium behaviour and its approach to equilibrium of the ordered spins within the droplet model. If we quench a spin glass in zero magnetic field from $T = \infty$ to $T \ll T_f$, a mixed of Γ and $\bar{\Gamma}$ (reversal) domains are formed. After the quench, the system will try to lower its free energy by decreasing the amount of interface between Γ and $\bar{\Gamma}$, thereby growing larger and larger domains of Γ and $\bar{\Gamma}$. For a spin glass this is a tediously slow procedure because of the randomness-induced free-energy barriers. Such barriers must be surmounted before sections of the wall between Γ and $\bar{\Gamma}$ can be moved. Since $B = \Delta(T)L^\psi$, in a time t after the quench, the characteristic length scale of the domains R_{t_a} grows as

$$R_{t_a} \sim \left[\frac{k_B T \ln (t_a/t_o)}{\Delta(T)} \right]^{\frac{1}{\psi}} \tag{5.89}$$

where t_a is the total age of the system after the quench and t_o is a microscopic unit of time ($t_o = 1$ is the usual basis). This growth of domains is the fundamental process of the long-time, non-equilibrium dynamics of the spin glasses below T_f. A corresponding 'quench' can be performed by turning off ($H = 0$) an infinite magnetic field at $T \ll T_f$. The magnetization is expected to decay with the formation of domains according to

$$m(t) \sim \frac{1}{R^\lambda} \sim \left[\frac{\Delta}{k_B T \ln t} \right]^{\frac{\lambda}{\psi}} \tag{5.90}$$

where λ is a new dynamic exponent unrelated to the equilibrium ones.

Now that we have established the length scale that determines the approach to equilibrium, we can consider waiting-time or ageing effects. These are the experiments described in section 3.3.1, where, for example, a sample can be cooled in zero field, and after waiting a time t_w a weak field is turned on and the magnetization measured as a function of time, $m(t)$. Or, the sample can be field cooled, and after waiting t_w the weak field is switched off and $m(t)$ tracked in time. In these experiments, relaxation processes are probed on a length scale L which begins at time t when the field is changed. Here

$$L_t \sim \left[\frac{k_B T \ln t}{\Delta(T)} \right]^{\frac{1}{\psi}} \tag{5.91}$$

Let us treat the processes which occur in the two limiting cases: $\ln t \ll \ln t_w$ and $\ln t \gg \ln t_w$. For the first case we have

$$L_t \sim \left[\frac{k_B T \ln t}{\Delta(T)} \right]^{\frac{1}{\psi}} \ll R_{t_a} \sim \left[\frac{k_B T \ln t_a}{\Delta(T)} \right]^{\frac{1}{\psi}} \tag{5.92}$$

5.10 Fractal cluster model

where now $t_a = t_w + t$. For all practical purposes R_{t_a} will be constant in this 'early epochs' regime, and since L_t, or experiment, probes length scales which are much smaller, we use the term quasi-equilibrium dynamics. In other words, our experimental scale is so small that it does not see the domain walls, only the pure states Γ and $\bar{\Gamma}$. So to a first approximation the relaxation processes are characteristic of equilibrium. And the magnetization should decay as

$$m(t) \sim \frac{H}{(\ln t)^{\theta/\psi}} \qquad (5.93)$$

with θ and ψ the equilibrium exponents. Figure 5.13 sketches the various functions R^ψ, L^ψ and the expected derivative of $m(t)$ with respect to $\ln t$. For the second case: $\ln t \gg \ln t_w$ ('late epochs'), and therefore $L_t \approx R_{t_a}$ since $t \approx t_a = t + t_w$. Now the experiments are clearly seeing the domain walls and thus the non-equilibrium dynamics are detected with

$$m(t) \sim H \left[\frac{\ln t_w}{\ln t} \right]^{\frac{\lambda}{\psi}} \qquad (5.94)$$

where naturally the non-equilibrium exponent λ enters the magnetization decay.

There are two crossover regimes of interest. The first is at $\ln t \approx \ln t_w$ or where $L_t \to R_{t_a}$, i.e., a transformation from quasi-equilibrium to non-equilibrium dynamics. Here the measured magnetization decay shows an 'S'-shape which, when plotted as $dm(t)/d(\ln t)$, reflects a peak at $t \approx t_w$; see Fig. 5.13. The second crossover occurs in the limit of infinite time which of course cannot be reached by experiment. At this point L and R have reached macroscopic length scales so the number of domains and their walls become insignificant. True equilibrium has finally been attained and the magnetization and its slope have decayed to zero. In general, based on this picture of the domains and their growth, we have a simple and qualitative model for interpreting the long-time and ageing behaviours of the spin glasses presented in section 3.3.1.

5.10 FRACTAL CLUSTER MODEL

A scaling theory, which makes particularly close contact with our experimentalist's interpretation (Chapter 3), is the fractal-cluster model of Malozemoff and Barbara (1985). They proposed a scaling theory of the spin glasses by considering clusters of correlation length ξ which diverge as $[(T-T_f)/T_f]^{-\nu}$. These coherent regions have a cluster size s_ξ on which all relevant physical quantities depend and whose volume is ξ^D, where D is the *fractal* dimension because the clusters are expected to be highly irregular and branched. The clusters have a magnetization $m_\xi \sim s^{\frac{1}{2}}$ or s, if the given cluster consists of

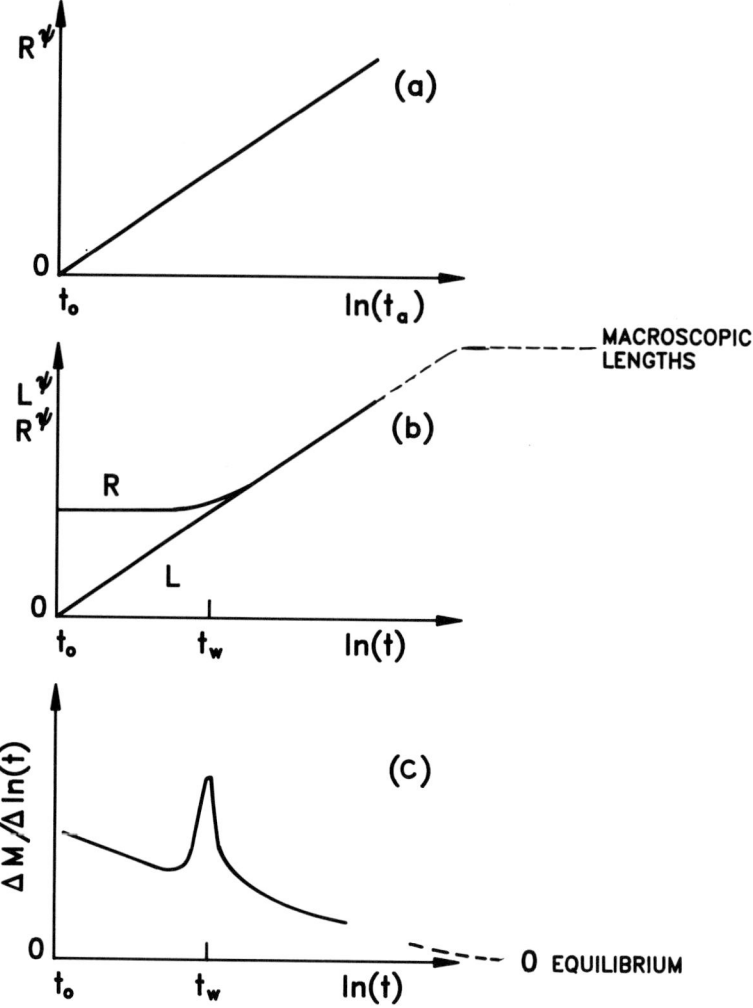

Fig. 5.13 Growth of domain size R and experimental-probing-length scale L versus logarithm of time (ψ is the barrier exponent); (*a*) $t_a = t+t_w$ the total age of the system; (*b*) t is the time of measurement, t_w, the waiting time, and t_o, a microscopic time; and (*c*) relaxation rate $\partial M/\partial \ln t$ versus $\ln t$. From Lundgren (1988).

a random distribution of spins or a ferromagnetic one. Above T_f the regions can be viewed as superparamagnetic clusters which rotate in response to the applied field H giving a net magnetization

$$M \propto n_s m_\xi \mathscr{B}\left(\frac{m_\xi H}{k_B T}\right) \tag{5.95}$$

where n_s is the number of clusters with spin s and $\mathscr{B}$ is the Brillouin

5.10 Fractal cluster model

function. The criterion $m_\xi H \approx k_B T$ corresponds to a crossover permitting the clusters to align along the field. Using

$$m_\xi \sim s^{\frac{1}{2}} \sim \xi^{\frac{D}{2}} \sim \left(\frac{T-T_f}{T_f}\right)^{-\frac{\nu D}{2}} \tag{5.96}$$

we can define the crossover exponent ϕ as $\phi = \nu D$, and since T is not critical, the crossover H–T line follows immediately

$$\left(\frac{T-T_f}{T_f}\right)^{-\frac{\phi}{2}} H = \text{const.} \quad \text{or} \quad \frac{T-T_f}{T_f} \propto H^{\frac{2}{\phi}} \tag{5.97}$$

Now if $\phi \approx 3$, as is suggested by experiment, we obtain a new, scaling derivation of the AT line without recourse to a mean-field instability. At T_f an infinite percolating cluster forms which we call the frozen matrix. Embedded in the matrix or exterior to it are the remaining correlated regions or clusters which govern the low-temperature behaviour.

Below T_f, these correlated regions can be visualized as rotating in a frozen matrix. Here the matrix imposes internal energy barriers which hinder the rotations, but which may be overcome via thermal activation. For further decrease of temperature the barrier heights stay fixed while the size of the remaining clusters grows or becomes part of the matrix. Using an Ising fractal-cluster model the net magnetization at T and H is an average over all cluster sizes:

$$M \propto \int_0^\infty n_s s^{\frac{1}{2}} \tanh\left(\frac{\mu_o s^{\frac{1}{2}} H}{k_B T}\right) ds \tag{5.98}$$

where we have applied a weak field so as to use the small-argument limit of the Brillouin function. According to percolation-scaling theory, the cluster-size distribution n_s is given by

$$n_s = s^{-\tau} f\left(\frac{s}{s_\xi}\right) \tag{5.99}$$

where s_ξ is the characteristic cluster size, τ is another critical exponent and f is the scaling function, i.e., $f(x) = \text{const.}$ as $x \to 0$ and $f(x) \to 0$ for $x > 1$.

We now calculate the time decay of the magnetization for $T < T_f$ when the weak field H is switched off at $t = 0$.

$$\frac{M(t)}{M(0)} = \frac{\int_0^\infty s^{\frac{1}{2}-\tau} f\left(\frac{s}{s_\xi}\right)\left(\frac{\mu_o s^{\frac{1}{2}} H}{k_B T}\right) \exp\left(-\frac{t}{t'}\right) ds}{\int_0^\infty s^{\frac{1}{2}-\tau} f\left(\frac{s}{s_\xi}\right)\left(\frac{\mu_o s^{\frac{1}{2}} H}{k_B T}\right) ds} \tag{5.100}$$

This equation is derived assuming that the characteristic relaxation time t of a given cluster is exponential and that $\tanh x \doteq x$. It is further assumed that t' depends on the size of the cluster as $t' = t_o s^x$ where $t_o = t_{oo} \exp(E_B/k_B T)$ and x is a *new* dynamical critical exponent. Here the effects of energy barrier E_B activation have been placed in the t_o-constant. Rewriting (5.100), we obtain

$$\frac{M(t)}{M(0)} = \frac{\int_0^\infty s^{1-\tau} \exp[-(s/s_\xi)^\nu] \exp[-(t/t_o)s^{-x}] \, ds}{\int_0^\infty s^{1-\tau} \exp[-(s/s_\xi)^\nu] \, ds} \quad (5.101)$$

The percolation-theory result has been substituted for

$$f\left(\frac{s}{s_\xi}\right) \approx \exp[-(s/s_\xi)^\nu] \quad (5.102)$$

with $T \leq T_f$ and sufficiently large clusters. In the long-time limit and ignoring algebraic prefactors, the integration of the numerator gives

$$M(t) \sim \exp[-(t/t_1)^{1-n}] \quad (5.103)$$

where $n = x/(x + \nu)$ and $t_1 \sim s_\xi^x$. This is our old friend, the stretched exponential, which has been used to describe the slow relaxation of real glasses and also certain regimes of spin-glass relaxation. If we repeat the above calculations for $M(t)$, but at $T = T_f$, we can use the percolation-theory result right at the percolation threshold: $n_s(T_f) \approx s^{-\tau}$. In this case the integral for $M(t)$ becomes

$$M(t) \propto \int_0^\infty s^{1-\tau} \exp[-(t/t_o)s^{-x}] \, ds \propto \left(\frac{t_o}{t}\right)^{\frac{D}{\delta z}} \quad (5.104)$$

i.e., a power-law behaviour is obtained with exponents $D/z = 1/x$ and $1/\delta = \tau - 2$. Comparatively, there is a similar expression found in the dynamic of the mean-field theory and some experimental indications for such intermediate behaviour have been reported.

We could proceed further with this model and determine the dynamical-scaling behaviour of the $H-T$ phase boundary, the so-called critical line. As the flavour is clear we stop. The utility of this fractal-cluster approach is that it begins very practically, employing the percolation picture demanded by experiments. Then using various scaling functions it attempts to calculate a number of physical quantities which seem to agree with measurement. With greater efforts the model could probably be extended to addition parameters both at and below T_f. Since the clusters and percolation are so helpful in visualizing the physical phenomenon in a spin glass, any theory done with them provides a viable alternative to the rather abstract mean-field approach.

5.11 NON-ISING SPIN GLASSES

Up until now we have been mainly considering Ising spin glasses. We will now relax this restriction and briefly deal with m-component vector spins $\mathbf{S}_i$, such that $\mathbf{S}_i \cdot \mathbf{S}_i = m$. The spin-glass order parameter is now a tensor in spin space which for $H = J_o = 0$ reduces to a simple scalar, i.e., the system will be isotropic in spin space. And here one can repeat the standard procedures of RSB and obtained modified solutions and properties for the static and dynamical behaviours.

A more interesting case is the isotropic spin glass in non-zero field with $J_o = 0$. Now $M \equiv \langle\langle S_i^{(1)}\rangle_T\rangle_c \neq 0$ and $q_{11} \equiv \langle\langle S_i^{(1)}\rangle_T\langle S_i^{(1)}\rangle_T\rangle_c \neq 0$ for the parallel (to the field) magnetization and spin-glass order parameters, respectively. In addition, there are various transverse order parameters which allow the possibility for transverse freezing

$$\delta_{kk'} q_\perp = \langle\langle S_i^{(k)}\rangle_T\langle S_i^{(k')}\rangle_T\rangle_c \tag{5.105}$$

with $k, k' = 2, 3, \ldots, m$. The line in the H–T plane where $q_\perp$ first becomes non-zero is called the Gabay–Toulouse (GT) line (1981) and it varies for small fields as

$$\frac{k_B[T_f - T_{GT}(H)]}{\Delta} = \frac{m+4}{2(m+2)}\left(\frac{H}{\Delta}\right)^2 \tag{5.106}$$

recalling that Δ is once again the variance of our gaussian $P(J_{ij})$ distribution. A sketch of the GT-line is shown in Fig. 5.14 along with the AT line. Notice the different $H(T)$ dependences between the two lines representing

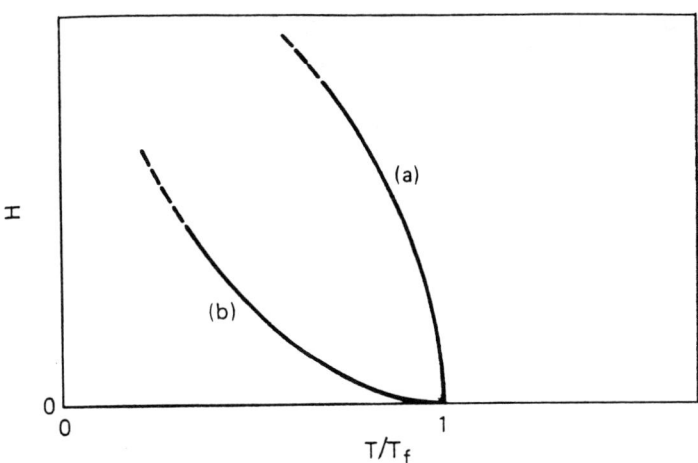

Fig. 5.14 Sketch of field-temperature phase diagram for the mean-field model with vector spins: (*a*) freezing of transverse spin components with weak irreversibilities (GT line); and (*b*) onset of strong irreversibilities with RSB (AT line).

the (GT) transverse freezing $\delta T_f \propto H^2$, followed by the (AT) longitudinal freezing $\delta T_f \propto H^{\frac{2}{3}}$. RSB occurs at both lines and the transverse spin-glass susceptibility is expected to diverge as $[T-T_{GT}(H)]^{-1}$ for $T \to T_{GT}^+(H)$. If we additionally let J_o (the mean of the gaussian $P(J_{ij})$ distribution) be non-zero, Gabay and Toulouse predict the more complicated 're-entry' $T-J_o$ phase diagram shown in Fig. 5.15. Here there is a set of transitions: paramagnetic to ferromagnetic to the first mixed phase M_1 where ferromagnetic and spin-glass (frozen transverse components) phases coexist. Finally $M_1 \to M_2$ signifies another transition where a crossover occurs to a new mixed state with strong irreversibilities as opposed to the weak irreversibilities in M_1.

Continuing with the vector spin glasses we next discuss systems with a uniaxial anisotropy. Here the hamiltonian becomes from equation (1.4)

$$\mathcal{H} = - \sum_{\{i,j\}} J_{ij} \mathbf{S}_i \cdot \mathbf{S}_j - D \sum_i (S_i^{(1)})^2 \qquad (5.107)$$

where the $S_i^{(1)}$ spin component is taken along D, the magnitude of the uniaxial anisotropy which has a fixed direction. If $D > 0$, ordering is preferred along this direction (longitudinal ordering), whereas transverse ordering is favoured for $D < 0$. The phase diagram can be calculated within

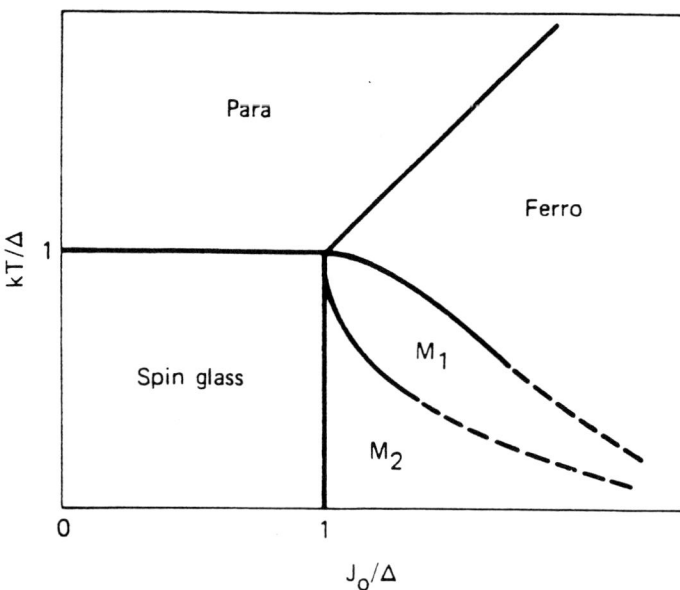

Fig. 5.15 Phase diagram for vector spins interacting with infinite-ranged gaussian distribution (J_o is the mean and Δ the width or variance). M_1 and M_2 represent mixed magnetic phases of the re-entry transition. From Gabay and Toulouse (1981).

5.11 Non-Ising spin glasses

RSB theory setting $J_o = 0$. The results are shown in Fig. 5.16. 'P' represents the paramagnetic phase at high temperature, 'T' the transverse phase with $q_{11} = 0$ and $q_\perp \neq 0$, 'L' the longitudinal phase with $q_{11} \neq 0$ and $q_\perp = 0$, and a mixed phase, 'LT' is also present, where both order parameters are not equal to zero. A variety of different phase transitions is predicted with two possible types of entry into the mixed LT phase. Experiments, mainly using crystalline anisotropy as the source of D, seem to exhibit the L or T transitions, however, values of D are not accessible in real materials to test for the LT phase.

We conclude this section by mentioning the theoretical conjecture that the Dzyaloshinskii–Moriya (DM) interaction (see section 1.4) and its corresponding random anisotropy in an isotropic (or Heisenberg) spin glass can be converted by an applied field into an Ising spin glass. The small field reorganizes the random DM directions to make a component of them point along the field. This then establishes an Ising-like (longitudinal) frozen alignment and changes the universality class from Heisenberg to Ising. With further increasing field the behaviour transforms to a new regime which exhibits both a transverse freezing with RSB at higher temperatures followed by a crossover to longitudinal freezing (strong irreversibility) as T is lowered. The two H–T lines are similar to the GT and AT lines – see Fig. 5.14 – and note the different $H(T)$ dependences. Experimental searches for these various lines and crossovers have had to rely on a precise determination of the freezing temperature over a wide range of magnetic fields. As we have

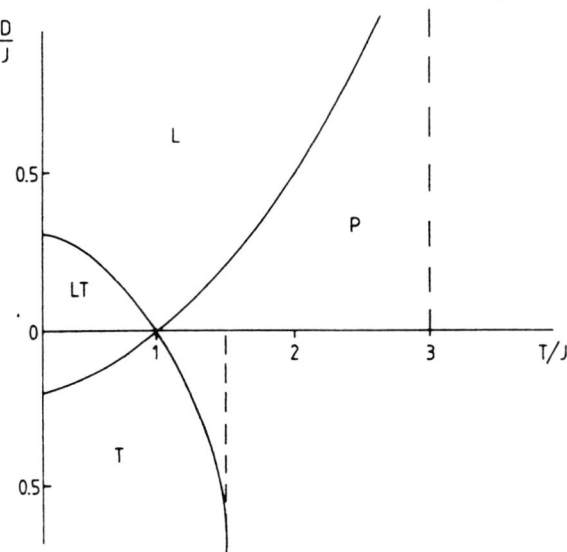

Fig. 5.16 Phase diagram of an infinite-range vector spin glass with uniaxial anisotropy D; from Roberts and Bray (1982) and Cragg and Sherrington (1982).

previously discussed (section 3.4) this is impossible. Furthermore, dynamical effects are playing an important role near T_f, hence the resolution of a so-called critical temperature is strongly contingent on the type and time of measurements. And the controversy remains that the different lines (AT or GT) are mostly dynamical in nature and not the result of a mean-field phase transition as predicted by RSB and the SK model.

5.12 COMPUTER SIMULATIONS

Spin glasses are a natural choice for numerical simulations and the Monte Carlo method lends itself readily for computational studies. These computer experiments serve to span the gap between the measurements done on non-ideal materials and the analytic calculations performed via approximate and oversimplified theoretical models. But first what is the Monte Carlo (MC) method as applied to a spin glass?

We pick a lattice in arbitrary dimension usually 2 or 3, but depending on computer power we can go even higher, e.g. $d = 5$. Sizes of the lattice or number of spins can reach 64^d using fast special purpose machines especially built for simulations of random spin systems. One then selects the number of spin components: $m = 1$ Ising, $m = 2$ x-y, and $m = 3$ Heisenberg. Finally, we need a probable distribution $P(J_{ij})$ for the random-bond configurations. This is customarily taken as the EA or SK gaussians or simply $\pm J$. Although a more realistic model, the canonical RKKY-random site '$CuMn$' situation, has also been MC simulated (see Chapter 8). Once we have decided upon the desired spin-glass 'system' via the above choices and with a given initial configuration of energy U_o, the game starts by flipping one spin. If the energy change $U_1 - U_o < 0$, the spin flip is accepted and we repeat the process for a second spin giving a U_2. However, if the energy change $U_1 - U_o > 0$, then we calculate the exponential $B = \exp[-(U_1-U_o)/k_BT]$ and compare the result with a random number R between 0 and 1. If $B \geq R$, then the spin flip is accepted (Metropolis acceptance criterion) and we go on to a U_2. If $B < R$, the spin flip is rejected and we return to our original initial condition U_o and start all over again by flipping another spin. Each time we attempt a spin flip, we set the time scale with MC step/spin (MCS). This for a good equilibrium ensemble average can reach 10^7 or 10^8 MCS. For better averaging we should repeat the calculations with 30 to 100 realizations $\{J_{ij}\}$ of the random-bond configuration. In addition, the initial condition should be checked for not influencing the results.

Such recursive procedures give a nearly exact calculation of the partition function $Z\{J_{ij}\}$ averaged over the various sets of $\{J_{ij}\}$. MC simulations have been employed from almost the beginning of the spin-glass problem. The early computations (late 1970s) focused upon 2D lattices with Ising spins (Glauber dynamics) and a gaussian distribution of near-neighbour bonds.

5.12 Computer simulations

The temperature was normalized to the distribution width. Interestingly, these pioneering efforts found susceptibility and specific-heat behaviours mimicking those of experiments on the canonical spin glasses. Furthermore, the dependences at $T < T_f$ for the remanent magnetizations (IRM and TRM) on field closely correspond to those measured on AuFe – see Fig. 3.32. These initial results were hampered by insufficient lattice size (finite-size effects) and not enough MCS to obtain true equilibrium (dynamical effects or relaxation times are longer than MC 'observational times'). Nevertheless, in recent years such difficulties have been surmounted and meaningful conclusions have been drawn and reasonable predictions can be made.

The time-dependent EA order parameter $q(t)$ may be directly obtained from the standard MC time averaging (t is measured in MCS)

$$q(t) = \frac{1}{N}\sum_{i=0}^{N}\left(\int_0^t S_i(t')\,dt'/t\right)^2 \tag{5.108}$$

Accordingly, the time-dependent (normalized) susceptibility becomes

$$\chi(t) = \frac{1}{k_B T}[1 - q(t)] \tag{5.109}$$

and can be interpreted in terms of the Sompolinsky/Parisi (section 5.8) dynamical susceptibility. Furthermore, we can proceed with the two-spin correlation function and define the time-dependent, non-linear susceptibility

$$\chi_{nl} = \sum_j \langle\langle S_i S_j\rangle_T^2\rangle_C = \sum_j \left(\int_0^t S_i(t')S_j(t')\,dt'/t\right)^2 \tag{5.110}$$

where

$$\langle\langle S_i S_j\rangle_T^2\rangle_C \sim \left(\frac{T-T_f}{T_f}\right)^{-\gamma} \sim \exp(-R_{ij}/\xi_{SG}) \tag{5.111}$$

and ξ_{SG} is the spin-glass correlation length

$$\xi_{SG} \sim \left(\frac{T-T_f}{T_f}\right)^{-\nu} \tag{5.112}$$

(γ and ν are the usual critical exponents associated with a possible spin-glass transition). Again using MC simulations these quantities may be numerically computed. And in a similar manner just about any quantity of physical or theoretical interest could be calculated for the various models of spin glasses.

Rather than delve into the many such MC experiments we limit ourselves to discussing the main and accepted (by the MC experts) results. Remember these numerical simulations are a middle route between theory and experiment, and they are most useful as a test or proving grounds for one or the other. Hopefully, as is the case in simpler problems, all three should agree.

In huge simulations on a specially built processor Ogielski and Morgenstern (1985) considered the Ising, $\pm J$, 3D spin glass. The static properties showed a sping-glass ordering transition at $T_f = 1.175J$ with critical exponents for the correlation length $\nu = 1.3$ and for the non-linear susceptibility $\gamma = 2.9$. Working within the mean-field RSB approach, Bhatt and Young (1985) compared order-parameter distributions $P(q)$ computed in different ways. A finite-size scaling function was derived from the moments of $P(q)$ which should be independent of size at T_f and then splay out again at lower temperatures. The various size curves do indeed cross at $T = 1.75J$ for $d = 4$. In the case of $d = 3$ they come together at $T \cong 1.2J$, but stay merged as T is reduced below T_f. This indicates that $d = 3$ is very close to the lower critical dimension d_l below which $T_c = 0$. Additional scaling fits yield $\gamma = 3.0$.

In another series of computer simulations called domain-wall renormalization-group techniques, Bray and Moore (1984) and McMillan (1984) further studied the dimensionality of the spin-glass transition. In this method one computes the variation of the ground-state energy δE_g by changing the boundary conditions from periodic to antiperiodic in one direction. For a spin glass $\langle \delta E_g \rangle = 0$, but $\langle \delta E_g^2 \rangle^{\frac{1}{2}}$ and $\langle |\delta E_g| \rangle$ are not zero. So if the energy change decreases with system size, then there is no order because at some length scale $\delta E_g \approx k_B T$, and this will disorder the system. The results indicate order for $d = 3$, but not for $d = 2$.

Naturally, these techniques can be repeated for isotropic (Heisenberg) vector spin glasses with short-range $\pm J$ or gaussian bond distributions in 2, 3 and 4 dimensions. Even the long-range RKKY interaction may be included in the simulations. The results suggest that $T_f = 0$ for $d = 2$ and 3 isotropic spin glasses and that $d_l \approx 4$ (above which T_f should be finite). Also the long-range RKKY interacting spin glasses are at, rather than below, their lower critical dimension.

MC simulations (Ogielski, 1985) have also been used to determine the spin-glass dynamics via the behaviour of $q(t)$, the dynamic correlation function defined above. Figure 5.17 exhibits how $q(t)$ decays in time (MCS) at various temperatures spanning T_f. Note the system size is $32^3 = 32\,768$ spins and about 5×10^7 MCS have been performed. Once one knows the $q(t)$ dependence various analytic fits can be made. For example, above T_f the functional form is

$$q(t) = C_o t^{-x} \exp[-(\omega t)^\beta] \qquad (5.113)$$

where both exponents x and β depend on the temperature. Above the Curie temperature of the *non-random* 3D Ising model ($T_c = 4.5J$) $x = \frac{1}{2}$ and $\beta = 1$. While from T_c down to T_f (the Griffiths phase) β is between 0.4 and 1 indicating that the time decay is a stretched exponential or Kohlrausch function. Below T_f it changes to power law

5.12 Computer simulations

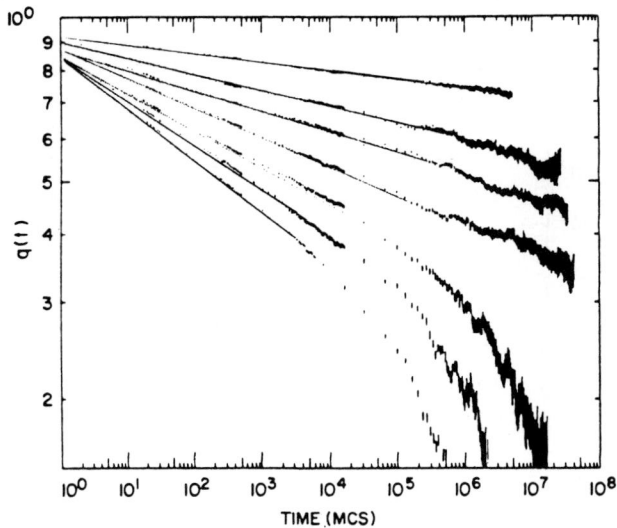

Fig. 5.17 Dynamic correlation function $q(t)$ calculated by Monte–Carlo simulation (lattice size 32^3) for temperatures around and below $T_f = 1$. From bottom to top $T = 1.30, 1.25, 1.20, 1.10, 1.00, 0.90$ and 0.70. From Ogielski (1985).

$$q(t) = C_o t^{-x} \quad T \le T_f \tag{5.114}$$

with x very small (≈ 0.1 at T_f decreasing to ≈ 0.01 at $\frac{1}{2}T_f$, the lowest temperature of the simulation). This is such a tiny exponent, i.e., the time dependence is exceedingly slow, that a $\ln t$ or even a $\ln(\ln t)$ behaviour may also be a suitable fit to the computer plots in Fig. 5.17.

If we extract from $q(t)$ the relaxation time τ used in the dynamical scaling of equation (5.73), then we can estimate the dynamical exponent z. This is accomplished by the proper integration of $q(t)$, namely

$$\tau = \frac{\int_0^\infty t\, q(t)\, dt}{\int_0^\infty q(t)\, dt} \tag{5.115}$$

By performing the above integrals at various temperatures approaching T_f, we can plot log τ versus log $(T - T_f)/T_f$. Figure 5.18 shows one such graph for a 64^3 system where the slope gives $zv = 7.9$ with the 'proper' choice of T_f ($=1.175J$). The determination of T_f (here it is a critical temperature) is very difficult also in the computer experiments as it is in a real experiment. Note that τ starts to exceed the maximum MCS (here $\approx 5 \times 10^8$) at about $T = 1.30J$. This means that a reduced temperature of only $(T - T_f)/T_f \approx 10^{-1}$ can be reached. So the fit in Fig. 5.18 is restricted to less than a decade. Not very good for the extraction of meaningful

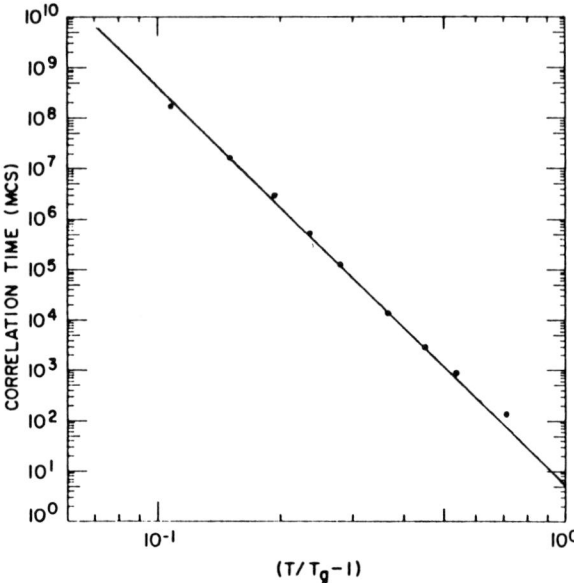

Fig. 5.18 Power-law fits for the relaxation time τ versus $(T-T_c)^{-z\nu}$ from the Monte–Carlo simulations for lattice size 64^3 and $T \geq 1.30$; from Ogielski (1985).

exponents, which also depend on the exact ascertainment of T_f. Nevertheless, the above simulations are the biggest and longest of the 1980s. And they show a rapid increase described by an unusually large exponent $(z\nu)$ power law as $T \to T_f^+$.

Various computer studies all agree that $T_f = 0$ for the 2D *Ising* spin glass. The d_l lies between 2 and 3, and the correlation length ξ_{SG} and non-linear susceptibility χ_{nl} should both diverge at $T = 0$ K. The critical exponents ν and γ are different depending upon which distribution is used. For the $\pm J$ distribution ν is about 2.6 and γ between 4 and 5.5. From the scaling relation $\gamma = (2 - \eta)\nu$ we can estimate η the length exponent of the correlation function. For the gaussian distribution ν is around 3.5 to 4.4 while γ can be above 7 even reaching 9. Substituting these 'value spreads' into the above scaling-equation results in an estimate for $\eta \approx 0$. Presumably, the large ground-state degeneracy of the $\pm J$ *discrete* model places it in a different universality class from models with a continuous J_{ij} distribution, if $T_f = 0$.

Now we proceed down into the frozen spin-glass state. Here the question is how to probe the equilibrium state by computer experiments which are always out of equilibrium. We rely on the two main theoretical models which have been previously proposed. The first is the mean-field SK model with Parisi RSB solution. This model leads to many thermodynamic states unrelated by symmetry and differing little in their ground-state free energy.

The distribution of order parameters $P(q)$ is an important property of the solution. Below T_c it should exhibit a peak at finite q and a tail with finite weight at $q = 0$. Using computer simulations of the SK model such a form for $P(q)$ has indeed been found. The alternative model is the droplet approach of Fisher and Huse where there is only a single thermodynamic state Γ whose excitations are droplets of globally reversed spin directions $\bar{\Gamma}$. For a finite system these states can produce a finite $P(q)$ down to $q = 0$. However, $P(0)$ should vanish as $L^{-\theta}$ as the linear size of the lattice L goes to infinite. Recent MC calculations for a $d = 4$ Ising spin glass displayed very little reduction in the $P(0)$ tail as L was increased. Therefore, at the present state of numerical simulation, the results seem to be favouring the SK-Parisi theory over the droplet model. The above examples have nicely illustrated the power and effectiveness of special-purpose computers in the study of spin glasses.

With this cursory review of the spin-glass models and theory we conclude the chapter. The reader is referred to the listings in Further Reading and References for more details and full derivations of the theoretical results. As mentioned earlier the spin-glass problem has not yet been solved and new developments should be expected, perhaps requiring a future addendum of the most recent progress. We now proceed to the ideal spin glasses and their comparison with theory.

FURTHER READING AND REFERENCES

Statisical mechanics

Tolman, R. C. (1962) *Principles of Statistical Mechanics* Oxford.
Huang, K. (1963) *Statistical Mechanics* Wiley, New York.
Landau, L. D. and Lifshitz, E. M. (1989) *Statistical Physics* Pergamon, London.
Ma, S.-K. (1985) *Statistical Mechanics* World Scientific, Singapore.
Ziman, J. M. (1979) *Models of Disorder* Cambridge.
Binder, K. (editor) (1986) *Monte Carlo Methods in Statistical Physics* Springer, Heidelberg; (1987) *Applications of the Monte Carlo Method in Statistical Physics* Springer, Heidelberg.

Phase transitions

Brout, R. H. (1965) *Phase Transitions* Benjamin, New York.
Stanley, H. E. (1971) *Introduction to Phase Transitions and Phenomena* Oxford.
Domb, C. and Green, M. S. (eds) (1972) *Phase Transitions and Critical Phenomena* Academic, New York.
Hohenberg, P. C. and Halperin, B. I. (1977) *Rev. Mod. Phys.*, **49**, 435.
Ma, S.-K. (1976) *Modern Theory of Critical Phenomena* Benjamin, New York.

Amit, D. J. (1984) *Field Theory, the Renormalization Group and Critical Phenomena* McGraw-Hill, New York.

Theoretical spin-glass reviews

Chowdhury, D. (1986) *Spin Glasses and Other Frustrated Systems* Princeton.
Binder, K. and Young, A. P. (1986) *Rev. Mod. Phys.*, **58**, 801.
Fisher, K. H. and Hertz, J. A. (1991) *Spin Glasses* Cambridge.
Mezard, M., Parisi, G. and Viraso, M. A. (1987) *Spin Glasses Theory and Beyond* World Scientific, Singapore.
Moorjani, K. and Coey, J. M. D. (1984) *Magnetic Glasses* Elsevier, Amsterdam.

References

Anderson, P. W., Halperin, B. I. and Varma, C. M. (1972) *Philos. Mag.*, **25**, 1.
Bhatt, R. N. and Young, A. P. (1985) *Phys. Rev. Lett.*, **54**, 924.
Blandin, A. (1961) Thesis, University of Paris (1961), unpublished; (1978) *J. Physique*, **39** C6, 1499.
Bray, A. J. and Moore, M. A. (1984) *J. Phys.*, **C17**, L463 and L613.
Cragg, D. M. and Sherrington, D. (1982) *Phys. Rev. Lett.*, **49**, 1190.
de Almeida, J. R. L. and Thouless, D. J. (1978) *J. Phys.*, **A11**, 983.
Edwards, S. F. and Anderson, P. W. (1975) *J. Phys.*, **F5**, 965; (1976) *ibid* **6**, 1927.
Fischer, K. H. (1975) *Phys. Rev. Lett.*, **34**, 1438.
Fisher, D. S. and Huse, D. A. (1986) *Phys. Rev. Lett.*, **56**, 1601; (1988) *Phys. Rev.*, **B38**, 386.
Gabay, M. and Toulouse, G. (1981) *Phys. Rev. Lett.*, **47**, 201.
Klein, M. W. and Brout, R. (1963) *Phys. Rev.*, **132**, 2412.
Larkin, A. I. and Khmel'nitskii, D. E. (1970) *Sov. Phys. – JETP*, **31**, 958; (1971) *ibid*, **33**, 458.
Lundgren, L. L. (1988) *J. Physique*, **49** – C8, 1001.
Malozemoff, A. P. and Barbara, B. (1985) *J. Appl. Phys.*, **57**, 3410.
Marshall, W. (1960) *Phys. Rev.*, **118**, 1519.
McMillian, W. L. (1984) *J. Phys.*, **C17**, 3179; (1980) *Phys. Rev.*, **B30**, 476.
Mezard, M., Parisi, G., and Viraso, M. A. (1987) *Spin Glasses Theory and Beyond* World Scientific, Singapore.
Ogielski, A. T. (1985) *Phys. Rev.*, **B32**, 7384.
Ogielski, A. T. and Morgenstern, I. (1985) *Phys. Rev. Lett.*, **54**, 928.
Parisi, G. (1979) *Phys. Rev. Lett.*, **43**, 1754; (1980) *J. Phys.*, **A13**, 1101; (1980) *ibid* **13**, 1887; (1980) **13**, L115; (1980) *Phys. Rep.*, **67**, 97.
Parisi, G. and Toulouse, G. (1980) *J. Physique Lett.*, **41**, L-361.
Phillips, W. A. (1972) *J. Low Temp. Phys.*, **7**, 351.
Roberts, S. A. and Bray, A. J. (1982) *J. Phys.*, **C15**, L527.
Sherrington, D. and Kirkpatrick, S. (1975) *Phys. Rev. Lett.*, **35**, 1792; (1978) Kirkpatrick, S. and Sherrington, D. (1978) *Phys. Rev.*, **B17**, 4384.
Sompolinsky, H. (1981) *Phys. Rev. Lett.*, **47**, 935; (1981) *Phys. Rev.*, **B23**, 1371.
Thouless, D. J., Anderson, P. W. and Palmer, R. G. (1977) *Philos. Mag.*, **35**, 593.
Walker, L. R. and Walstedt, R. E. (1977) *Phys. Rev. Lett.*, **38**, 514.

6

Ideal spin glasses: comparisons with theory

Now that we have toured the main spin-glass experiments, listed the various materials and scanned the highlights of the theory, how close can we relate theory to experiment on the most ideal systems? We do not wish to make a detailed comparison of every aspect, but chiefly to show which models can be mimicked by experiment and whether their predictions are valid especially regarding the freezing transition. Remember we have our experimentalist's rendition of the basic phenomenon: growing and freezing clusters with a broad and shifting distribution of relaxation times. Theory, if it is correct, can take us much further with a quantitative description of experiment and the predictions of new effects. Once we gain a sufficient rapport between the two, experimentation and theory, the problem is solved and the physics is said to be understood. However, for the spin glasses the full solution has not yet been attained. There is no complete quantitative comparison or detailed set of predictions directly relating to experiment. Great progress has indeed been made and is continuing with the various models and calculations, particularly the computer simulations. On the other hand, the majority of the measurements have been performed on systems which are too complicated and greatly differ from the simplified theoretical models. So how can you expect to compare experiments on a complex material with an idealized theory?

In this chapter we adopt the point of view that real spin-glass systems must be made or chosen to resemble intimately the theoretical model. And these special materials we call the *ideal spin glass*. We distinguish an ideal spin glass from a canonical (RKKY) or good spin glass – these generic terms were frequently used in the previous chapters – as follows. An ideal spin glass *completely* conforms to the theoretical model under consideration. In contrast, a canonical or good spin glass exhibits the main experimental features of a spin glass exposed in Chapter 3. There is no question about the latter being a spin glass with a *sharp* freezing transition. However, complications are present due to its universality class, deviations from randomness, departures from pure RKKY behaviour, higher spin values, orbital moments, undefined anisotropy, etc. Such aberrations add to the complexity and prevent a meaningful comparison with the simple theoretical models.

First, we explain what is an ideal spin glass in the abstract. Then, we try to come as close as possible with some recently discovered mixed compounds. Here space dimensionality will play an important role as we are only interested in Ising spin glasses. For the 2D and 3D cases we examine the experiments and draw a comparison with the existing theoretical results which are mainly related to the static and dynamic scaling behaviours of the freezing transition. As we shall see there are weaknesses from both sides, but such is the present state of the science of spin glasses. Finally, some future possibilities are offered for the fabrication of new ideal spin glasses using the many sophisticated techniques of modern materials research.

6.1 WHAT IS AN 'IDEAL' SPIN GLASS?

The many canonical or good spin glasses, which exist in nature and possess the prerequisite experimental criteria, are not at all 'ideal', i.e., closely conform to the basic theoretical model (see below). For example, most of the natural spin glasses are Heisenberg systems with site randomness. Very many, especially the metallic alloys, are severely complicated with local chemical clustering that leads to atomic, and in turn, to magnetic short-range order. While the RKKY interaction is long range, it is greatly modified by the magnetic clusters and may even taken on different distance dependences for real metals. Another possibility is that a truncated spin density wave will form. Molecular-field theory is inappropriate to treat these effects.

Other spin glasses do not have fully localized moments and are intricated with spin fluctuations or the Kondo effect. Still more materials are unknown regarding their exchange interactions and, in the case of metals, are arbitrarily defined as proper RKKY systems. Nevertheless, despite these difficulties, such systems as (canonical) CuMn and AuFe or (good) (EuSn)S and a-GdAl$_2$ are often assumed to be ideal and compared in great detail to Ising-model, random bond, mean-field theory. Insulating mixed compounds also have their difficulties with only antiferromagnetic exchanges or complicated crystal structures and Ising systems are usually hard to find.

We now propose to create an artificially structured spin glass which will fully satisfy the theoretical models and their computer simulations. Figure 6.1 shows a 2D projection of our 'ideal' spin glass. The 'material', which also could be 3D, is an insulating Ising system with good local moments ($S = \frac{1}{2}$) occupying regular lattice sites of a simple crystal structure. In Fig. 6.1 the magnetic sub-lattice is square and the spins can point only up or down (Ising). An interpenetrating sub-lattice is available for non-magnetic elements. Here we randomly distribute two types of ligands which control the superexchange coupling. The exact ligand configuration between two of

6.1 What is an 'ideal' spin glass?

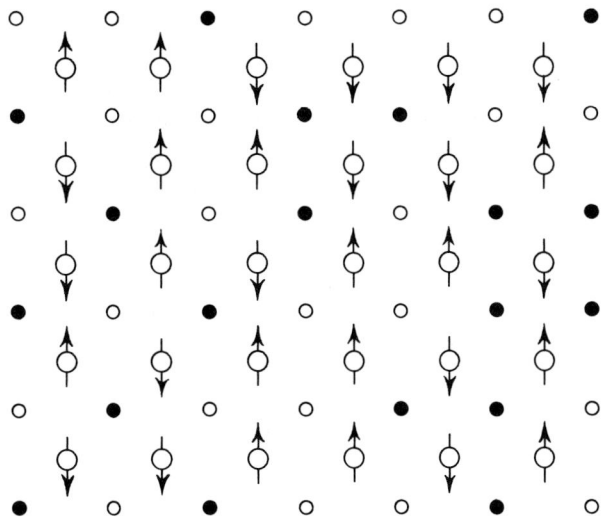

Fig. 6.1 Schematic in 2D of a 3D Ising *ideal* spin glass formed by $\pm J$ distribution of bonds. Note the randomly distributed ligands (small open and filled circles).

the moments will govern their alignment: parallel or antiparallel. This results in a short-range, well defined $\pm J$ exchange interaction which can be adjusted by varying the relative concentrations of the two different ligands represented by the small *open* and *filled* circles in Fig. 6.1. With this abstract creation we have constructed an Ising, spin $\frac{1}{2}$, random-bond (Edwards–Anderson), $\pm J$ spin glass. Such we dub *ideal*, as it closely mimics the simplest of theoretical models and therefore merits experimental study.

It is all well and good to fantasize with sketches, but how close can we come in practice? Presently, there are no artificially made spin glasses that satisfy all of the above model criteria. The full power of modern material science, e.g. sputtering deposition, molecular-beam epitaxy, etc., has not yet been brought to bear on fabricating an ideal spin glass. We should add that some of the above techniques have been used to examine the effects of a reduced dimension, 3D → 2D, using thin films of mainly *Cu*Mn and *Au*Fe. However, nature and our chemist friends have provided us with two insulating compounds that are reasonable approximations of the ideal spin glass defined above. In the subsequent two sections we shall examine in some detail these materials, 2D-$Rb_2Cu_{1-x}Co_xF_4$ and 3D-$Fe_{1-x}Mn_xTiO_3$, and see what comparisons and conclusions may be drawn from the theoretical predictions. At the present moment this is the best we can do, but hopefully in the future better 'real-artificial' materials will be concocted that do an even better job of being ideal. For, only then can the question of a true phase transition be fully resolved.

6.2 $Rb_2Cu_{1-x}Co_xF_4$, AN IDEAL 2D SPIN GLASS

The crystal structures of both Rb_2CuF_4 and Rb_2CoF_4 are the K_2NiF_4-type, meaning a body-centred tetragonal sub-lattice of magnetic Cu or Co atoms. Since the spins are separated by such a large distance along the long or c-axis, the compound is magnetically quasi-2D. In the decoupled (a–b) planes a simple square lattice of strongly interacting moments is formed. Rb_2CoF_4 is an archetypal Ising (up/down spins along the c-axis) antiferromagnet with $T_N = 103$ K. Rb_2CuF_4 possesses a small x–y anisotropy which causes the spin to lie ferromagnetically aligned in the a–b plane, $T_C = 6$ K. The magnetic interactions are superexchanged via the F-ligands and fall off very rapidly such that the next nearest-neighbour coupling is only 1% of the nearest neighbour one. Thus a nearest-neighbour exchange is fully sufficient to describe the magnetic behaviour of this system.

Now instead of mixing ligands to create random bonds as in Fig. 6.1, we substitute Cu for Co which effectively mimics bond randomness. The reasons for this are as follows – recall section 1.3.5. The Cu^{2+} ion has two types of ground state orbitals whose lobes alternate along the x and y axes. As nearest neighbours two Cu ions are ferromagnetically coupled and their lobes are always orthogonal. But a Cu lobe pointing towards a no-lobe Co neighbour (remember the square-lattice symmetry) generates an antiparallel alignment. For a Cu lobe perpendicular to its Co neighbour, a parallel spin pair results. Finally, Co–Co nearest neighbours are always ferromagnetic. A sketch of the lattice with its 'random' and competing bonds is given in Fig. 6.2. Notice we have two ferro (↑ ↑) and two antiferromagnetic (↑ ↓) interactions. These can be varied by changing the concentration x, and for a particular $x = 0.29$ the resulting $P(J)$ was already shown as part of the introduction to spin glasses in Fig. 1.8. In essence we are now repeating what has been previously stated in section 1.3.5 for a *random-bond* spin glass.

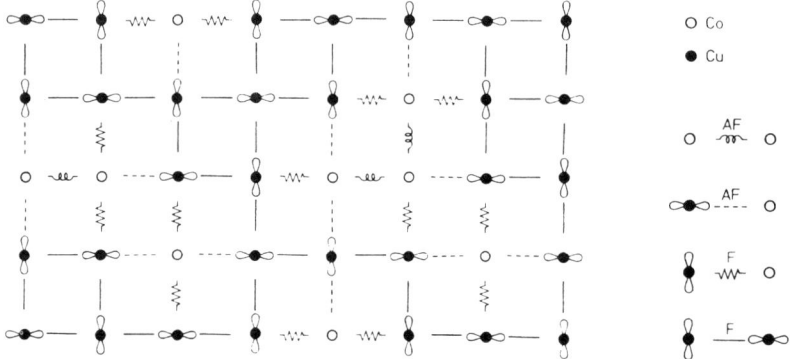

Fig. 6.2 Representative distribution of ferromagnetic (F): Cu–Cu and Cu–Co and antiferromagnetic (AF): Co–Co and Cu–Co exchange interactions in $Rb_2Cu_{1-x}Co_xF_4$. Also indicated are the in-layer d-orbital lobes of Cu^{2+}. From Dekker (1988).

6.2 $Rb_2Cu_{1-x}Co_xF_4$, an ideal 2d spin glass

The Co, although in the minority, plays a very interesting role. Because it experiences a tetragonal crystal field, Co^{2+}-ions have only their lowest Kramers doublet occupied. Thus its effective spin is $\frac{1}{2}$, exactly the same as the Cu. Furthermore, the single-ion anisotropy of the Co^{2+} is such as to create a good Ising, uniaxial-oriented system. Last but not least, when single crystals are grown, and this procedure is not without difficulty, they form without short-range chemical order. So there is a true distribution of random bonds unhampered by any atomic clustering. Consequently, the $Rb_2Cu_{1-x}Co_xF_4$ compounds represent an Ising, spin $\frac{1}{2}$, random bond, short-range ($\pm J$), square-lattice spin glass. This set of adjectives closely conforms to our ideal sketch in Fig. 6.1 with the exception of the symmetric $\pm J$ distribution. For the real case there are 2+Js and 2−Js, and, while the relative probabilities $P(J)$ (see Fig. 1.8) may be adjusted by varying x, the asymmetric magnitude of the four Js is an aberration from our idealization.

The next step is to optimize the x-value and to demonstrate that the system is not long-range ordered but indeed a random freezing occurs at some temperature. In the present 2D case we expect $T_f = 0$. Experiment establishes the spin-glass regime to exist for $0.18 < x < 0.40$, with a good working concentration $x = 0.22$. The usual measurements of ac- and dc-susceptibility, magnetization and specific heat are carried out to prove the absence of long-range order, i.e., a featureless specific heat, and the beginning of a random freezing via the time (frequency) and field (FC versus ZFC) dependences of the other quantities.

Once the spin-glass properties have been firmly established and a T–x phase diagram mapped out, we assume a phase transition transpires at some T_c. Since we are being guided by theory which would predict a critical temperature T_c, we have changed our notation from T_f to T_c. The *static* critical behaviour can be studied through the non-linear susceptibility χ_{nl}. Here a dc method in different applied fields (vibrating sample magnetometer) was used to determine χ_{nl} taking advantage of the magnetization expansion indicated in equation (3.41).

$$M = a_1\left(\frac{H}{T}\right) - a_3\left(\frac{H}{T}\right)^3 + a_5\left(\frac{H}{T}\right)^5 - \cdots \tag{6.1}$$

The definition of the non-linear susceptibility was already given in (3.49). Remember theory predicts the linear susceptibility $\chi_0 = a_1/T$ to remain non-singular at T_c, below which a cusp appears for fast times and a plateau for very long times (equilibrium). The other a_n-coefficients should diverge and depend on the critical exponents according to theory as

$$a_3 \propto (T-T_c)^{-\gamma} \tag{6.2}$$

$$a_5 \propto (T-T_c)^{-(2\gamma+\beta)} \tag{6.3}$$

etc.

Assuming that χ_0 for $Rb_2Cu_{0.78}Co_{0.22}F_4$ obeys a simple Curie law, i.e.,

possesses a symmetry distribution of exchange couplings, we can rewrite (3.42) as

$$\chi_{nl} = a_3(\chi_o H)^2 - a_5(\chi_o H)^4 + \cdots \qquad (6.4)$$

with the coefficients reflecting the critical divergence. A typical plot for the experimental data is given in Fig. 6.3, where the solid lines are a fit to the above equation. The net result of this analysis is a resolution of the temperature dependences of a_3 and a_5, and thereby the γ and β exponents, once T_c is established.

However, a severe problem remains with the exact determination of T_c, the critical temperature. From the ac-susceptibility (see below) a cusp is visible at about 3 K and is strongly frequency dependent, shifting downwards in T with decreasing ω. What then is the true T_c in the limit $\omega \to 0$? Theory has already prognosticated the answer, namely, $T_c(\omega=0) = 0$. But we want to test this conclusion via experiment. So, since we cannot let

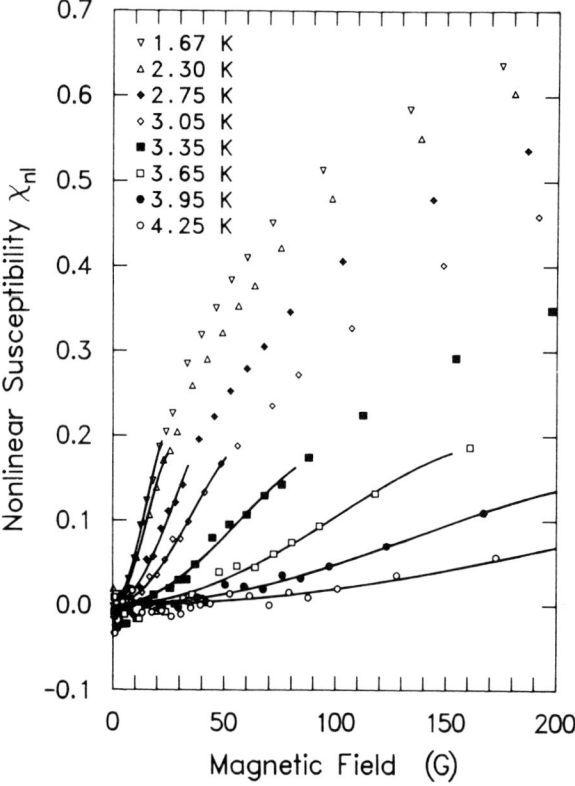

Fig. 6.3 Non-linear susceptibility $\chi_{nl} = 1 - (M/\chi_o H)$ versus the internal field for a selection of temperatures. Solid lines are fits to determine the a-coefficients. From Dekker et al. (1988).

6.2 $Rb_2Cu_{1-x}Co_xF_4$, an ideal 2d spin glass

$\omega = 0$ which may be less than 10^{-10} Hz or some geological time scale, we analyse the χ_{nl} data with T_c as a free parameter. Figure 6.4 shows the results. Especially important is the statistical analysis performed on the data using the χ^2-test. Note how χ^2 decreases and converges nicely to 1 as $T_c \to 0$. We judge our best fit to be with $T_c = 0$, $\gamma = 4.5$ and $\beta = 0.0$ (see Fig. 6.4).

In order to check further this delicate point we can perform a scaling analysis of the χ_{nl}-data. The predicted scaling relation for $T_c = 0$ is

$$\chi_{nl} = T^\beta G\left(\frac{H}{T^\Delta}\right) \tag{6.5}$$

where the scaling function $G(x)$ goes to $c_o x^2$ in the limit $x \to 0$ and to $1 - c_\infty x^{-1/\Delta}$ in the limit $x \to \infty$, with $\Delta = 1 + \frac{1}{2}(\gamma+\beta)$. By plotting $\chi_{nl} T^{-\beta}$ versus H/T^Δ at a variety of temperatures, excellent overlap is obtained –

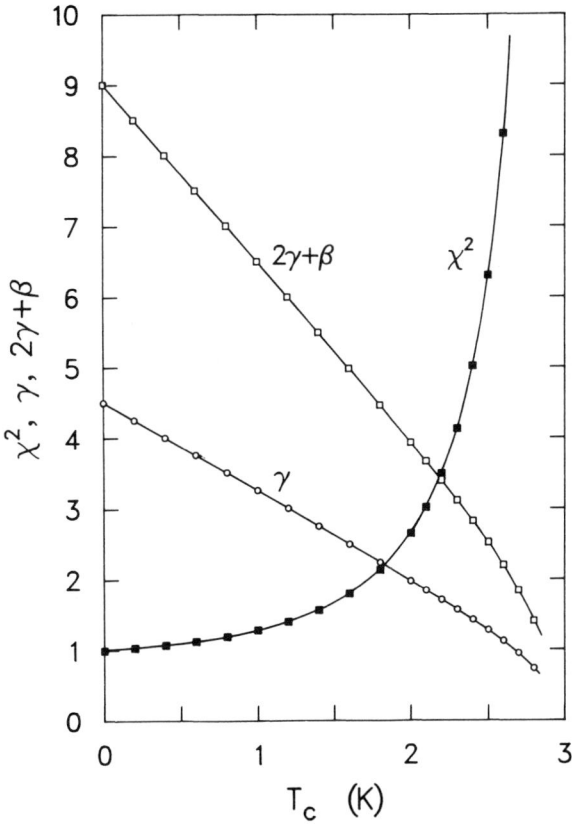

Fig. 6.4 Results for χ^2 test (best-fit for T_c), γ, and $2\gamma + \beta$ versus T_c, obtained from fitting the non-linear susceptibility with various values of 'T_c'. Solid lines are guides to the eye. From Dekker et al. (1988).

see Fig. 6.5. Superimposing data at different values of an experimental parameter (T, H, etc.) is called a scaling plot and the best overlap is used to determine the values of T_c and the critical exponents. For Fig. 6.5 a log–log plot is utilized and the superposition looks good except at the lower values of the abscissa and ordinate. Accordingly, the best overlap gives $T_c = 0$ K, $\Delta = 3.2$ and $\beta = 0.0$. Furthermore, if we used the scaling relation appropriate to a 3D, $T_c \neq 0$ system, fits or data overlap can be obtained, but at the expense of large and unphysical values of the critical exponents. Here there appears a definite trend, the lower T_c, the better and more meaningful the fits. This once again confirms our conclusion that $T_c = 0$ in accord with the theoretical prediction. At this point a very disturbing question arises: how do we treat the cusp in the ac-susceptibility which is usually taken not only as a measure of T_f but also of T_c – the critical temperature of the supposed phase transition? This should serve as a warning to experimentalists – a cusp in χ_{ac} does not automatically mean a phase transition, or in other words $T_f \neq T_c$! One must really perform a

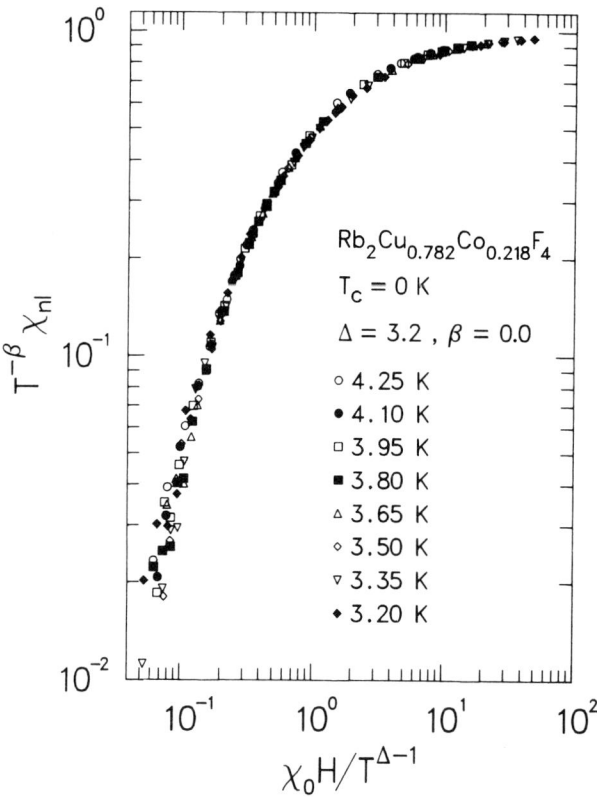

Fig. 6.5 $T_c = 0$ scaling plot of the non-linear susceptibility data. External fields range from 1.0 G to 12.5 kG. From Dekker et al. (1988).

6.2 $Rb_2Cu_{1-x}Co_xF_4$, an ideal 2d spin glass

detailed analysis of χ_{nl} on an ideal spin glass before jumping to conclusions about phase transitions. To shed more light on this problem let us consider the dynamical behaviour of $Rb_2Cu_{0.78}Co_{0.22}F_4$.

Now we measure the ac-susceptibility (χ' and χ'') over as wide frequency and temperature ranges as possible. In practice five or six decades of ω can easily be generated and T must span the region of maximum response in both χ' and χ''. Figure 6.6 exhibits these two susceptibilities from 0.3 Hz to 50 kHz. Note the 'cusp' in both χ' and χ'' which becomes sharper and shifts to lower temperature. There are three distinguishing features of this cusp behaviour (see Fig. 6.6) compared to that usually found in the 3D spin glasses. Firstly, the maximum value (height) of χ' is very strongly dependent on the frequency; a factor of three change is noted in the above frequency range. Secondly, the susceptibility deviates from isothermal already at T_{max}, meaning a χ'' contribution appears at this 'high' temperature. And thirdly, χ'' strongly decreases at the lower temperatures instead of remaining constant. Well how can we interpret this different behaviour within the contexts of a dynamical phase transition? Answer: analysis of the susceptibility data is necessary to extract a relaxation time.

As revealed in section 3.2.1, an average relaxation time may be extracted from a broad distribution $g(\tau)$ by employing the Cole–Cole phenomenological approach. Here the basic equation is

$$\chi(\omega) = \chi'(\omega) - i\chi''(\omega) = \chi_s + \frac{\chi_o - \chi_s}{1 + (i\omega\tau_c)^{1-\alpha}} \qquad (6.6)$$

where χ_s and χ_o are the adiabatic and isothermal susceptibilities, respectively,

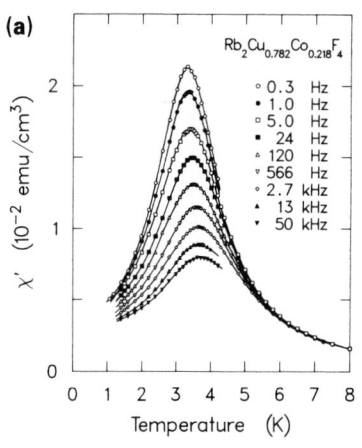
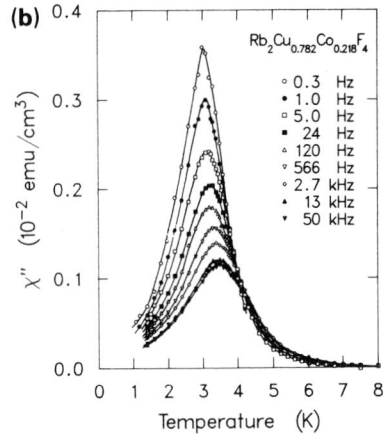

Fig. 6.6 (a) In-phase linear susceptibility $\chi'(\omega, T)$ for $Rb_2Cu_{0.782}Co_{0.218}F_4$ versus the temperature. Data were taken at frequencies $\omega/2\pi$ ranging from 0.3 Hz to 50 kHz, and with the driving field along the c axis. Solid lines are guides to the eye. (b) Same as (a), but out-of-phase linear susceptibility $\chi''(\omega, T)$. From Dekker *et al.* (1989).

and τ_c is an average relaxation time. The fitting parameter α determines the width of the distribution such that $\alpha = 1$ corresponds to an infinite-width distribution while $\alpha = 0$ reverts back to the Debye form of a single relaxation time. The trick now is to plot χ'' (χ') for the various frequencies at each temperature (Argand diagram – section 3.2.1). At the maximum of χ'', $\omega\tau_c = 1$, thereby establishing the value of τ_c for the given T. A collection of such τ_cs is shown in Fig. 6.7 as a function of temperature. The special feature of the Cole–Cole approach is that it allows average relaxation times to be derived even though they lie far outside the measurement window. The experiments spanned a few seconds to a few microseconds. Yet the analysis determines τ_cs from 10^6 down to 10^{-11} seconds. Once we have $\tau_c(T)$ plot as in Fig. 6.7, theory can be invoked.

A prediction from the droplet model (section 5.9) is that for a 2D Ising $\pm J$ spin glass, activated dynamics governs the dynamical critical behaviour. Activated dynamical scaling leads to the following equation, derived from (5.85), for the characteristic relaxation time,

$$\ln\left(\frac{\tau_c}{\tau_0}\right) = \frac{B}{T} \propto \frac{\xi^\psi}{T} \propto \frac{1}{T^{1+\psi\nu}} \tag{6.7}$$

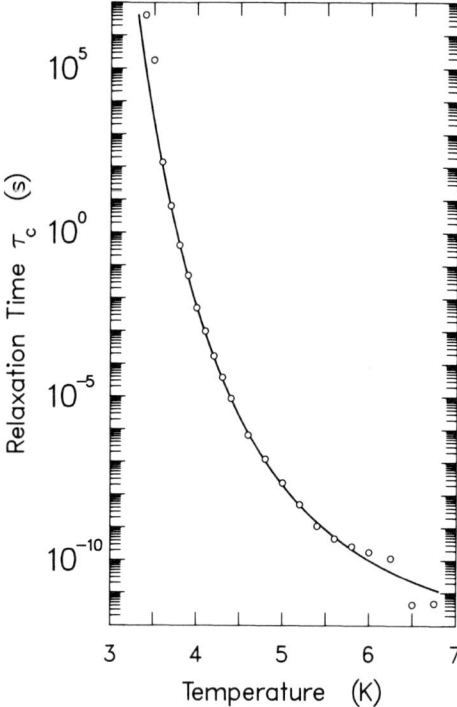

Fig. 6.7 Relaxation time τ_c versus the temperature. Solid line is a fit to the activated-dynamics equation. From Dekker et al. (1989).

6.2 $Rb_2Cu_{1-x}Co_xF_4$, an ideal 2d spin glass

where B is the barrier height (ψ is its exponent) and $\xi \propto t^{-\nu}$ for a zero temperature critical point. Thus, according to theory τ_c diverges much faster with temperature than a power-law divergence. Returning to Fig. 6.7, we have displayed the fit to the above equation by the solid lines with $\psi\nu = 2.2$, $\tau_o = 2 \times 10^{-13}$ s and $B = 11$ K – a physical set of values. Other fitting procedures were attempted, e.g. power laws with $T_c = 0$ or T_c as a parameter. But in all cases poor χ^2-tests and unphysical values of the parameters resulted. The best agreement with experiment is clearly via the activated dynamics of the droplet model with critical exponent $\psi\nu = 2.2$.

One can also use the Cole–Cole analysis to generate the form of the relaxation-time distribution $g(\tau)$. Figure 6.8 exhibits the evolution of $g(\tau)$ as a function of temperature. Notice the many decades shift with T especially in the long-time tails. These enormous displacements and broadening are all occurring far above $T_c = 0$, at a few kelvin. Once again the dynamics is dominant and such huge changes within the measurement window between 5.8 K and 3.4 K could easily and incorrectly be taken for finite temperature phase transition.

Alternative methods exist to derive τ_c from the spin-correlation function and by use of the fluctuation-dissipation theorem:

$$q(t) = \langle S_z(0) S_z(t) \rangle \tag{6.8}$$

and

$$\chi(\omega) = -\chi_o \int_0^\infty [\exp(-i\omega t) \frac{dq(t)}{dt}] \tag{6.9}$$

Two choices exist for the form of $q(t)$:

(i) an exponential-logarithmic decay

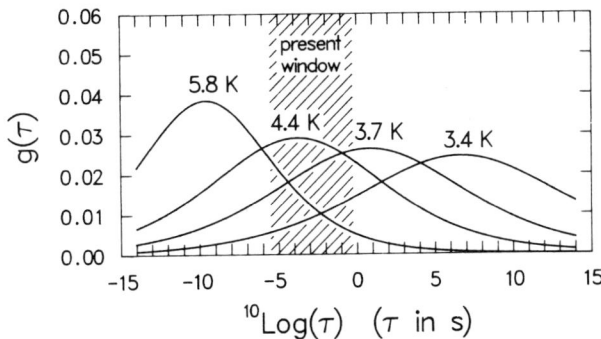

Fig. 6.8 Distribution of relaxation times $g(\tau)$ according to the Cole–Cole analysis for a selection of temperatures. From Dekker et al. (1989).

$$q(t) = \exp\left[-\left[\frac{\ln(t/\tau_o)}{\ln(\tau_c/\tau_o)}\right]^\delta\right] \quad (6.10)$$

and

(ii) a stretched exponential

$$q(t) = \exp[-(t/\tau_c)^\beta] \quad (6.11)$$

When the numerical integrations are performed to determine optimum values of τ_c (and also χ_o and δ or β keeping $\tau_o = 10^{-13}$ s), the temperature can be varied to generate a plot similar to Fig. 6.7. Best fitting leads to slightly different values for the $\psi\nu$ exponent. For (i) the result is $\psi\nu = 1.6$, while for (ii) $\psi\nu$ is 1.9, not too bad results considering the small experimental window compared to the enormous range (15 to 20 decades) of $\tau_c(T)$.

In order to double check the above approaches a dynamical scaling analysis was carried out using the scaling function suggested by the droplet model

$$\chi'' T^{-p} = F[-\ln(\omega\tau_o)T^{\psi\nu}] \quad (6.12)$$

where $p = -1 - \gamma + \psi\nu$. Excellent scaling exists over the entire range, except near $\omega\tau_c \approx 1$, with $p = -3$, $\psi\nu = 2.2$, $\tau_o = 10^{-13}$ and most important $T_c = 0$. Also the data overlaps remain valid even for temperatures where the system is clearly out of equilibrium. So the Cole–Cole approach and similar χ' and χ'' analyses give quantitative conclusions which are consistent with the dynamical scaling method.

Let us summarize these results and their meaning for this 'ideal' 2D spin glass. First of all, $Rb_2Cu_{1-x}Co_xF_4$ has been shown to possess practically all the specific characteristics of an Edwards–Anderson model spin glass. Table 6.1 lists and compares these characteristics for the system and for the

Table 6.1 Comparison of $Rb_2Cu_{1-x}Co_xF_4$ and the Edwards–Anderson spin-glass model for $d = 2$

	$Rb_2Cu_{1-x}Co_xF_4$	Edwards–Anderson SG model for $d = 2$
dimensionality	$d = 2$	$d = 2$
lattice	square	square
spin value	$S = \frac{1}{2}$	$S = \frac{1}{2}$
type of randomness	bond randomness	bond randomness
type of interactions	Ising	Ising
range of interactions	nearest neighbours	nearest neighbours
distribution of interactions	discrete (four competing interactions shown in Figure 1.8)	Gaussian (continuous) or $\pm J$ (discrete)

6.2 $Rb_2Cu_{1-x}Co_xF_4$, an ideal 2d spin glass

model, only with the distribution of interactions is there a deviation. Then, for a limit x-range this system was shown to obey all the experimental criteria related to spin-glass freezing. With these properties firmly established we can delve more systematically into the freezing phenomenon. Our experiments are non-linear susceptibility and ac-susceptibility. The former determines the static critical behaviour, the latter the dynamics. Measurement, using different forms of analysis to confirm the results, demonstrates that $T_c = 0$; $\gamma = 4.5$, $\beta = 0.0$ and $\Delta = 1+\frac{1}{2}(\gamma+\beta) = 3.2$ for the static critical exponents. And for the dynamical behaviour a new type of activated dynamics was established with exponents $\psi\nu = 2.2$ and $p \equiv -1-\nu(z-\eta-\psi) = -3.0$. Although various scaling relations exist between the different exponents, $\gamma = \nu(z-\eta)$, $z\beta = \nu(d-z+\eta)$ and $\nu = -1/\theta$, we still have one unknown too many. Here we revert to theory for the most innocuous choice, namely the prediction of a very small positive η ranging from 0 to 0.28. So taking $\eta = 0.2$, we then proceed to calculate the remaining critical exponents. Table 6.2 collects the results for the six exponents, the first four are the usual static ones, the last two are those based on the droplet model and its activated dynamics. The theoretical predictions, which are also listed, have been gained mainly through computer simulations (see section 5.12). A study of Table 6.2 will reveal good agreement with the $\pm J$ model and a self-consistent set of exponents, at least within the rather large error bars of both the experimental and the simulation, results. Very important, T_c was found from various best fits to be zero, nicely confirming the theoretical conclusion drawn from the 2D computer calculations. Furthermore, the exponent θ being negative corroborates that there is no finite-temperature phase transition for a 2D Ising spin glass. In addition, the dynamics is activated as forecast by the droplet model. All in all, we have a rather

Table 6.2 Comparison of critical exponents obtained for $Rb_2Cu_{0.782}Co_{0.218}F_4$ with those of the $\pm J$ and Gaussian Edwards–Anderson Ising SG models

Critical exponent	$Rb_2Cu_{0.782}Co_{0.218}F_4$	$\pm J$ $d = 2$ Ising SG	Gaussian $d = 2$ Ising SG
ν	2.3 ± 0.4	2.59 ± 0.13	3.4 ± 0.1
	2.5 ± 0.3	2.6 ± 0.2	3.5 ± 0.3
		2.64 ± 0.23	3.6 ± 0.1
γ	4.2 ± 0.6	4.5 ± 0.5	7.0 ± 0.2
	4.5 ± 0.2	5.3 ± 0.3	
β	0.0 ± 0.1	0.2 ± 0.1	0
η	0.2 ± 0.2	0.20 ± 0.02	0
		0.28 ± 0.04	
ψ	0.9 ± 0.2	$0 \leq \psi \leq 1$	$0 \leq \psi \leq 1$
θ	−0.42 ± 0.06	−0.38 ± 0.03	−0.29 ± 0.01

complete description for this ideal spin glass and a most favourable experimental-theoretical agreement. Certainly more work is needed to reduce the error bars to justify that the four real Js of $Rb_2Cu_{0.78}Co_{0.22}F_4$ are definitely in the same universality class as the simple $\pm J$ model. And finally upon viewing Fig. 6.7 again, the warning is clear. Far above a possible T_c the relaxation shifts outside the realm of feasible experiment. Here the spin glass is out of equilibrium and we must live with these disturbing non-equilibrium effects.

With this in mind, let us move onto the available 3D materials where the question of a finite T_c is paramount and, if so does $T_c = T_f$? And what about the behaviour below T_c, once a phase transition has been established? Keep in mind we are using T_c to denote the critical temperature of a phase transition as predicted by theory. In contrast we reserve T_f for the freezing temperature as is clearly characterized by experiment. The latter may or may not be related to the former.

6.3 IDEAL 3D SPIN GLASSES

Enormous effort has gone into attempts to compare experiments of χ_{nl}, $\chi''(\omega)$, etc. on various 3D spin-glass materials with theory using different types of scaling functions and analyses which give the critical exponents. The approach is similar to that followed in the previous section for the ideal 2D spin glass. The hope is to gain some insights into the true nature of a phase transition, if there is one. For example, CuMn, AgMn and AuFe have all been leading candidates for such endeavours and once the scaling analysis is completed, the values of the critical exponents are likened to those obtained from Monte-Carlo simulations of specific model (mainly Ising) spin glasses. Certain difficulties exist with this prescription. Firstly, it is based on the assumption that a standard second-order phase transition occurs with conventional static or dynamic critical behaviour. We have already experienced in section 6.2 the unconventional activated dynamics for 2D $Rb_2Cu_{1-x}Co_xF_4$. So would a 3D spin glass exhibit a completely normal second-order phase transition? Probably not! The above assumption assumes too much and begs the question of a phase transition. Secondly, for the many different and definitely non-ideal or in some cases not even good spin glasses, there is a vast range of critical exponents which have been determined from the scaling analysis. These depend strongly on the range and quality of the experimental data and the particular analytic procedure. One jestingly speaks of the 'right order of magnitude' for a critical exponent. Thirdly, the manifold values usually do not agree with the Monte-Carlo results. And fourthly, more refined and recent scaling analysis using linear plots instead of log–log ones has demonstrated the ambiguity of even the finest of the older work. The latest scaling usually

6.3 Ideal 3D spin glasses

shifts the exponents upwards and T_c, which is taken as a fitting parameter, downwards. Furthermore, the 'best' exponent values obtained to date are anomalous, i.e., much too large when compared to those observed in normal phase transitions. These points when taken collectively, indicate that a 3D spin glass is close to or just above its lower critical dimension. In addition, we have all the complications of non-equilibrium setting in as we approach T_c, so that even if there is a phase transition, we could not measure its equilibrium properties.

Rather than belabour the above obstacles (or shall we say challenges), let us consider the present situation regarding a possible *ideal*, or model-conforming (see Table 6.1 and dilate the dimensionality from 2 to 3), 3D spin glass. To date our 'best choice' lies with the $Fe_{0.5}Mn_{0.5}TiO_3$ mixed compound, although strictly speaking it is not as ideal in 3D as the Rb_2CuCoF_4 is in 2D. For the former compound we have a random substitution of Fe and Mn spins situated in a hexagonal lattice; see Fig. 6.9. In the basal plane the Fe–Fe nearest neighbour superexchange is ferromagnetic, while the Mn–Mn one is antiferromagnetic. However, it has recently been verified that the Fe–Mn coupling is also antiparallel. Between the planes, the interplanar coupling is always weakly antiferromagnetic as shown in Fig. 6.9, and this gives the system its 3D character. So for a 50–50, Fe–Mn composition, there are three $-J$s with large weight and one $+J$ creating the short-range interactions. This represents a most asymmetric distribution $P(\pm J)$ of exchanges along with a large amount of frustration due to the 'missing-centre' hexagonal (see Fig. 6.9) crystal structure. Moreover, the Fe and Mn ions have different values of spin, both not $\frac{1}{2}$, which makes the system seem random-site-like due to the distinction between Fe and Mn ions. In spite of these difficulties a strong unaxial anisotropy keeps the spins aligned (up/down) along the c (or hexagonal) axis, so $Fe_{0.5}Mn_{0.5}TiO_3$ is indeed a 3D Ising short-range spin glass. And this conclusion has naturally been substantiated by the usual measurements of ac- and dc-susceptibility, FC versus ZFC magnetization, time dependences, etc. These experiments have nicely mapped out the $(T$–$x)$ phase diagram. Equal amounts of Fe and Mn lies right in the middle of the spin-glass regime.

Let us proceed with the scaling analysis of this quasi-ideal 3D spin glass. By defining the non-linear susceptibility in a similar manner as in section 6.2, we expect the coefficient a_3 to diverge as $(T-T_c)^{-\gamma}$. One then plots χ_{nl} versus H_o^2 (the small applied dc field) and in the limit of $H_o \to 0$, i.e., the initial slope a_3 is acquired at the various temperatures. Then a_3 may be plotted on a log–log scale against $t \equiv (T-T_c)/T_c$ for various choices of T_c. Figure 6.10 shows such dependences where $\chi_{nl} \propto a_3$, and $T_g \equiv T_c$. Note there are two parameters here T_c and γ. A 'best fit' is curve B, $T_c = 20.70$ K and $\gamma = 4.0$. The reduced temperature interval is very small because, according to FC-magnetization, the system loses equilibrium below ≈ 22.0

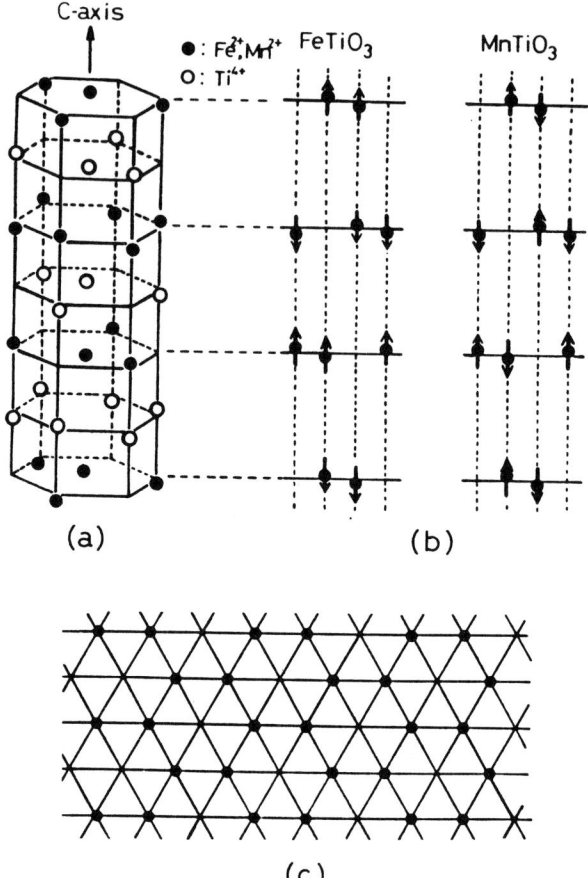

Fig. 6.9 (a) Crystalline structure of ilmenite-type compounds in the hexagonal lattice. (b) Magnetic structure of FeTiO$_3$ and MnTiO$_3$. (c) Arrangement of magnetic ions in the hexagonal c-layer. From Ito et al. (1988).

or $\log_{10}(T/T_c - 1) = -1.2$, and at high T the non-linear response is very weak and the critical region may be exceeded. Approximately a decade of reduced temperature is present in Fig. 6.10, which is not a scaling analysis.

The usual spin-glass scaling equation for the non-linear susceptibility is

$$\chi_{nl} = t^\beta F\left(\frac{H_o^2}{t^{\beta+\gamma}}\right) \qquad (6.13)$$

where $F(x) \propto x$ for small x and $F(x \to \infty) \propto x^{1/\delta}$ with δ another critical exponent. This scaling function is displayed in Fig. 6.11 as a log–log plot. The data 'collapsing' is obtained using the previously determined values for T_c and γ. β was the only remaining parameter and a value of 0.54 was optimized from the scaling overlap. It should be noted, however, the

6.3 Ideal 3D spin glasses

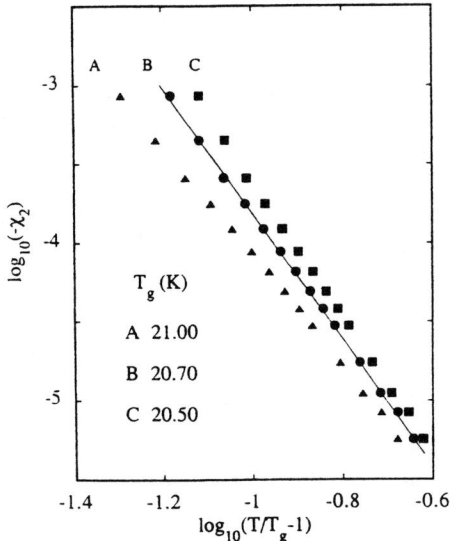

Fig. 6.10 $\text{Log}_{10}(-\chi_2)$ versus $\log_{10}(t)$ for three different values of $T_g = T_c$. The best fit, indicated by a solid line, is obtained for $T_g = 20.70$ K, yielding $\gamma = 4.0$. From Gunnarsson et al. (1991).

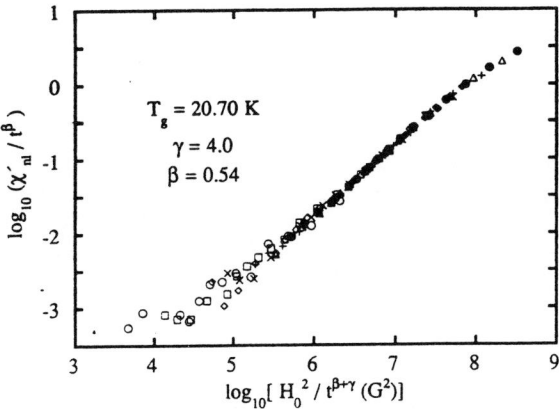

Fig. 6.11 $\text{Log}_{10}(\chi'_{nl}/t^\beta)$ versus $\log_{10}(H_0^2/t^{\beta+\gamma})$. The points represent the data collapsing obtained using $T_g = 20.70$ K, $\gamma = 4.0$, and $\beta = 0.54$. From Gunnarsson et al. (1991).

superposition of data point is not very good for lower magnitudes of the coordinates.

Via ac-susceptibility experiments the dynamical behaviour can be analysed using the conventional divergence of $\tau_{av} = \tau_o(T-T_c)^{-z\nu}$. Here for $Fe_{0.5}Mn_{0.5}TiO_3$ eight decades of frequency are available from $\chi'(T)$ and

$\chi''(T)$ measurements. Defining an τ_{av} according to a ratio of $\chi''(\omega)/\chi'(\omega)$, we can make the standard plots of log τ_{av} versus log $(T-T_c)$ and try for a unique determination of T_c, τ_o and zv. The results are $T_c = 20.95$ K, $\tau_o \approx 10^{-13}$ s and $zv = 9.3$. However, once again this is not a scaling analysis, but the conventional dynamical critical-phenomenon relation for a second-order phase transition.

A new improved scaling function has recently been suggested which takes the form

$$\frac{\chi''T}{\omega^{\beta/zv}} = \tilde{F}\left(\frac{T-T_c}{\omega^{1/zv}}\right) \qquad (6.14)$$

Since we expect $1/zv$ and β/zv to be much less than one, the above equation considerably weakens the frequency dependences. Now a *linear* scaling plot of $\chi''T/\omega^{\beta/zv}$ versus $(T-T_c)/\omega^{1/zv}$ can be made and this permits a much better judgement of the scaling quality. Such refined analyses result in $T_c = 20.5$ K, $zv = 11.5$ and $\beta = 0.58$. Significant here is that as always when a more sophisticated scaling analysis is made T_c is reduced and the critical exponents are increased.

As a final comparison we use the static form of this improved scaling analysis to treat the non-linear susceptibility for $Fe_{0.5}Mn_{0.5}TiO_3$. With the same rationale as above in mind, we write

$$\chi_{nl} \approx H^{\frac{2\beta}{\gamma+\beta}} \tilde{G}\left(\frac{T-T_c}{H_o^{2/\beta+\gamma}}\right) \qquad (6.15)$$

where the scaling function $\tilde{G}(x) \to 1$ for $x \to 0$ and $\tilde{G}(x) \approx x^{-\gamma}$ for small H_o. As before scaling plots are constructed by plotting on linear scales $\chi_{nl}/H_o^{2\beta/\gamma+\beta}$ versus $(T-T_c)T_c/H_o^{2/\gamma+\beta}$. Figure 6.12 displays the scaling analysis for a particular choice of parameters. The inset is an enlargement of the scaled data around the upturn. Unfortunately, it is impossible to find a uniquely optimum set of parameters (here T_c, γ and β) that gives a perfect collapse of the data. Various sets of the three parameters produce equally good 'collapsibility'. T_c can vary between 20.5 and 21.0, γ between 3.6 and 4.3 and $\beta \approx 0.5$ to 0.7.

Now that we have summarized the latest attempts at scaling the most ideal 3D spin glass, we should collect the results and relate them to the theoretical predictions from the Monte-Carlo simulations of a 3D Ising short-range spin glass (see section 5.12). Tables 6.3 and 6.4 list the salient features and results. Regarding the 'idealization', notice in Table 6.3 the inconsistency with respect to the different spin values and the distribution of interactions. The unlike crystal structures should not be of great significance, except that the hexagonal lattice will increase the amount of frustration and strongly reduce the magnitude of superexchange coupling $-J_4$ along the c-axis.

Table 6.4 requires a close examination. Here we make a direct comparison

6.3 Ideal 3D spin glasses

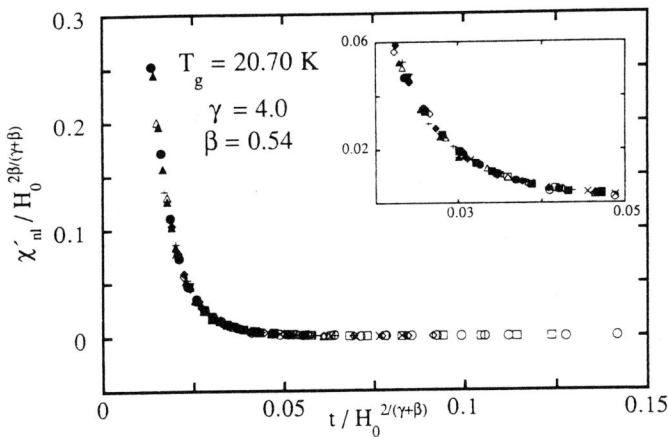

Fig. 6.12 $\chi'_{nl}/H_o^{2\beta/(\gamma+\beta)}$ versus $t/H_o^{2/(\gamma+\beta)}$. The points represent the data collapsing obtained using $T_g = 20.70$ K, $\gamma = 4.0$, and $\beta = 0.54$. The insert figure displays a magnified part of the plot. From Gunnarsson et al. (1991).

Table 6.3 Comparison of $Fe_{1-x}Mn_xTiO_3$ and the Edwards–Anderson spin-glass model for $d = 3$

	$Fe_{1-x}Mn_xTiO_3$	EA Model
dimensionality	$d = 3$	$d = 3$
lattice	hexagonal	cubic
spin value	$S(Fe) = 2$, $S(Mn) = \frac{5}{2}$	$S = \frac{1}{2}$
type of randomness	bond randomness	bond randomness
type of interaction	Ising	Ising
range of interaction	nearest neighbours only in plane	nearest neighbours
distribution of interactions	discrete and anisotropic ($+J_1$, $-J_2$; $-J_3$ in plane, plus weak $-J_4$ interplane)	Gaussian or $\pm J$

between the experiments on $Fe_{0.5}Mn_{0.5}TiO_3$ and the large-scale Monte-Carlo (MC) simulations for a 3D (cubic), Ising, $\pm J$ spin glass. Table 6.3 has already suggested that the similarity (real system with MC model) is reasonable and besides it is the best we have available at present. Both (real and MC) experiments indicate a finite T_c, i.e., a phase transition is occurring at a non-zero temperature. The MC computations have determined the full range of exponents, except for α the specific-heat 'divergence' which must be calculated from a scaling relation and β, the order-parameter exponent (see below). Since the simulations cannot reach below or even too close to T_c (loss of equilibrium), they give a 'best-fit' function for the order parameter $q(t)$ which contains a temperature-dependent exponent

Table 6.4 Comparison of critical exponents obtained for $Fe_{0.5}Mn_{0.5}TiO_3$ with those of the 3D Ising $\pm J$ Monte-Carlo simulations

Critical exponent		$Fe_{0.5}Mn_{0.5}TiO_3$	MC simulations
directly determined from experiment	γ	4.0 ± 0.4	2.9 ± 0.1
	β	0.5 ± 0.2	≈ 0.5
	$z\nu$	10.5 ± 1.0	7.9 ± 1.0
	$T_c(K)$	20.7 ± 0.3	finite
	δ	$= \gamma/\beta + 1 = 9.0$	6.8 ± 1.2
	α	$= 2 - 2\beta - \gamma = -3.0$	-1.9 ± 0.3
	ν	$= (2-\alpha)/d = 1.7$	1.3 ± 0.1
	η	$= 2 - \gamma/\nu\ \ -0.35$	-0.22 ± 0.05
	z	$= 10.5/\nu = 6.2$	6.0 ± 0.5

$\beta(T)$. Around T_c, where the simulations may not even be valid, $\beta \approx 0.5$. So this is a deficiency of the MC technique, namely that β cannot be accurately determined. Returning to the real experiments only 3 exponents can be directly evaluated via the diverse scaling analyses of the measurements: γ, the susceptibility exponent, β, and $z\nu$, a product of the dynamical exponent with the correlation-length exponent. These are shown with the appropriate error bars in Table 6.4 In order to continue the comparison, different scaling relations (equations connecting the various critical exponents) must be used to calculate δ (the magnetic field exponent at T_c), α, ν, η (the correlation function exponent), and z. With this set of exponents we should be able to completely characterize a second-order phase transition. In general, the agreement is at best fair. Particularly disturbing are the too large values of γ and ν for our real (ideal) 3D spin glass; also the inability of fixing T_c to within 1% weakens the analysis. While it seems that there is a phase transition, it is not of the conventional type. Either the lower critical dimension is close to 3, e.g. as some theories predict 2.7, so that a marginal phase transition occurs, or the non-equilibrium behaviour which is certainly present hinders a meaningful characterization of the critical phenomenon. At this time we cannot proceed and resolve these difficulties. One may study Table 6.4 and draw his or her own conclusions about the nature of the spin-glass freezing. To be sure we must await future experiments and theory we can offer a definite and physical settlement. Perhaps such will be possible in the second edition of this book.

6.4 FUTURE POSSIBILITIES

Never before has so much effort been put into the study of a possible phase transition as with the spin glasses. The previous sections have only scratched the surface of huge amounts of work done on the problem. For we have

6.4 Future possibilities

intentionally omitted the many experiments and comparisons with theory regarding the abundance of *non-ideal* (canonical and good) spin-glass materials. Tables exist for their critical exponents and temperature, but we have maintained the viewpoint: first to understand the most ideal and simple systems before proceeding with imperfect and complex examples.

For the case of a 2D Ising short-range ($\pm J$) spin glass (section 6.2) we have a reasonable material which closely matches the constraints of the EA model. A series of accurate and systematic experiments have been performed and firmly compared to theory which predicts two dramatic features: $T_c = 0$ and activated dynamics. Both of these highlights were borne out by the measurements and a relatively consistent set of exponents was found to agree (within the large error bars) with the Monte-Carlo simulations. Our life is made quite easy in all the fitting procedures by setting $T_c = 0$. Hence, we seem to now possess a basic understanding of the 2D situation.

Clearly, for the 3D Ising short-range spin glass (section 6.3), the present state of the science is less satisfactory. Our best-choice system, while not perfectly ideal, does meet many of the constraints of the EA model. Here theory is more mundane: a finite T_c and conventional dynamics are inferred. However, these predictions cause a lot of trouble, T_c cannot be determined accurately from experiments and becomes a fitting parameter. And the conventional dynamics results in an exceptionally large value of zv which makes one suspect that something is wrong with the fitting function itself. In addition, the static critical exponents do not agree with those found from the Monte-Carlo simulation. Since $T_c = 0$ for $d = 2$, we would expect $T_c \neq 0$ for $d = 3$ with both systems in the Ising universality class. While a phase transition in 3D spin glass seems likely, it probably requires a more sophisticated (beyond-the-conventional) type of critical-phenomenon theory to describe its properties. Mean-field-type theories do not help us here because they always predict a phase transition with a standard set of exponents. Now, finally, we have arrived at the 'cutting-edge' of present-day spin-glass research. What will the future bring? Let's try to glance a bit into the uncertain prospects for future progress.

Above all, an *ideal* 3D Ising $\pm J$ spin glass system must be discovered or deliberately fabricated. The latter could be intentionally created using the modern techniques of materials research. A material resembling that shown in Fig. 6.1 and discussed in section 6.1 would then be measured completely: specific heat, ac- and dc-susceptibility, magnetization, neutron scattering, etc. The accumulated data with temperature, field and concentration dependences could be fully compared with existing theory and the results on $Fe_{0.5}Mn_{0.5}TiO_3$. Primarily, we would more severely test the theory and hopefully there would be a better agreement than now appears. The key question here is how to obtain the 3D sample. Again it lies in the hands of our materials or chemist friends.

A hint along these lines comes from the recent work to fabricate thin films of metallic spin glass, e.g. *Cu*Mn, *Ag*Mn and *Au*Fe. Various deposition techniques have been employed to make samples of these alloys thinner and thinner. To increase the sample mass usually a multilayer or superlattice is constructed with non-magnetic interlayers (Cu or Ag) separating the thin spin-glass films. Different experimental methods try to detect the freezing temperature as a function of thickness, e.g. SQUID magnetometry or anomalous Hall effect to determine the magnetization. The problem is to get enough sample mass so that clear signs of the freezing, not simply relaxational effects related to the loss of equilibrium, may be detected. Remember it is rather difficult and subtle to distinguish the 3D spin-glass cusps from the 2D spin-glass peaks caused by non-equilibrium effects. This impediment becomes even more severe, if signal sensitivity is low due to small sample size. The initial results of such experiments are ambiguous. One group sees a strong reduction of $T_f \to 0$ with d (the sample thickness) $\to 0$. Another claims T_f decreases but remains non-zero down to thicknesses of 1 and 2 monolayers. Also $T_f(d)$ depends strongly on what the interlayer is in the superlattices. The experiments are hard to perform and the interpretation of data is even more troublesome when it comes to establishing T_f or T_c. So far, accurate ac- or non-linear susceptibility has not yet been obtained although attempts are presently being made. Thus it remains a pending task to overcome these difficulties and settle the freezing behaviour of truly 2D (Heisenberg) spin glass. How thin is 2D? What is the length scale governing the crossover from 3D to 2D? At what thickness will we observe a change in the dynamics? These and many other questions can be posed concerning the artificial reduction of spin glass dimensionality. Future research will certainly provide clear answers to these questions. And the same sophisticated deposition techniques as used for the multilayers could be upgraded to produce the artificially-structured ideal 3D spin glass shown in Fig. 6.1.

The theory is far from complete and improvements are necessary. Tremendous exertion has gone into the mean-field theory and obtaining its solution. And we have tried to outline the physical interpretation of its concepts and calculations (section 5.7) which are not obvious and took a number of years to develop. Moreover, for this most-simple starting point of theoretical approaches, enormous complexities lie in the continuous order parameter and its probability distribution of overlaps. Even here experiment is left behind. The first suggestions for mesoscopic systems are now beginning to be made on how to measure the overlap distance or to estimate the order-parameter probability distribution function based upon fluctuations (noise), time dependences and ageing effects in the frozen spin-glass state. But the basis is only mean-field theory – a first order approximation. What happens when we advance to include fluctuations, or short-range interactions, or remove the Ising up/down restriction? We call all this future exertion: beyond the mean-field theory. Also the droplet and fractal cluster models need to be pursued and extended with more and closer experimental contact and comparisons. There is still a vast amount to be accomplished with the theory of spin glasses and despite the past successes new approaches, not so heavily based on second-order phase transitions, and models from other areas should be tried ensuring future progress not just in spin glass, but more extensively with random and glassy systems in general.

FURTHER READING AND REFERENCES

Bass, J. and Cowen, J. A. (1992) *Recent Progress in Random Magnets* (Edited by D. H. Ryan) World Scientific, Singapore.
Dekker, C. (1988) Ph.D. Thesis University of Utrecht.
Dekker, C., Arts, A. F. M. and de Wijn, H. W. (1988) *Phys Rev.*, **B38**, 8985.
Dekker, C., Arts, A. F. M., de Wijn, H. W., van Duyneveldt, A. J. and Mydosh, J. A. (1988) *Phys. Rev. Lett.*, **61** 1780; (1989) *Phys. Rev.*, **B40**, 11243.
Gunnarsson, K., Svedlindh, P., Nordblad, P., Lundgren, L., Aruga, H. and Ito, A. (1991) *Phys. Rev.*, **B43**, 8199.
Ito, A., Aruga, H., Kikuchi, M., Syono, Y. and Takei, H. (1988) *Solid State Commun.*, **66**, 475.

7
Spin-glass analogues

The previous six chapters have dwelt upon the spin-glass phenomenon with emphasis on the word 'spin'. Except for a few cursory remarks pertaining to real glasses, we have remained fully on the track of magnetic effects and spin models. Naturally, magnetism does have its advantages since we can easily couple into such systems with a magnetic field, the conjugate field of the problem, and there exists a vast range of sensitive experimental techniques to measure this coupling under various conditions. However, now that we have outlined the basic *spin*-glass behaviour and theory, and hopefully have gained an understanding of the rudimentary physics, the obvious question arises: are there non-magnetic analogues? Answer: yes, many! So let's at this point take a look at some of these and venture beyond the spin and even beyond the glass into life sciences.

We begin with the electrical analogues where electric-dipole moments giving rise to a polarization **P** replace the spin moment causing the magnetization **M**. Higher-order moments are also mentioned in the form of electrical quadrupoles. By substitution of a moment bearing species for a non-moment or a different-interacting moment, we introduce the elements of randomness and competition. Thus, we can create electric dipolar and quadrupolar glasses.

Our analogues are continued with various superconducting phases. First, there are granular superconductors which behave like a network of randomly coupled tunnel junctions. Then, the high-T_c oxide superconductors have been modelled on the spin glasses, initially because of their ceramic nature and later due to the unusual field penetration ($H_{c1} < H < H_{c2}$) creating a flux-line 'lattice' which becomes glassy.

In order to contrast the spin-glass phenomenon with a random magnetic system which does exhibit a *good* second-order phase transition, we briefly discuss the random-field Ising model. Here we focus upon the critical behaviour and the H–T phase diagram whose critical line must be traversed in a certain way in order to observe the phase transition.

Finally, we conclude the chapter with three non-physical analogues to the spin glasses. These 'spin-offs' are combinatorial optimization, popularly known as the travelling-salesman problem, neural networks or how our

7.1 ELECTRIC DIPOLAR AND QUADRUPOLAR GLASSES

brain can be modelled by a spin glass, and finally biological evolution. Is the origin of life a spin glass?

7.1 ELECTRIC DIPOLAR AND QUADRUPOLAR GLASSES

The electrical analogue of a magnetic spin is a small displacement d of two unlike charges e creating a dipole moment $p = ed$. Such dipoles can be formed by the displacement of dopant atoms in certain compounds from the centrosymmetric host sites or by molecules with their bonds not having a spherical symmetry. Along these lines we can make direct contact with our magnetic systems via the long-range order of the ferroelectrets (all the electric-dipole moments are aligned parallel) or the antiferroelectrets (antiparallel sublattices). In mixed (ferro/antiferro) or diluted (ferro) electric materials we have a close analogue of a spin glass which we call a dipolar glass. Here the electric field serves as the conjugate field. Theoretically, we can use a 'pseudospin' formalism to represent the electric dipoles and their interactions, measure the dielectric polarization and susceptibility in place of the magnetic ones, and apply an electric field for the FC or ZFC procedures.

A random-site example of a dipolar glass is $K_{1-x}(Li_x, Na_x$ or $Nb_x) TaO_3$ where the displacement of the x-dopant (Li, Na or Nb) forms an electric-dipole moment. For sufficiently large Li, Na or Nb concentrations, collective effects occur via dipole–dipole or lattice-stress-field interactions, and through experiment we can investigate the freezing behaviour. Figure 7.1 illustrates

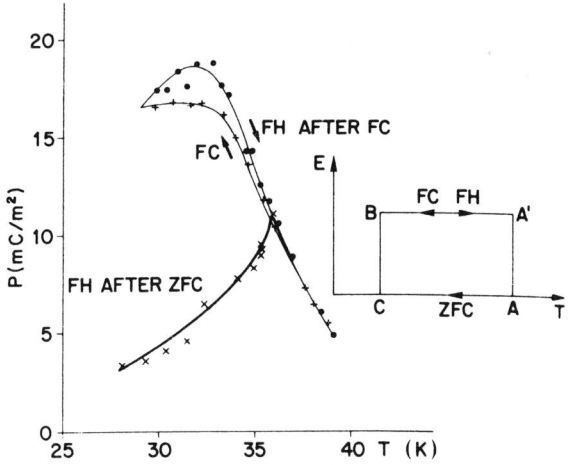

Fig. 7.1 Polarization versus temperature in a T-sweep experiment for $K_{1-x}Li_xTaO_3$. Note that the temperature where P becomes independent of time, $\partial P/\partial t = 0$ (not shown), coincides with the departure of the FC curve from the ZFC curve. $x = 0.016$, $dT/dt = 3$ mK/sec, and $E = 30$ kV/m. Inset: Field-temperature cycle. From Höchli et al. (1985).

the typical static, electric polarization P (analogue of M) versus temperature curves for $K_{0.984}Li_{0.016}TaO_3$ according to the different (electric) E-field-cooling procedures shown in the inset. Note the ZFC-field heating (FH) cusp found at the freezing temperature $T_f = 37$ K. This nicely corresponds to the ZFC magnetization for a spin glass. However, in the electric system the FC plateau is missing and there is noticeable hysteresis below T_f between FC and FC-FH. Another attempt at comparison is displayed in Fig. 7.2 where the non-linear part of the 'static' (1 kHz) dielectric susceptibility ϵ_{nl} (obtained from the higher-order E-field dependence) is plotted against the temperature for $K_{0.8}Na_{0.2}TaO_3$. The raw data are shown as ϵ_{nl}, but the 1 kHz measurement frequency prohibits their use below ≈ 20 K (where a dispersive ϵ''-component begins to appear). A special analysis tries to extract the divergent coefficient of the non-linear susceptibility: $a(T) \propto [(T-T_f)/T_f]^{-\gamma}$. This is exhibited as the curve (a) in Fig. 7.2 and yields T_f

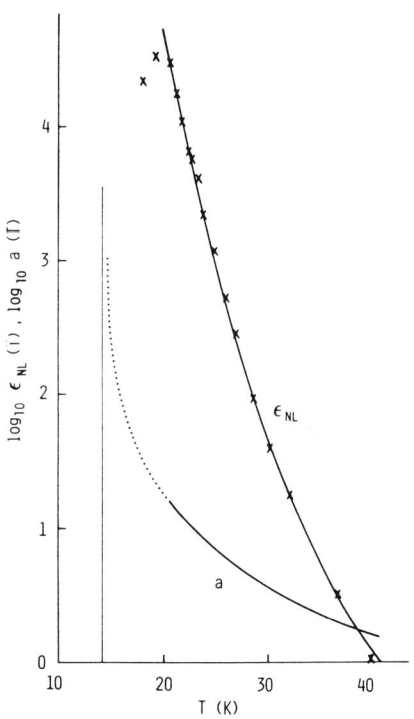

Fig. 7.2 Non-linear dielectric susceptibility $\epsilon_{nl}(T)$ and critical coefficient $a(T)$ in $K_{1-x}Na_xTaO_3$. Solid curves: best fits to power law. Dotted curve: extrapolation of $a(T)$ to show criticality. From Maglione et al. (1986).

7.1 Electric dipolar and quadrupolar glasses

$\simeq 14.5$ K and $\gamma \simeq 2.7$. Unfortunately, better fitting closer to T_f could not be accomplished due to the loss of equilibrium represented by the peak and decrease in ϵ_{nl}.

Another dipolar glass is the mixed system K_{1-x} or $Rb_{1-x}(NH_4)_xH_2PO_4$. These polar materials are characterized for sufficiently high temperature by a disordered network of O–H--O bonds corresponding to randomly rotating dipole moments. Here we have the paraelectretic phase. At low temperatures the ordering of these H-bonds or dipoles leads to a ferroelectric polarization in pure K or RbH_2PO_4, but in pure $NH_4H_2PO_4$ an antiferroelectret develops. These long-range transitions are suppressed by the random substitution of K or Rb by $(NH_4)_x$, i.e., competing or mixed interactions. A glass freezing results below $\simeq 30$ K for $x = 0.35$. The quantity to measure once again is the dielectric constant, real and imaginary (dielectric loss) parts, ϵ' and ϵ''. The data are shown in Fig. 7.3. While there is no cusp in the real part, the dielectric loss exhibits a frequency-dependent peak which can be related

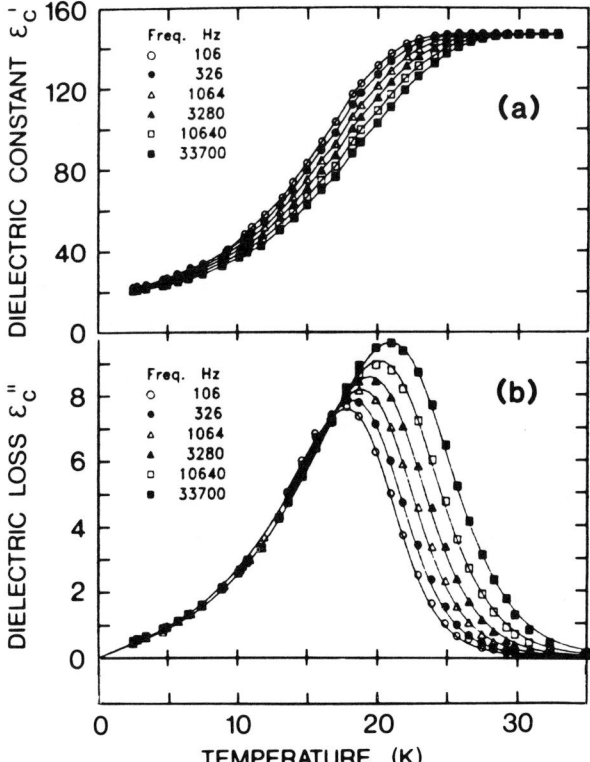

Fig. 7.3 (a) The real part of the dielectric constant ϵ'_c for $Rb_{1-x}(NH_4)_xH_2PO_4$. The lines are guides to the eye. (b) The imaginary part of the dielectric constants ϵ''_c fitted via the lines to a broad distribution of relaxation times. From Courtens (1984).

to a broad distribution of relaxation times $g(\tau, T)$. This function, similar to that of a spin glass, extends to longer and longer times as the temperature is reduced. So at a particular T, the characteristic or average relaxation time becomes much greater than any time of measurement (loss of equilibrium) and freezing results with the associated glassy dynamics. Nevertheless, from the curves in Fig. 7.3, it does not seem that the freezing transition displays sharp co-operative effects.

Quadrupolar glasses are formed when the elastic quadrupole–quadrupole interaction is more dominant than the electric dipole–dipole interactions. A good example is the mixed molecular crystal $(KCN)_x(KBr)_{1-x}$. In this case the CN molecules possess not only a small dipole moment, but also a quadrupolar one since they are rod- or cigar-shaped compared to the spherical (meaning no moments of any order) Br-ions. Both the dipoles and the quadrupoles couple *individually* via the quadrupolar stress field generated by the lattice and its elastic constant. Pure KCN undergoes a transition at 168 K from a completely disordered (paraelectret) cubic phase to an orthorhombic phase where the *axes* of the CN molecules become oriented along a given (b-axis) direction without any associated dipolar ordering, i.e., the arrows or heads of the dipoles point randomly up or down along the b-axis. At a lower temperature 83 K, there is an order-disorder transition which creates an antiferroelectric structure of the dipole heads along the b-axis. Now we mix $(KBr)_{1-x}$ into $(KCN)_x$ and around $x \approx 0.5$ for $T \leq 60$ K, a *random* isotropic freezing of both the molecular orientation (the axes) and the dipoles (the arrows) simultaneously occurs. Figure 7.4 sketches this glassy state for $(KCN)_{0.5}(KBr)_{0.5}$. The experimental proofs for such a phase come from measurements of specific heat (a linear low-temperature dependence), neutron scattering (appearance of a central peak) and dielectric susceptibility (low-frequency peak in ϵ' and ϵ'' as a function of T). Figure 7.5 shows the latter for various frequencies. At least around the peak temperature the behaviour of the (dielectric) susceptibility is standard spin-glass-like, except for broadness of the maximum in ϵ'. Moreover, the shift of the peak temperature with frequency seems to follow an Arrhenius law: $\omega = \omega_o \exp(-E_B/k_B T)$, for the accumulated data. So once again it would seem that with the mixed molecular crystals a reasonable analogue reveals itself to a spin glass, albeit without the sharpness or criticality of a good phase transition.

We could mention other examples of dipolar or quadrupolar glasses $[(N_2)_{1-x}Ar_x$, etc.] with the same general properties. But instead, let us conclude this section by reviewing the interesting case of solid, molecular (ortho/para) hydrogen, which is a quadrupolar system without a dipole moment.

In molecular hydrogen because of the fermion statistics of the protons, two forms are possible: (ortho o-H_2) and para (p-H_2) hydrogen. o-H_2 has a total nuclear spin of unity and odd angular-momentum values. Oppositely, p-H_2 has zero total nuclear spin and even values of the angular momentum,

7.1 Electric dipolar and quadrupolar glasses

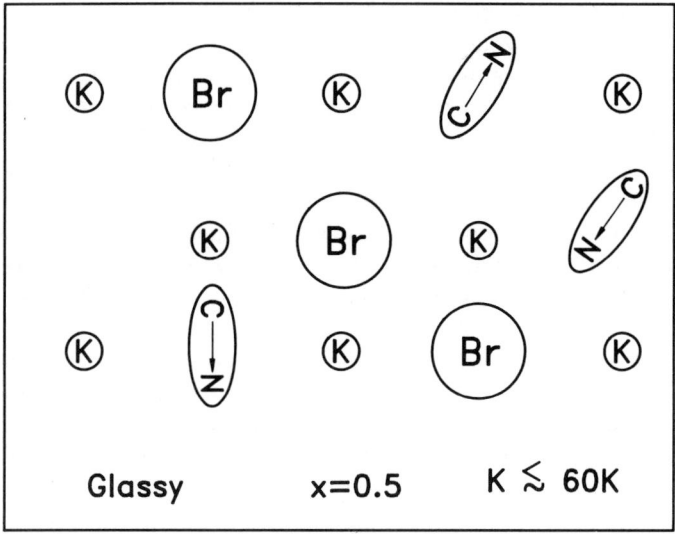

Fig. 7.4 Sketch of the various molecules in a (1, 0, 0) plane for 50% $(KCN)_x(KBr)_{1-x}$. The arrows on the cyanide molecules indicate the random freezing of the dipoles. From Sethna and Chow (1985).

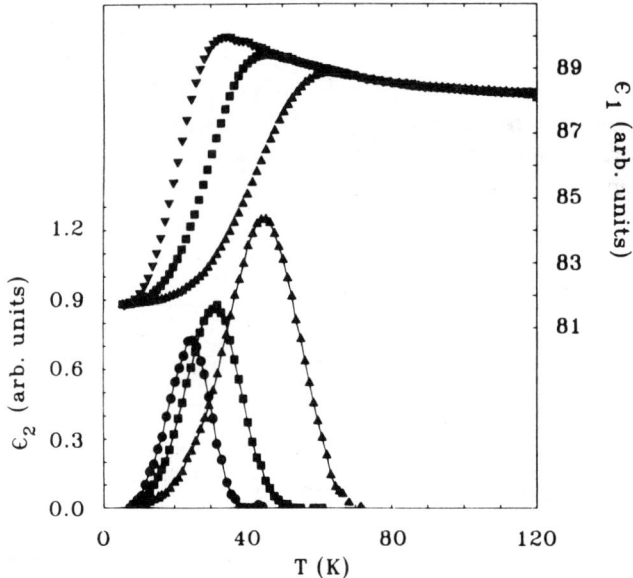

Fig. 7.5 ϵ_1 (=ϵ') and ϵ_2 (=ϵ'') versus temperature at several frequencies for $(KBr)_{0.5}(KCN)_{0.5}$: $\omega = 6\times10^7 s^{-1}$ (▲), $6\times10^4 s^{-1}$ (■), $60 s^{-1}$ (●), and $12 s^{-1}$ (▼). From Birge et al. (1984).

which vanishes in the ground state of the solid phase. As a result of this difference, o-H_2 molecules are non-spherical, while p-H_2 molecules have perfect spherical symmetry. Therefore, the asymmetry of the o-H_2 offers the possibility for various types of quadrupolar orientational order (dipole moments are negligible) whereas the no-moment p-H_2 can be used as an inert host. Figure 7.6(a) illustrates the situation for the long-range-ordered phase of pure o-H_2. A planar projection of the fcc-like crystal structure is shown with the novel type of periodic long-range orientational ordering at low temperatures. More important for us is the mixed crystal where sufficient amounts p-H_2 have been substituted for o-H_2 to form the mixed phase. This is displayed in Fig. 7.6(b) where the p-H_2 molecules are indicated by the 'dotted' spheres and there are various shapes and orientations for the o-H_2 molecules. The latter is caused by a probability distribution for different molecular shapes which occurs in the mixed phase. Here molecules can be orientated preferentially along an axis ('cigars') or preferentially in a plane ('discuses') with a continuum of intermediate shapes. At high enough temperatures these dissimilar-shaped quadrupoles can freely rotate,

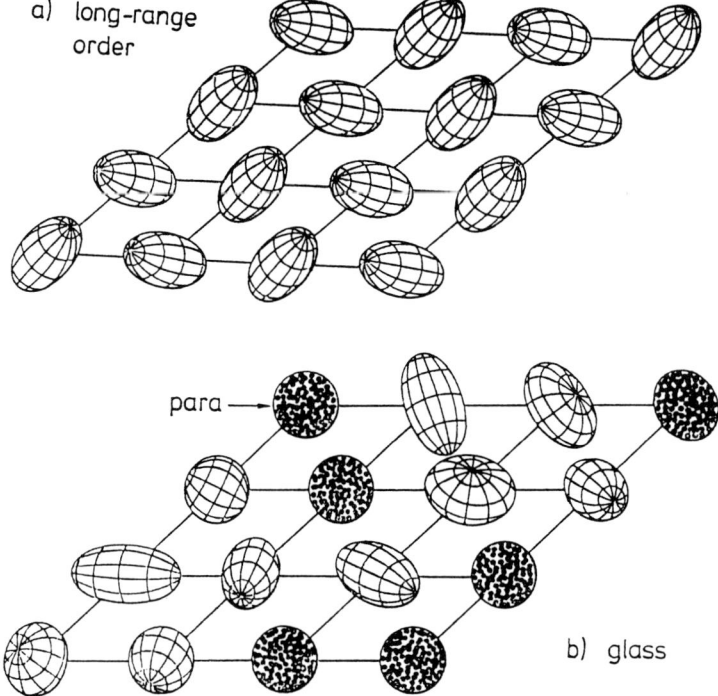

Fig. 7.6 Comparison of the orientational ordering for (a) a plane section of the ordered structure and (b) the quadrupolar glass phase; from Sullivan (1983).

7.1 Electric dipolar and quadrupolar glasses

but as the temperature is lower they seem to gradually freeze out into a random static alignment. Such glassy state is sketched in Fig. 7.6(b).

The main experimental technique used to measure these effects has been NMR on the protons; resonance-line shapes and nuclear-relaxation times are determined as a function of temperature and concentration. Based upon the results a phase diagram ($T-x$) is proposed in Fig. 7.7 for the mixed (o-H$_2$)$_x$ (p-H$_2$)$_{1-x}$ system. Notice at the very low temperatures (below 0.3 K) a frozen (orientational glass) phase appears to form for a restricted range of x. The coupling between the o-H$_2$ molecules is the weak electric quadrupole–quadrupole interaction. Its conjugate field is an electric-field gradient which would be required for the measurement of the quadrupolar susceptibility. However, it is quite difficult experimentally to produce a sufficiently large electric-field gradient in order to perform accurate susceptibility measurements. Consequently, most of the information pertaining to o-H$_2$/p-H$_2$ mixed crystals has been obtained from NMR which should be complemented by other techniques. The generally agreed interpretation of the NMR results is a gradual freezing of quadrupole moments with a distribution of relaxation times which seems to continuously evolve as the temperature is lowered. The quadrupoles appear frozen when the time of measurement is smaller than the average relaxation time, whereas no freezing is observed for experiments with longer observation times. As above we have a spin-glass analogue, but without any indications of a sharp and critical phase transition which seems to be generally missing in these 'electrical' systems.

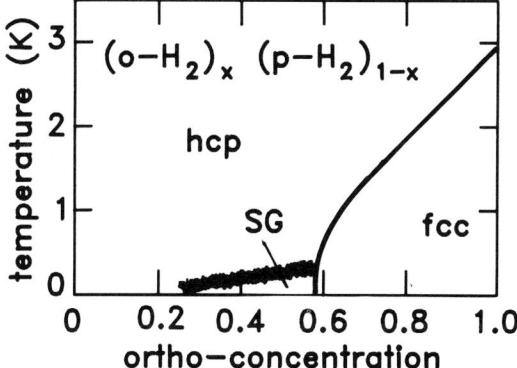

Fig. 7.7 Phase diagram for solid solutions of o-H$_2$ and p-H$_2$. For large concentrations of the orientationally anisotropic component (o-H$_2$) the orientationally ordered phase has face-centred-cubic structure, the disordered one is hexagonal close packed. The orientational glass phase (denoted by SG) forms a narrow low-temperature wedge. From Sullivan *et al.* (1978).

7.2 SUPERCONDUCTING GLASS PHASES

Consider a granular superconductor that consists of superconducting islands randomly embedded in a non-superconducting host. The host can be an insulator, semiconductor, metal or even another superconductor with a lower T_c. The various superconducting grains or islands interact via Josephson tunnelling of Cooper pairs, i.e., a phase coherence of the electron pair across a narrow insulator or the proximity effect, i.e., an induced but weakened superconductivity in the host material. There are many experimental realizations of such systems. Examples are In or Pb with insulating Ge forming a random continuous medium, Pb ($T_c = 7.2$ K) spheres randomly dispersed in Zn ($T_c = 0.8$ K) and Nb composites with the normal metal Cu. In addition, a more modern technique is to artificially fabricate a network of Josephson junctions out of Al/Al-oxide with different size 'weak links' connected to each other. And indications exist that in a magnetic field these disordered superconducting systems may freeze into a state which exhibits spin-glass-like order: a random orientation of the supercurrents in the various grains. Here we need to define our order parameter which for a superconductor is the complex energy gap (wavefunction of the Cooper pairs) in the ith grain ψ_i,

$$\psi_i = \Delta_i(T) \exp(i\phi_i) \tag{7.1}$$

where $\Delta_i(T)$ is the temperature-dependent real energy gap in the density of states and ϕ_i is the phase factor.

The hamiltonian describing the collection of grains and their interactions can be written as

$$\mathcal{H} = -\sum_{\{ij\}} J_{ij} \cos(\phi_i - \phi_j - A_{ij}) \tag{7.2}$$

where

$$A_{ij} = \frac{2e}{hc} \int_i^j \mathbf{A} \cdot d\mathbf{l} \tag{7.3}$$

$\mathbf{A}$ is the vector potential and the integration is along a path connecting the centres of the ith and jth grains. J_{ij} varies between 0 and J without reversing sign, but it falls off with distance between the grains as $\propto 1/r_{ij}$ for a Josephson coupling or $\propto \exp(-r_{ij})$ for the proximity effect. A *random* $\pm$ (or ferro/antiferro) interaction is caused by the variable A_{ij} as part of the argument of the cosine in equation (7.2). Let us position the applied magnetic field along the z-axis, $\mathbf{B} = B\hat{z}$ and use the gauge for the vector potential, $\mathbf{A} = Bx\hat{y}$. Then

$$A_{ij} = \frac{2e}{hc} B \left(\frac{x_i + x_j}{2}\right)(y_j - y_i) \tag{7.4}$$

where $x_i y_i$ and $x_j y_j$ are the centre coordinates of the ith and jth grains,

7.2 Superconducting glass phases

respectively. Disorder is introduced by an amorphous structure of the grains such that the x and y become a set of random variables. This creates a variation in A_{ij} which may be further tuned by increasing the field B. Another way is to vary the Josephson-tunnelling coupling between the grains such that $J_{ij} = J$ with occupation probability p, and $J_{ij} = 0$ with probability $1-p$. Because of these random, yet weak, interactions, frustration results and causes random orientations of the supercurrents among the collection of coupled grains.

The thermodynamic properties of this disordered systems are governed by the free energy and associated partition function. In order to calculate these quantities from the above hamiltonian both thermal and configurational averaging is required, necessitating the replica trick. Thus, the order parameter for the set of superconducting grains becomes analogous to the Edward–Anderson one

$$q = \langle\langle\psi_i\rangle^2_T\rangle_C = \Delta^2(T) \langle\langle\exp i\phi_i\rangle^2_T\rangle_C \tag{7.5}$$

And the state it represents is a randomly oriented distribution of frozen-in supercurrents instead of the EA spins.

The mean-field phase diagram based upon these model calculations is shown in Fig. 7.8. Note the figure has three coordinates (T, p, H) with applied magnetic field $(B = H)$ as the third axis. For a granular superconductor just above its percolation threshold $(p \geq p_c)$, there are the collective Meissner phase – a complete expulsion of the flux up to H_{c1}, and the Abrikosov one – above H_{c1} a penetration of the field in a regular triangular lattice which has a uniform orientation from grain to grain. Really

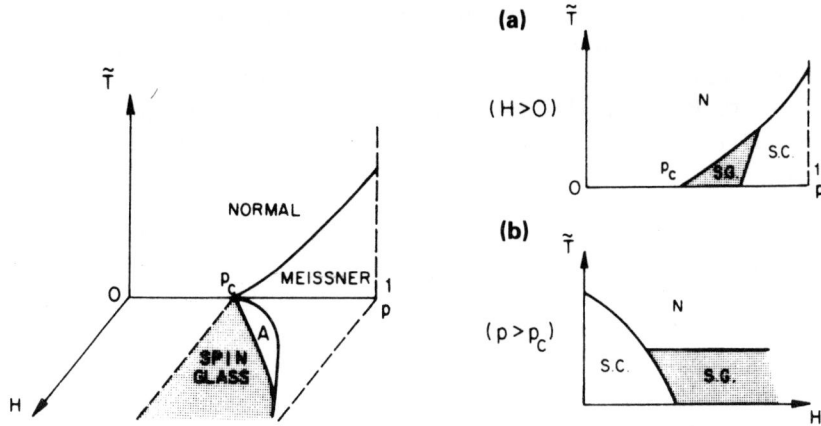

Fig. 7.8 Left-hand side: mean-field phase diagram as a function of temperature T, applied magnetic field H, and Josephson bond-occupation probability p near percolation threshold p_c. Four distinct phases are exhibited: normal, Meissner, glass, and Abrikosov (A) phases. Right-hand side: phase diagrams for fixed applied magnetic field H, showing spin-glass (S.G.), superconducting (S.C.), and normal (N) phases for (a) $H > 0$ and (b) $p > p_c$. From John and Lubensky (1985).

new here is the prediction of a glassy phase just beyond p_c for sufficiently large fields. Glassy means a penetration of the magnetic field into the superconductor via a completely random alignment of frozen-in supercurrents among the various grains. This leads to long-range fluctuations in the magnetic-field profile which decay as a power law with distance. So we have a highly inhomogeneous field distribution within the superconductor which yet remains to be detected. Unfortunately, the superconducting order parameter is experimentally inaccessible. However, the differences between FC and ZFC of the magnetization and the time dependences (dynamics) of the glassy state should be measurable characteristics.

Recently there has been renewed interest in the superconducting/glass concept due to the high-T_c oxide superconductors and their inability to carry large currents in a field. Firstly, the ceramic form of these materials is a natural granular superconductor. Thus, the model of the preceding paragraphs may be directly applied. And, indeed, the experiments just mentioned have borne out the spin-glass-like behaviour. Secondly, for the oxide superconductors there are an enormous number of microscopic defects, which, when combined with the unusually small superconducting coherence length and large anisotropy, make even bulk *single crystals* a candidate for glassy behaviour. The name 'vortex glass' has been given to describe the random penetration of magnetic flux in such single-crystal materials.

It is instructive to reconsider our superconducting order parameter in order to elaborate our magnetic analogue. Recall that ψ is a complex quantity with two components: amplitude and phase. This corresponds to the order parameter of a two-components (x–y) magnetic system. Returning to Fig. 7.8 in the Meissner phase ($H < H_{c1}$), ψ is uniform as in a ferromagnetic (uniform magnetization). In the Abrikosov phase ($H_{c1} < H < H_{c2}$) a periodic flux-line lattice is formed without any disorder in which $\psi(\mathbf{r}) = \psi(\mathbf{r} + \mathbf{R})$ where $|\mathbf{R}| = a_o \cong (h/2eH)$. However, if some quenched disorder exists, as it does in the high-T_c materials through the assortment of defects, the flux-line lattice will be destroyed on long-length scales. Figure 7.9 illustrates the difference: long-range ordered Abrikosov phase versus short-range ordered domains of flux lines and finally complete disorder (glass). It is expected that the latter case is more vulnerable to displacements from the Lorentz force, $\mathbf{I} \times \mathbf{H}$, if the superconductor carries a current I. When such motions of the previously-rigid flux-line lattice occur, a resistance appears and the superconductivity will effectively disappear.

As explained before it is the magnetic field via its random vector potential $\mathbf{A} \rightarrow A_{ij}$, which arises in our hamiltonian that creates the 'competing interactions' and thereby frustrations. Such are the basic ingredients for forming an analogue spin glass. In a type-II superconductor a vortex glass (see Fig. 7.9) may, therefore, exist and, since the flux-line lattice is rigidly frozen, have no resistance at least for small currents. Although our initial discussions have dwelt upon granular superconductors, now with the advent

7.2 Superconducting glass phases

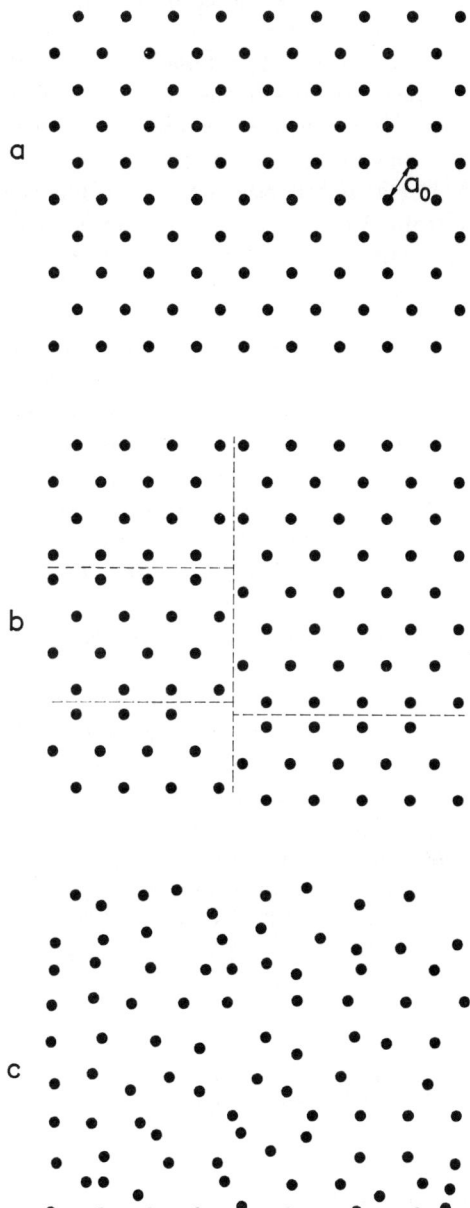

Fig. 7.9 Superconducting flux-line lattice for $H_{c1} < H < H_{c2}$: (*a*) long-range ordered, Abrikosov triangular lattice; (*b*) short-range ordered lattice with domains; and (*c*) vortex glass. For the magnetic field into the page the supercurrents circulate clockwise around each dot.

of the high-T_c oxide materials, we can apply the model to a distinct phase of bulk, single-crystal superconductors. We continue by drawing a direct analogue with our spin glasses. The hamiltonian, if the A_{ij} values are constrained to the range between 0 and π, is similar to the EA x–y $\pm J$ spin glass. Based upon our previous review of the theory, especially the computer simulations, this universality class is not expected to have a phase transition in $d = 3$. For a superconductor a more realistic model would be to let the A_{ij} vary from 0 to 2π. Such a model is called a 'gauge glass'. This means that the frustration is a serious disorder which is invariant under the gauge transformations $S_i \rightarrow -S_i$ and $J_{ij} \rightarrow -J_{ij}$. The disorder cannot be 'gauged away' using the word gauge in analogy with the invariant transformations of electro-dynamics. A gauge glass does not have reflection $\phi_i \rightarrow -\phi_i$ as one of its symmetries because the magnetic field breaks the time reversal invariance. And this lack of symmetry distinguishes it from the x–y (0 to π) model. So present theoretical thinking is that the gauge glass belongs to a different universality class and may very well have a phase transition in 3D. Extensive Monte–Carlo computations indicate that the gauge-glass model in 3D has behaviour comparable to an Ising spin glass ($d = 3$) which should have a finite T_c rather than to an x–y spin glass which should not.

Experiment has tried to determine the various properties of the flux-line lattice, especially to establish the existence of an intermediate phase with zero resistance corresponding to the vortex glass (see Fig. 7.9). Initial indications seem to be favourable for its formation accompanied by a sort of phase transition and there is experimental agreement with other effects predicted by the model. Since a flux-line lattice is a mesoscopic phenomenon with larger than atomic dimensions (typical lattice constants are of order microns), it can be directly observed via scanning tunnelling microscopy or decoration techniques. To find such a glassy structure as sketched in Fig. 7.9 would be a challenging experiment and complete confirmation of the model, as would the observation of the melting behaviour at higher fields and temperature.

7.3 RANDOM-FIELD ISING MODEL

Before continuing with our spin-glass analogues and our transversal into areas far different from our spin systems, let us pause to examine another type of random magnet which is not a spin glass. The model, first proposed by theorists (Imry and Ma, 1975), is called the random-field Ising model (RFIM). Here an Ising ferromagnet of various dimensions is placed in a site-random magnetic field. The model hamiltonian becomes

$$\mathcal{H} = -\sum_{ij} J_{ij} S_i S_j - \sum_i H_i S_i \qquad (7.6)$$

7.3 Random-field Ising model

where the spins are all z-components, J_{ij} is a constant (J), representing positive ferromagnetic interactions between nearest-neighbour spins, and the H_i are independent, local, random variables with

$$[H_i]_{AV} = 0 \quad \text{(mean)} \quad [H_i^2]_{AV} = H_R^2 \quad \text{(variance)} \tag{7.7}$$

Using the following simple argument Imry and Ma concluded that such RFIM systems would exhibit a phase transition for dimension $d > 2$ ($d_l = 2$, the lower critical dimension). The net-energy cost for the formation of an oppositely-oriented domain of volume R^d out of the ferromagnetic state is

$$E(R) = JR^{d-1} - H_R R^{d/2} \tag{7.8}$$

The first term represents the domain-wall energy where the spin bonds are broken along a surface R^{d-1}. The ferromagnet in the presence of random fields has a mean-square energy of $H_R^2 R^d$. For a particular choice of an oppositely-oriented domain, this energy (after taking the square root) can lower the total energy of the domain formation by a more favourable alignment with the random fields. And thus we arrive at the second term which is of opposite sign to the first. For $H_R \ll J$ the total energy is positive for $d > 2$ (domains cannot form), but for $d < 2$ it becomes negative for sufficiently large R. This means that domains are energetically favourable and will be created thereby destroying the long-range order phase and its phase transition. The $d = 2$ case is unresolved, hence giving rise to the lower critical dimension.

Therefore, in 3D a special phase transition and critical phenomenon are expected for the RFIM but not in 2D. And the H–T mean-field phase diagram for this transition should be as sketched in Fig. 7.10. However, in order to verify such behaviour we need a real system to measure. How do

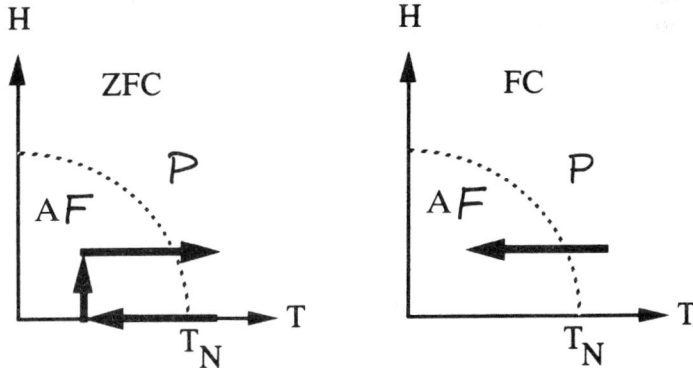

Fig. 7.10 Zero-field-cooling (ZFC) and field-cooling (FC) procedures for the RFIM. The dotted line indicate the AF phase boundary in the H versus T phase diagram. The arrows show the direction of the field-temperature cycling in each procedure. P represents paramagnet and AF the RFIM-state which develops at sufficient field. T_N is the REIM Néel temperature ($H = 0$).

we in the laboratory turn on a site-random field in a ferromagnet? Answer: we do not! Instead, we use another theoretical result that maps the above RFIM onto an Ising antiferromagnet with *random* exchange (site dilution of the spins staying more concentrated than the percolation threshold) in a *uniform* magnetic field H. These systems exist, for example, in 3D: $Fe_xZn_{1-x}F_2$, $Mn_xZn_{1-x}F_2$, and $Fe_xMg_{1-x}Cl_2$, and in 2D: $Rb_2Co_xMg_{1-x}F_4$ are well-studied representations. The theory has demonstrated that such dilute antiferromagnets in a uniform field lie in the same universality class as RFIM, i.e., they both exhibit identical critical phenomena. Careful experimentation has established that the above materials are, in single-crystal form, *ideal* RFIM systems (closely corresponding to the theoretical-model requirements).

At constant field there are two ways to cross the phase-transition line as indicated in Fig. 7.10. One path is a direct field cooling (FC); the second is ZFC and then applying the field and heating. These paths are illustrated by the bold lines with arrows. Dramatic differences occur in the static experimental properties between the two crossings. For ZFC all measurements (specific heat, Mössbauer effect, neutron scattering, etc.) evidence a sharp phase transition to a long-range ordered antiferromagnetic state with unique critical exponents which are distinct (crossover effects have occurred) from those of the random-exchange Ising model (REIM) or the zero-field behaviour. Of particular importance are the resolution-limited Bragg-peaks of the neutron scattering which track the long-range order right up to $T_c(H)$. In stark contrast, FC experiments exhibit broadened transitions characteristic of domains (short-range order) being frozen in below $T_c(H)$. Here the FC procedure causes the system to lose equilibrium and enter a metastable, frozen, domain state without long-range order. Above an 'equilibrium' boundary $T_{eq}(H) > T_c(H)$, the two procedures (ZFC and FC) yield the same experimental properties. So the domain formation and loss of equilibrium with its peculiar dynamics begin already at $T_{eq}(H)$ for FC and this masks the determination of $T_c(H)$. Figure 7.11 collects the various results into a schematic H–T phase diagram for $d = 3$. Now $H^{2/\phi}$ is plotted as ordinate since this gives a straight line to the scaling behaviour of T_c with $H^{2/\phi}$, where ϕ ($= 1.42$) is the crossover exponent. At sufficiently large H for ZFC there is a crossover from REIM critical phenomenon to RFIM critical phenomenon. The RFIM region of H–T space is delineated by the two dashed lines labelled T_{cr}^- and T_{cr}^+. Within this region critical behaviour and exponents corresponding to the RFIM will be observed. As we approach the equilibrium phase-transition line $[T_c(H)]$ the critical slowing down associated with the RFIM transition is encountered. The $T_c(H)$ line (see Fig. 7.11) is weakly time or frequency dependent as indicated by the shaded region surrounding it (T_-^* to T_+^*) with the various measurement times given. Outside these narrow triangles of the experimental time window the system is always in equilibrium on all accessible time scales. If a *FC* procedure is used to approach the $T_c(H)$ line, the system will *first* confront an equilibrium

7.3 Random-field Ising model

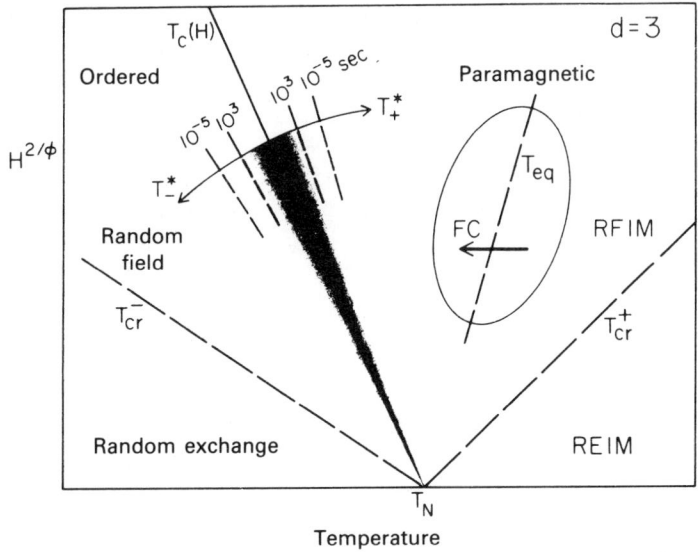

Fig. 7.11 Schematic $H-T$ phase diagram for the diluted AF, $d = 3$, RFIM system at a fixed concentration, following a ZFC procedure. New RFIM critical behaviour is observed for $T_{cr}^- < T < T_{cr}^+$. All boundaries, both static and dynamic, obey REIM to RFIM crossover scaling $H^{2/\phi}$, with $\phi \simeq 1.42$. $T^*(H)$ indicates the onset of the dynamic rounding of the critical behaviour associated with the extreme slowing down as $T \to T_c(H)$. It is only weakly time (frequency) dependent, as indicated by the characteristic times associated with particular measurements. Outside of the shaded region, the system is in equilibrium on all accessible time scales. In an FC route (inset), the system falls off equilibrium at $T_{eq}(H)$, preventing the correlation length ξ from diverging further for $T < T_{eq}(H)$ and ultimately leading to a frozen-in (non-equilibrium) domain state below $T_c(H)$. From Jaccarino and King (1990).

line labelled $T_{eq}(H)$ in the ellipsoidal inset of Fig. 7.11. Now the system falls out of equilibrium for $T < T_{eq}(H)$, thereby preventing the correlation length ξ from further diverging and ultimately leading to a non-equilibrium frozen-in domain state which forms below $T_c(H)$.

In 2D RFIM specific-heat and neutron-scattering measurements of $Rb_2Co_xMg_{1-x}F_4$ display a systematic rounding of the critical behaviour, independent of the cooling procedure. There is no hysteresis in the vicinity of the 'destroyed' phase transition at '$T_c(H)$'. From these data it was concluded that $d_1 \geq 2$. By utilizing the field scaling of the non-equilibrium properties well below '$T_c(H)$', the crossover exponent ϕ is determined to be 1.74, indistinguishable from the pure Ising value of 1.75. Figure 7.12 gives the schematic $H-T$ phase diagram for the diluted antiferromagnetic $d = 2$ case. Since hysteresis only occurs below '$T_c(H)$', the FC and ZFC procedures yield identical results in the surrounding '$T_c(H)$' region which represents an *equilibrium* domain state, i.e., no long-range order or ξ remains finite. When hysteresis finally begins (see ellipsoidal inset of Fig. 7.12), it is to a non-equilibrium, metastable, frozen state that depends on

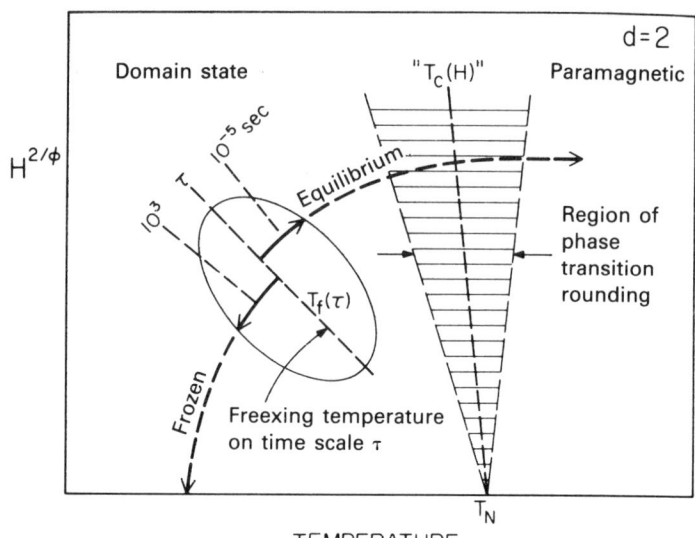

Fig. 7.12 Schematic H–T phase diagram for the diluted AF, $d = 2 = d_l$ RFIM system at a fixed concentration. The destroyed transition '$T_c(H)$' broadens and shifts with $H^{2/\phi}$ crossover scaling; it separates a domain state below '$T_c(H)$' from a paramagnetic one above. Here $\phi = 1.75$. A freezing (metastability) boundary $T_f(H)$ exists well below '$T_c(H)$' and also scales as $H^{2/\phi}$; its position depends on the measurement time scale τ. Notice that destruction of the phase transition is an equilibrium event. Unlike the $d = 3$ case, non-equilibrium behaviour at $d = 2$ has nothing to do with extreme critical slowing down. From Jaccarino and King (1990).

the time scale of the measurement as indicated in the figure. Here $T_f(\tau)$ is the time-dependent freezing temperature which tracks this loss of equilibrium. Note the distinction between this freezing (non-equilibrium) behaviour and the extreme critical slowing down in equilibrium of the rounded or destroyed phase transition at '$T_c(H)$'.

Returning to the 3D RFIM systems, we briefly examine the dynamics of the phase transition. Theory (Villian, 1985; Fisher, 1986) predicts an activated-dynamics model where now the relaxation time becomes

$$\frac{\tau}{\tau_o} = \exp(\xi^\theta) = \exp\left[\frac{C_o}{[T - T_c(H)]^{\nu\theta}}\right] \quad (7.9)$$

here we have used the standard result for the RFIM correlation length

$$\xi = \xi_o \left[\frac{T - T_c(H)}{T_N}\right]^{-\nu} \quad (7.10)$$

and θ is a new 'violation of hyperscaling' critical exponent which controls the anomalous growth of the free energy in a correlation volume via the scaling relation $2 - \alpha = (d - \theta)\nu$. The above application of activated dynamics

7.3 Random-field Ising model

should be compared to our previous usage in the 2D, Ising spin glass ($T_c = 0$) of section 6.2. According to the RFIM theory, the free energy and the associated thermodynamic quantities vary logarithmically with time or frequency rather than linearly. For example, the frequency-dependent free energy scales as

$$F = \left(\frac{T - T_c(H)}{T_N}\right)^{2-\alpha} f\left[\frac{\ln(\omega/\omega_o)}{\xi^\theta}\right] \quad (7.11)$$

and the ac-susceptibility as

$$\chi(\omega, T) = \xi^{2-\eta} X\left[\frac{\ln(\omega/\omega_o)}{\xi^\theta}\right] \quad (7.12)$$

Hence, there exists a dynamical rounding temperature which limits the divergences of the experimental quantities to a region of reduced temperature

$$\frac{T - T_c(H)}{T_N} > t^*(\omega) \propto \left|\ln\frac{\omega}{\omega_o}\right|^{\frac{-1}{\theta\nu}} \quad (7.13)$$

since dynamical effects (rounding) should start appearing for $\ln(\omega/\omega_o)/\xi^\theta \approx 1$. In addition, the peak height of the ac-susceptibility under activated scaling becomes a function of frequency

$$[\chi'(\omega)]_p \approx \left|\ln\left(\frac{\omega}{\omega_o}\right)\right|^{\frac{\alpha}{\nu\theta}} \quad (7.14)$$

which in the limit α (the specific-heat exponent) $\to 0$,

$$[\chi'(\omega)]_p \approx \ln[\ln(\omega/\omega_o)] \quad (7.15)$$

Measurements on (FeZn)F$_2$ have been performed employing ac-susceptibility for the higher frequencies (kHz) and Faraday rotation for the lower (mHz). Some typical low-frequency results are displayed in Fig. 7.13 where the critical part of $\chi'(\omega)$ is extracted as shown in the inset. Here it is important to note, in contradiction to the spin glasses, the peak position does *not* shift in temperature as a function of frequency. This critical $\chi'_c(\omega)$ begins to diverge at the lower reduced temperatures (solid straight line) but breaks away or rounds at $t^*(\omega)$ given by the arrows. Also the peak height $[\chi'(\omega)]_p$ may be determined from the figure in arbitrary units. For the lowest reduced temperatures (shaded area) the concentration gradient of even the best single-crystal sample of Fe$_{0.46}$Zn$_{0.54}$F$_2$ prevents any meaningful interpretation. When the above data are combined with the Faraday-rotation susceptibility, six decades of frequency can be spanned. Comparison may then be made between the predictions of conventional dynamic scaling: $t^*(\omega) \approx \omega^{1/z\nu}$ and $[\chi'(\omega)]_p \approx \omega^{-\alpha/z\nu}$, and the above logarithmic behaviours of the activated dynamic scaling. The latest experimental results suggest the activated-dynamics model yields a better fit to the above

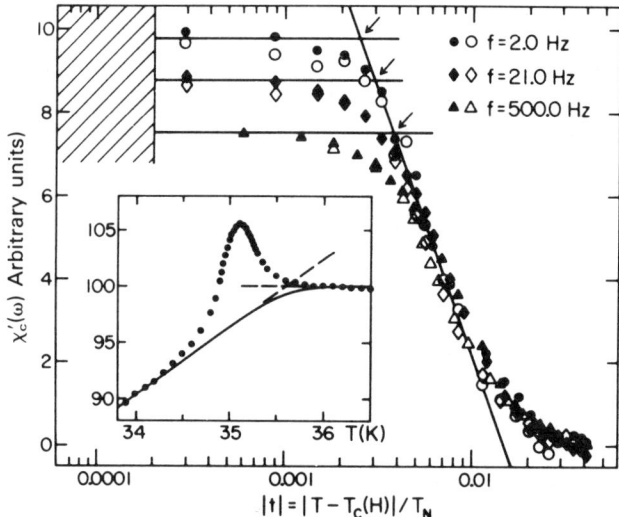

Fig. 7.13 $\chi'_c(\omega)$ versus $|t|$, the reduced temperature, at three frequencies after background subtraction, see inset. The open and closed symbols refer to $T < T_c(H)$ and $T < T_c(H)$, respectively. Rounding of the transition due to the concentration gradient occurs only within the shaded region, i.e. $|t| < 2 \times 10^{-4}$. The expected ln $|t|$ critical behaviour of $\chi'_c(\omega)$ is used to determine a dynamic rounding temperature $t^*(\omega)$ as indicated. From King et al. (1986).

two parameters with $\theta = 1.0$ and $\omega_o \approx 10^7$ rad/s. Since ν has also been found from neutron scattering to be equal to one, the product $\theta\nu = 1.0$ and hence the relaxation form is a simple Vogel–Fulcher type:

$$\tau = \tau_o \exp\left[\frac{E_a/k_B}{T - T_c(H)}\right] \tag{7.16}$$

Further effects of the 'extreme' critical slowing down can be observed in the neutron-scattering experiments. For example, there is a rounding of scattering intensity at the antiferromagnetic Bragg peak. The width of the peak scales with reduced temperature as $H^{2/\phi}$ as expected from RFIM dynamical crossover scaling. And a time dependence in the height of the peak appears near $T_c(H)$ in ZFC experiments. Here a logarithmic increase in peak-height occurs for times 10 to 10^4 s upon warming through $T_c(H)$. The good agreement between the various experiments and the theoretical models has led to a basic understanding of both the 2D and 3D RFIM and their 'destroyed' and real phase transitions.

7.4 OTHER 'SPIN-OFFS'

In this section we cull three *non-physics* analogues of the spin-glass problem. Our choices for discussion while arbitrary are governed by our search for

7.4 Other 'spin-offs'

interesting and practical examples of how the spin-glass concepts and mathematical methods can be applied to totally different types of complex systems. Here the *complexity* of the problem plays a major role and has even forged open a new area of physics called 'complexity'. We consider the three spin-offs: combinatorial optimization, neural networks and biological evolution. While our treatment will be cursory we, nevertheless, hope to illustrate the usefulness and direct contact of the spin glasses. We begin with the combinatorial-optimization topic, or for us the 'travelling salesman' problem: What is the shortest distance a salesman must travel to visit a given number of cities and return to the origin city?

Given a set of N points (cities) and the 'distance' between them – $\frac{1}{2}N(N-1)$ numbers – find the shortest path which traverses all of the points and returns to its starting point (origin). The correspondence with a travelling salesman is clear when we define the *cost function* as the total length of the 'tour'. The cheapest tour (optimization) has the shortest length, and the basic problem is to construct an algorithm which is able to find the configuration of lowest cost for any instance (the N cities and their various distances) with as little computing effort as possible. A useful classification of the various optimization problems can be made according to the time it takes to solve them on the computer. The 'easy' problems are those that can be solved by a polynomial algorithm. This P-class employs an algorithm which finds an optimal configuration in a computing time growing like a power n of size N. Algorithms, where time complexity functions are not bounded by such polynomials, generally require exponential time functions. The problems in this exponential category (no algorithm is known that provides an exact solution using a computation time that is polynomial in the size of the input) are usually called NP-complete problems where the NP refers to nondeterministic polynomial. If the running time necessary to construct an optimal solution increases faster than any power n of the problem size N, i.e., the computing time increases exponentially with size, the problem is intractable. The travelling salesman is one such problem since it becomes completely intractable (computing time $\to \infty$) for N of the order a few hundred. Other non-physics cases of NP-complete optimization problems are the placement and wiring of computer chips, image processing, code design, digital-signal processing, seismic-data analysis, neural networks, and many more.

In any event, connection of optimization and statistical (spin-glasses) physics is obvious. The cost function corresponds to the energy of a configuration. The exercise is to find the lowest cost or ground-state energy. This is the optimum configuration or the lowest valley in the multi-valley energy-configurational-space landscape of Fig. 5.11. In NP-complete problems we deal with disorder and frustration, so that it is not possible to satisfy simultaneously all the constraints which would locally minimize the energy. To see the analogue more clearly let us consider the following simple computer-design example. Suppose there are N circuits to be divided between two chips placed one above the other. The number of signals that pass between the ith and jth circuits is denoted by K_{ij}. Next we attach a two-value variable s_i to each circuit i, such that $s_i = +1$ for the upper chip and -1 for the lower one. The number of signals crossing from one chip to the other is

$$\sum_{\langle ij \rangle} K_{ij} \tfrac{1}{4}(s_i - s_j)^2 \qquad (7.17)$$

The difference between the number of circuits on the two chips is Σs_i. In order to optimize the 'eventual performance' of the computer, e.g. minimize the wiring length, reduce the noise and pick up, etc., the quantity

$$\sum_{\langle ij \rangle} (\lambda - K_{ij}/2) s_i s_j \qquad (7.18)$$

should be minimized. Here λ (a constant) takes into account the imbalance in the number of circuits on the two chips and in the number of boundary crossings. The above quantity to be minimized resembles an Ising, spin-glass hamiltonian where the 'interactions' are the short-range ferromagnetic-like K_{ij} and the long-range antiferromagnetic-like λ. Due to the competing interactions and the intrinsic randomness, frustration results, and thus, we have all the basic ingredients of a spin glass.

In order to secure the analogy of combinatorial optimization with the spin glasses let us re-examine the SK-model in the former language. An instance of this problem is a sample consisting of a given number of spins N and a set of couplings J_{ij} where $1 < i < j < N$. The domain of possible instances is defined by a probability distribution which reproduces the choice of couplings. In the SK-model each J_{ij} is taken as an independent random variable with a gaussian distribution: $\exp[-(N/2)J_{ij}^2]$. A configuration is characterized by the values of the N Ising spins, $S_i = \pm 1$. And finally the cost function is the energy given by the hamiltonian

$$\mathcal{H} = - \sum_{1 < i < j < N} J_{ij} S_i S_j \qquad (7.19)$$

Barahona (1982) has demonstrated that finding the ground-state energy (lowest or optimal cost function) is an NP-complete problem. In addition, Parisi using the RSB solution has determined what the ground-state energy should be in the limit of large N. Thus, the SK model has become one of the most studied of NP-complete problems and the knowledge gained here is very useful in treating other situations in combinatorial optimization.

For such NP-complete problems there are two main ways of attempting a near-optimum solution beyond that of simply trying clever algorithms. In the first, the so-called 'iterative improvement technique', a plausible configuration of spin orientations, the circuit locations for the above 'chip' example or the cities route for the travelling salesman, is chosen and various rearrangements of the bonds are tried. Only those rearrangements which

7.4 Other 'spin-offs'

lower the energy are accepted. This procedure is equivalent to a rapid quench of the system down to $T = 0$. Accordingly, it can often become stuck in one of the local minima and the spin, bond, et al. rearrangements are not sufficient to eliminate this trap in a metastable configuration. Hence, we are back to employing special algorithms and tricks to free the system and continue the search for the true ground state.

A second and more appealing general method is that of stimulated annealing (Kirkpatrick et al., 1983). Here one introduces a fictitious temperature as a parameter into classical optimization problems. This artificial parameter permits a slow cooling Monte–Carlo simulation to be carried out. Remember from section 5.12 such MC-simulations use a Metropolis acceptance criterion $\exp[-\Delta E/k_B T]$ which has the temperature as a natural variable. The stimulated-annealing algorithm begins by creating a high-temperature state where the system can probe every possible region of configurational space. Then, we *slow cool* the system so that it settles into a state of lower energy. Visualize the rough multi-valley landscape of Fig. 5.11 and slowly move downwards a horizontal line or bar corresponding to a reduction in $E = k_B T$. If, at an early stage, the system becomes stuck in a high-lying valley, then by slightly heating and slowly recooling it has a good chance of escaping over the nearest 'mountain pass' to seek a deeper valley or lower energy state. After many cycles of reheating and slow cooling, the simulated annealing algorithm has a high probability to yield a good solution, namely, a low-energy state which comes reasonably close to the very lowest one. To find the latter is extremely unlikely because of the huge amount of configuration space and the many nearly degenerate ground states.

The annealing schedule is the most difficult part of the procedure. One must choose a proper cooling rate $r = \Delta T/t$ for decreasing the T (t is the number of Monte–Carlo steps, MCS). If the cooling rate is too rapid, the final state will give too high an energy. Oppositely, too slow a rate is wasteful of computer time. Grest et al. (1987) have performed a detailed computer analysis on the ground-state properties for six model spin-glass systems and the travelling salesman problem using simulated annealing. The results showed the ground-state energy or cost function to be sensitive function of the cooling rate r. For 1D nearest-neighbour gaussian and 2D nearest-neighbour $\pm J$ and gaussian models

$$E(r) = E_o + C_1 r^n \tag{7.20}$$

In contrast for the 3D $\pm J$, a two-layer $\pm J$ and the infinite-range SK models along with the N-city travelling salesman problem, the dependence on r is much slower

$$E(r) \simeq E_o - C_2 \left(\frac{1}{\ln r}\right) \tag{7.21}$$

The conclusion is that for the latter four models the slow logarithmic dependence of finding the ground-state energy E_o makes the problem intractable, i.e., an exponentially large amount of computer time is needed to offset the very weak $(\ln r)^{-1}$ dependence of $E(r)$, thus it is NP-complete. Conversely, the polynomial r-dependence for the former three models would place them in P-class, even though they are not expected to exhibit a finite-temperature phase transition.

Let us examine the simulated-annealing results for the travelling-salesman problem given in Fig. 7.14. The plot shows the various distances travelled, l, of a 100-city and a 400-city tour as a function of $1/\ln r$ $(r = \Delta T/t)$. l is defined as the total distance traversed between the N-cities randomly distributed on a unit square divided by the square root of N:

$$l = \sum_{\langle ij \rangle} \frac{d_{ij}}{\sqrt{N}} \qquad (7.22)$$

This is the quantity to be optimized by finding the shortest length l_{min}. Δls are equivalent to ΔEs and the 'temperature' equivalent generates the acceptance criterion for a given configuration. Initially, at high temperature, the algorithm accepts deteriorations in the cost function with a high probability. However, in the course of the algorithm's running time the probability is gradually decreased to zero. This is accomplished by a control parameter or temperature c which is a function of t, the number of MCS, whereby $c \to 0$ at the end of the computer run. Notice that we are still far away from l_{min} with the finite rs used in the algorithm. So we must extrapolate to r (or $1/\ln r$) $\to 0$ as done in Fig. 7.14. The extrapolated results of l_{min} agree nicely with the 'exact' value for $N = 100$ and the 'expected' value for $N = 400$.

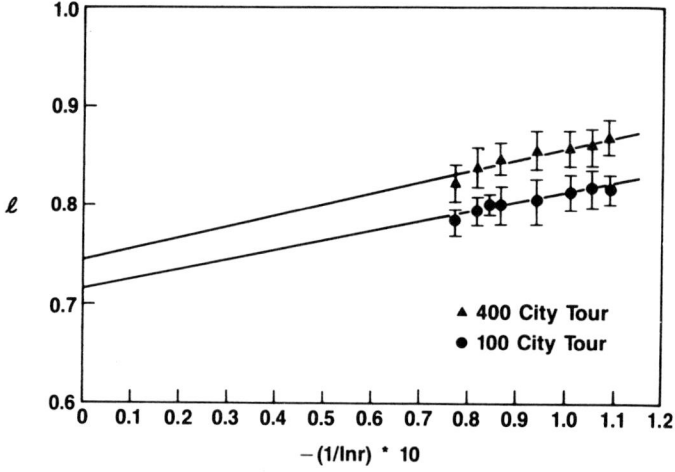

Fig. 7.14 Optimal tour length l versus $-(\ln r)^{-1}$ for the N-city 'travelling salesman' problem for $N = 100$ and 400. From Grest *et al.* (1987).

7.4 Other 'spin-offs'

In 1982 Hopfield proposed a neural-network model which relied heavily on spin-glass mathematics. He realized that his system of interacting neurons was similar to a spin-glass freezing and could perform computations and store information provided it had the appropriate dynamic rules. This model is especially interesting since it simulates the operation of the brain more accurately than a standard digital computer could. Let us first consider a little of the background to neurons and their functions, and then move on to draw the analogue with such networks and spin glasses. Finally, we shall illustrate one of the unique properties of these neural networks related to associative memory. In recent years the problem of neural-network models has evolved into a rapidly developing field of its own.

The neurons are nerve cells which for simplicity have two discrete states, namely active (fires or passes on an electrical impulse to another neuron) and inactive or dormant. The process of firing is called synapses and is accomplished via dendrites which reach out from the neuron or cell body to thin axons which connect the various neurons. Whether a given neuron remains in its present state or switches to the opposite one depends on the states of all the other neurons connected to it. If the sum total of the external stimuli exceeds a certain threshold, the neuron will change its state. The nature of the network's function, e.g. its computational task, is determined by the pattern of neural connections or interactions. We put this firing process of a neuron a bit more mathematically as follows. Denote the state of the ith neuron at time t by variable v_i which takes two values: 1 if the neuron is active and 0 if it is inactive. We let the strength of the coupling or impulse connection (synaptic efficacies) between the ith and jth neurons be C_{ij}. Then, the sum total of the stimuli or impulses at the ith neuron from all the other neurons is $\Sigma C_{ij} v_j$. T_i represents the threshold for the firing of neuron i. Accordingly,

$$v_i = 1 \quad \text{if } \Sigma C_{ij} v_j > T_i \qquad (7.23)$$

$$v_i = 0 \quad \text{if } \Sigma C_{ij} v_j < T_i \qquad (7.24)$$

To facilitate the analogy with the magnetic spin glasses we define

$$S_i = 2v_i - 1 \quad J_{ij} = \tfrac{1}{2} C_{ij} \quad \text{and} \quad H_i = \sum_j \tfrac{1}{2} C_{ij} - T_i \qquad (7.25)$$

This given the Ising-like pseudo-spins

$$S_i = +1 \quad \text{if } \Sigma J_{ij} S_j - H_i > 0 \qquad (7.26)$$

$$S_i = -1 \quad \text{if } \Sigma J_{ij} S_j - H_i < 0 \qquad (7.27)$$

which can be expressed by the single formula

$$S_i (\Sigma J_{ij} S_j - H_i) > 0 \qquad (7.28)$$

Therefore we can write a hamiltonian for the neural network as

$$\mathcal{H} = -\sum_{\{ij\}} J_{ij} S_i S_j - \Sigma H_i S_i \qquad (7.29)$$

Here the J_{ij} need not be symmetric, yet both $+$ and $-$ (competing ferro- and antiferromagnetic) pseudo-exchange interactions do exist since the neuron synopses can be 'excitatory' or 'inhibitory'. Also the various couplings J_{ij} may change with time, i.e., the system can learn in addition to store. The set of all the spin values represents the collective state of the system and such a situation closely resembles our SK, Ising hamiltonian in a random external field. The solution of this hamiltonian, based upon our previous spin-glass experience is expected to be represented by the multi-valley, phase-space landscape of the many nearly-degenerate ground states. In the present case these local energy minima of the collective system correspond to patterns to be recognized or memories to be recalled or other types of mental behaviour. It is the set of neural connections that fixes the number location and meaning of the valleys. A given external stimulus (input) determines the initial state of which neurons are firing and which are not. Afterwards the system via the particular set of couplings evolves to a special low valley (memorized pattern) that represents the class to which the input stimulus belongs. Thus, the memorized patterns are retrievable by their association with a given input pattern or addressable by the content of the input pattern. A 'basin of attraction' surrounds the lowest point in a valley such that if the system finds itself anywhere near the valley it will be attracted to the lowest point – a non-ergodic process. Again stimuli from the external world generate the selection of a special solution or ground state. Note that with N-neurons there are 2^N different patterns for the network.

The choice of the J_{ij} usually takes the form

$$J_{ij} = \frac{1}{p} \sum_{\alpha=1}^{p} \xi_i^\alpha \xi_i^\alpha \qquad (7.30)$$

where the ξ_i^α are quenched random variables assuming the values $+1$ and -1 with prescribed probabilities. p corresponds to the p-learned patterns of N-bit words which are fixed by the p sets of $\{\xi_i^\alpha\}$. From a straightforward stability analysis, the stored patterns would only be stable at $T = 0$ K for negligibly small values of p/N. However, we must recompute the stability criterion using an upper limit of retrieval error of learned patterns. This greatly increases the maximum number of patterns able to be stored in an N-neuron network. As a first step towards the calculation of retrieval error, we define a so-called Hamming distance d between a pure learned pattern S^k and the pattern S' as

$$d(S', S^k) = \tfrac{1}{2}[1 - q(S', S^k)] \qquad (7.31)$$

The Hamming distance represents the overlap between the two patterns

7.4 Other 'spin-offs'

and is the number of common bits. Compare it to the overlap distance of pure states in a spin glass defined in section 5.7 where

$$q(S^\alpha, S^\beta) = \frac{1}{N} S^\alpha \cdot S^\beta \qquad (7.32)$$

It is exactly the same concept resulting from RSB and ultrametricity. The average number of errors is given by the number of spins which do not align with the learned or embedded pattern, and mathematically this becomes

$$N_e = \frac{N}{2}(1-q) \qquad (7.33)$$

Recent calculations have shown that for $N_e/N \approx 1\%$, $p/N = 0.14$ and for smaller error limits, the ratio p/N goes rapidly to zero. In fact, if reasonable retrieval errors are permitted, the maximum storage capacity of learned patterns may be increased to $p_{\max} = 0.14N$, a rather high percent.

Relaxation of the neutron or spin system from the initial stimulus or state to the final steady-state pattern or valley state is governed by the dynamics. Any model of associated memory requires a well-defined dynamical process. The Hopfield model assumes a single spin-flip Glauber dynamics which was discussed in section 5.12 on Monte–Carlo spin-glass simulations. Here, after every updating of a spin, it is the new configuration which is used to update the next one. This is nothing but the heat-bath version of the Monte–Carlo process required for the Ising (up/down) spin glasses. The other model, called Little dynamics, updates all N-spins independently of each other. Now at each time step all the spins simultaneously check their states against the corresponding local fields. Hence, such an evolution of parallel relaxations is called synchronous whereas the Hopfield ones are asynchronous or series relaxations. Different forms are derived for the firing probability of the neuron (or spin) at site i and $t+\Delta t$ given the configuration $\{S_i\}$ at time t. But it must be remembered that the dynamics are part of the model and not a fundamental stochastic process. Various neuron-like electronic circuits with the above properties have been built to study collective and complicated computations and this is just the beginning of an exciting new field.

We conclude our brief treatment of neural networks with a simple illustration of associative memory whereby pieces of the stored information serve to retrieve the entire memory. Consider the six-neuron network of Fig. 7.15 and let the neurons have the $+1$ or -1 features corresponding to the description of a person according to name, height, age, weight, hair and eyes. See Table (*a*) as inset to Fig. 7.15. We wish to store three sets of characteristics for three different people as memories A, B and C. See Table (*b*) as inset to Fig. 7.15. Table (*b*) determines the wiring diagram of the 6-neuron network. For example, the link between neurons (1) and (2) must be for A $\rightarrow$ (+1)(+1), for B $\rightarrow$ (+1)(−1), and for C $\rightarrow$ (+1) (+1).

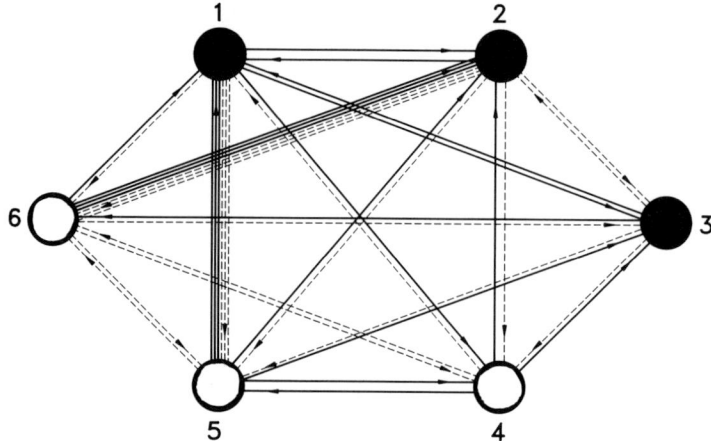

Table (a) Features assigned to nodes

	1 name	2 height	3 age	4 weight	5 hair	6 eyes
-1	Smith	tall	old	thin	brown	blue
+1	Jones	short	young	fat	blond	brown

Table (b) Nodes

		1	2	3	4	5	6
memories	A	+1	+1	+1	-1	-1	-1
	B	+1	-1	+1	+1	-1	+1
	C	+1	+1	-1	+1	-1	-1

Fig. 7.15 Associative memory with six nodes, or 'neurons', is linked by excitatory (solid line) and inhibitory (broken line) connections. The number of lines in each link indicates the strength of the connection; each solid line represents a connection strength of +1 and each broken line represents a strength of −1. Each node might denote a characteristic of a person, as is shown in the Table (*a*). Suppose one wants to store three memories, or sets of characteristics as given in Table (*b*). The nodes that are supposed to be in the +1 state are given an excitatory link to the other +1 nodes and an inhibitory link to the −1 nodes. To store information about all three memories one simply adds up the connections. From Tank and Hopfield (1987).

7.4 Other 'spin-offs'

Since the sum of these products is $+1$ we place one excitatory wire on a solid line to connect (1) and (2). Repeating this procedure for neurons (2) and (4), we obtain the sum equal to -1 which means one inhibitory wire or a dashed line should connect (2) and (4). Note for connections (1) and (5) and (2) and (6), the sum is equal to -3. Thus, 3 inhibitory wires must be used for these connections and by following the same routine we can establish the full wiring diagram for the six neurons of Fig. 7.15 according to the characteristics of Table (*b*). Now when the circuit is turned on for person or memory A, the network produces the correct pattern of current flows to ensure that neurons (1), (2) and (3) have their $+1$ states (solid circles) while (4), (5) and (6) are in their -1 states (open circles). The pattern is self-consistent, for, at each neuron the positive (solid arrow) currents and negative (dashed arrow) currents always add up to have the same *sign* as the neuron itself. Check this in Fig. 7.15! Note the associative-memory character. If the network is given only partial data, e.g. a thin, short Jones, it will immediately fall into a stable state from which one can retrieve the entire memory, namely A of a thin, short, young Jones with brown hair and blue eyes.

We conclude this section by mentioning another link between the spin glasses and biology. Naturally, we could use the ultrametric tree of RSB (see Fig. 5.7) as a classification scheme for the evolution of a species. However, we wish to go further and consider the central question of biological evolution: how does a 'soup' of small molecules evolve into highly-organized, information-carrying macro-molecules? Here attempts have been made to construct mathematical models of the first critical stages of molecular evolution where an information transition occurs from *little* to *much* information. Two basic elements, stability and diversity of the species, are required for their evolution. Without stability one species would quickly mutate into another. If there is one only stable state, thus no diversity, no possibility would exist to create more complex structures with greater amounts of information. A survival function selects the molecule's chance of formation and is equivalent to a random interaction between the components. Once again we have the basic elements of a spin glass: competing interactions, frustration, complexity, etc. Simple mathematical models are made to account for these factors statistically. While the words are different the calculations and pictures of these models closely resemble those of the spin-glass problem.

One important common denominator constantly appears in all these spin-offs. This is the rugged mountainous landscape in phase space. For the spin glasses the vertical axis is the energy states of the peaks and valleys, for the combinational optimization it is the cost function, for the neural network it is the memories to be recalled or patterns to be recognized. And finally for the biological evolution we have put into the model the survival-probability function with its many peaks and valleys in the 'state space' of the molecules. So perhaps we should all return to Fig. 5.11 for another glance because a great deal of our modern science of complexity depends upon it. And as mentioned at the beginning of this chapter we have progressed far beyond the spin and the glass and entered into totally foreign areas of life and evolution. Hence we stop.

FURTHER READING AND REFERENCES

Spin glass analogues

van Hemmen, J. L. and Morgenstern, I. (eds) (1983) *Heidelberg Colloquium on Spin Glasses*, Lecture Notes in Physics 192, Springer, Heidelberg; (1987) *Heidelberg Colloquium on Glassy Dynamics*, Lectures Notes in Physics 275, Springer, Heidelberg.
Höchli, U. T., Knorr, K. and Loidl, A. (1990) *Orientational Glasses, Advances in Physics*, **39**.
Stein, D. L. (ed.) (1989) *Complex Systems*, Addison-Wesley, New York.
Kauffmann, S. A. (1989) *Origins of Order: Self Organization and Selection in Evolution*.
Kandel, E.R. and Schwartz, J. H. (1985) *Principles of Neural Science*, Elsevier, Amsterdam.
Stein, D. L. (ed.) (1992) *Spin Glasses and Biology*, Directions in Condensed Matter Physics, Vol. 6, World Scientific, Singapore.
Zallen, R. (1983) *The Physics of Amorphous Solids*, Wiley, New York.

References

Barahona, F. (1982) *J. Phys.*, **A15**, 3241.
Birge, N. O., Jeong, Y. H., Nagel, S. R., Bhattacharya, S. and Susman, S. (1984) Phys. Rev. **B30**, 2306.
Courtens, E. (1984) *Phys. Rev. Lett.*, **52**, 69.
Fisher, D. (1986) *Phys. Rev. Lett.*, **56**, 416.
Grest, G. S., Soukoulis, C. M., Levin, K. and Randelman, R. E. (1987) *Heidelberg Colloquium on Glassy Dynamics*: Lecture Notes in Physics 275 (edited by J. L. van Hemmen and I. Morgenstern) Springer, Heidelberg.
Höchli, U. T., Kofel, P. and Maglione, M. (1985) *Phys. Rev.*, **B32**, 4546.
Hopfield, J. J. (1982) *Proc. Nat. Acad. Sci. (USA)*, **79**, 2554.
Imry, Y. and Ma, S. K. (1975) *Phys. Rev. Lett.*, **35**, 1399.
Jaccarino, V. and King, A. R. (1990) *Physica*, **A163**, 291.
John, S. and Lubensky, T. C. (1985) *Phys. Rev. Lett.*, **55**, 1014.
King, A. R., Mydosh, J. A. and Jaccarino, V. (1986) *Phys. Rev. Lett.*, **56**, 2525.
Kirkpatrick, S., Gelatt, Jr., C. D. and Vecchi, M. P. (1983) *Science*, **220**, 671.
Maglione, M., Höchli, U. T. and Joffrin, J. (1986) *Phys. Rev. Lett.*, **57**, 436.
Sethna, J. P. and Chow, K. S. (1985) *Phase Transitions*, **5**, 315.
Sullivan, N. S., Devoret, M., Cowan, B. P. and Urbina, C. (1978) *Phys. Rev.*, **B17**, 5016.
Sullivan, N. S. (1983) *Proceedings of the Symposium on Quantum Fluids and Solids* (edited by E. Adams and G. G. Ihas) American Institute of Physics, New York.
Tank, D. W. and Hopfield,, J. J. (1987) *Scientific American*, **257**, 62.
Villain, J. (1985) *J. Phys* (Paris), **46**, 1843.

8

The end (for now)

If you have come this far you must either be an expert on the spin glasses or slightly disappointed and still confused. In order to improve and solidify the first and to alleviate the second possibilities we offer this concluding chapter. It will not be a succinct recapitulation of the preceding seven chapters. Hopefully after a rereading they will become sufficiently clear and to the point. The phenomenon and physics of 'spin glass' have been explained via an experimental approach – the purpose of this monograph.

In this final chapter we wish to reiterate briefly, using other words, the basic and key concept of the spin glasses, namely, symmetry breaking resulting in the rough multi-valley landscape of the system's energy in phase space. Then, we proceed to one of the latest (1992) developments in the experimental area that of mesoscopic or nanostructured spin glasses and their special fluctuations. And afterwards we take a last quick look at the theory: what does it say about dimensional crossover (3D → 2D) and how does the freezing temperature depend on the length scale? This is followed by some of the most recent computer simulations. Here there are appearing some novel and altered interpretations. These final considerations will clearly illustrate to us the ongoing stream of progress in a relatively dormant field of physics. At last we end the book with a few compositional thoughts about our journey through the spin glasses and where to find what.

8.1 SYMMETRY BREAKING AND PHASE SPACE

All second-order phase transitions, especially ferro- and antiferromagnetic ones, break the symmetry of their respective hamiltonians, and thereby, are characterized by an order parameter. For the above two possibilities the spontaneous magnetization and the staggered magnetization, M, represent the order parameters. This means an enormous reduction in the system's possible spin configurations to simply one (with the help of an infinitesimal symmetry-breaking field in an Ising ferromagnet, see below). Such a situation is illustrated in Fig. 8.1 where we plot the free energy versus magnetization as a function of temperature. The figure symbolizes a Landau-

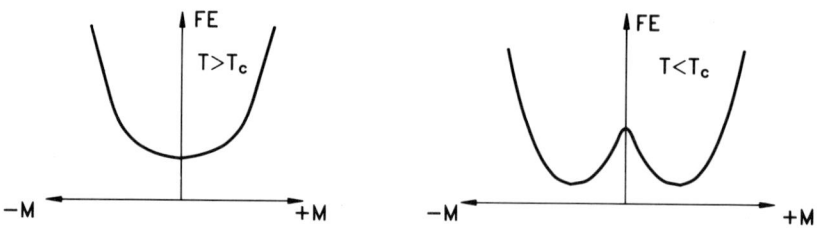

Fig. 8.1 Free energy as a function of the magnetization (order parameter) for a ferromagnetic above and below its Curie temperature T_C.

theory approach to a second-order phase transition. Above T_c the free energy has its minimum at $M = 0$ and all possible microscopic spin configurations are available which satisfy this 'null' macroscopic condition. Here the symmetry of the magnetic (Ising) hamiltonian

$$\mathcal{H} = -J \sum_{i,j} S_i S_j - H \sum_i S_i \qquad (8.1)$$

is preserved along with the ergodic hypothesis. However, below T_C there is the ferromagnetic transition and the spins must either all point up or all down and thus finite values of $\pm M$ appear. With our tiny symmetry-breaking field one of these two possibilities is chosen. Hence, only a *single* configuration of spins is allowed as lowest energy or ground state. Such a drastic constraint breaks the symmetry of the hamiltonian (8.1) and further violates the ergodicity by limiting the phase space to either the $+M$ or the $-M$ valley. Remember restriction of phase space is called broken ergodicity, in this case it is 'trivial'. The above description has now become standard in the theory of second-order phase-transition. Here the order is long-range and periodic. And we can calculate the ground-state energy, for there is only one, and further how the order grows with temperature $M(T)$. The critical-phenomenon problem has also been solved and the various critical exponents are known for the different universality classes – end of story.

Using this powerful theoretical framework let us consider the spin glasses. Attempts at devising a single order parameter, e.g. the Edwards–Anderson order parameter, have failed. In addition, to calculate the true ground-state energy is impossible or better said intractable. And all of these difficulties occur already within the mean-field or simplest theoretical approximation. The net result after many years of work was a continuous order parameter $q(x)$ and the corresponding multi-degenerate ground-states. So the transitional behaviour is very much different in a spin glass than in a ferromagnet.

We again look at our multi-valley picture which is now repeated in Fig. 8.2 as a function of temperature. First there is the case of $T > T_f$ where the paramagnetic state is modified by the various *clusters* that introduce small, local minima into the otherwise symmetric behaviour situated about

8.1 Symmetry breaking and phase space

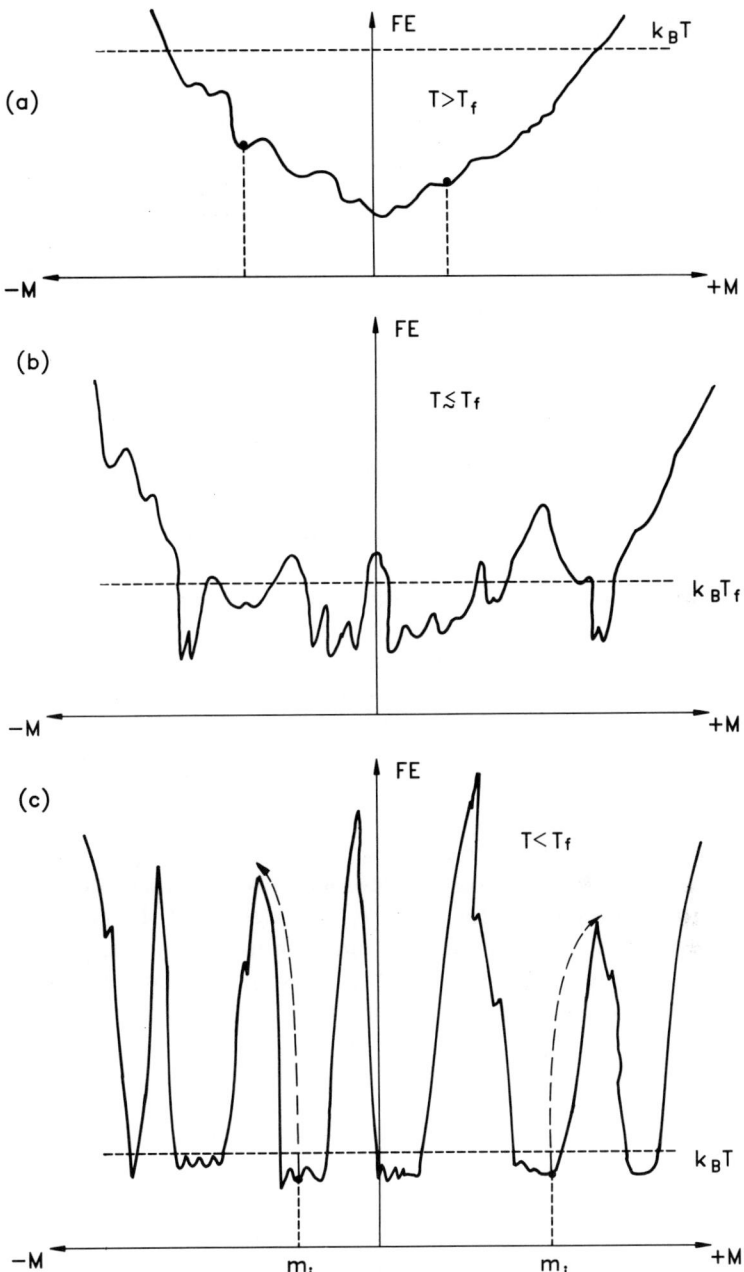

Fig. 8.2 Sketch of the multi-valley landscape as a function of temperature for a spin glass.

$M = 0$. The local minima represent the finite values of the cluster magnetization, however, because of the high temperature they are metastable and give only a fluctuation contribution to the free energy. It is the large temperature which causes the various energy barriers to be surmounted, and thus, the equilibrium thermodynamic state will be an average over a vast region of phase space with zero net magnetization. Equilibrium is now attained because the cluster barriers are much too small compared to the temperature.

But what happens as we lower the temperature to just under T_f? Here the variations in the free-energy have increased, i.e., the magnitudes of the energy barriers have become larger but remain finite. This is a direct consequence of our *interacting* system of spins and their co-operative freezing. Note now the appearance of many low-lying valleys possessing different internal structures of sub-valleys with various magnetizations. Remember the figure is only a schematic 'bare-bones' illustration, the real *Cu*Mn alloy is much more complex. Yet for this case the temperature is sufficient and the barrier heights not too intimidating so that the local metastabilities are quickly removed and the macroscopic free energy can span a reasonable region phase space within the available time. We would say because of the relatively rapid relaxation at T_f a quasi-equilibrium is maintained with, however, an increased chance for the system to become stuck in one of the low-lying valleys. The problem here is we do not know which one or even if it possesses the *lowest* possible energy. Furthermore, an external field will greatly influence the barrier-height distribution causing certain magnetization values to the preferred. All these processes are difficult to calculate since we do not know the free-energy landscape, or to which state the system will migrate, or how it depends on field.

If the number of these local magnetizations is large (multidegenerate) as it will be at low temperatures ($T \ll T_f$), then some of the energy barriers separating the different local m_is will tend towards infinite height – see Fig. 8.2(c). Accordingly, we can partition the landscape into mutually inaccessible valleys where each of the valleys corresponds to a thermodynamic phase similar to the two ferromagnetic minima in Fig. 8.1. Within each of the inaccessible valleys there will be many sub-valleys or smaller structures separated by finite (small) barriers. Since these sub-valleys have various non-equal energy minima, they represent the metastable states of the frozen spin-glass and are responsible for the unique experimental effects at $T \ll T_f$. When a system becomes trapped in one such valley, it will exhibit properties specific to that particular valley, i.e., non-trivial broken ergodicity. Such properties differ from those in true equilibrium, the latter involve averages scanning all valleys with the appropriate thermal weights. It is the EA order parameter q_{EA} which measures the mean-square *single-valley* local spontaneous magnetization averaged over all possible single-valleys. Note how q_{EA} will be non-zero for the distribution of inaccessible valleys given in Fig. 8.2(c) since the $\pm m_i$s are all squared eliminating the possibility of ± cancellations. q_{EA} is not, as already noted in Chapter 5, related to the equilibrium magnetization. Such requires a *new* order parameter q which takes into account the *intervalley* contributions. This necessitates the accumulation of a large region of phase space with the incorporation of many low-lying 'almost inaccessible' valleys. A dynamic picture is useful here. We let the barrier heights be large but not infinite so that low-probability transitions over the barriers are allowed as illustrated in Fig. 8.2(c). This will simply take a very long time according to $\tau = \tau_0$ exp $(+E_B/k_BT)$. If we use a time-scale long enough for the system to surmount the barriers and to pass statistically many times through all the valleys with a significant thermal weight, then true equilibrium is attained by including these intervalley contributions. Granted it would take a much longer time than is accessible to experiment, but that's the price we pay for our complex multi-valley landscape. These concepts were alluded to in Chapter 5 and can be phrased in mathematical terms. For example, a third order parameter Δ can be introduced which measures the degree of broken ergodicity:

$$\Delta = q_{EA} - q \qquad (8.2)$$

Notice that $\Delta \to 0$ when we return to our single equilibrium phase of Fig. 8.1 ($T < T_c$ with a small $+H$), even though one half of the phase space has been removed.

As stated previously the key to the spin glass and its many analogues is this peculiar and undeterminable multi-layer landscape – a unique but necessary conception. This is an important bit of new physics with significant consequences that can be tracked experimentally. For example, if we measure the ac-susceptibility we are effectively probing q_{EA} and its short-time quasi-equilibrium state. However, if we try to 'mimic' the very long-time behaviour by field cooling, we obtain a measure of q which gives the average equilibrium susceptibility. Can you remember this discussion in section 5.8? More important is not to forget the picture and underlying concepts of the many valleys, so you should look once again at Figs. 8.1 and 8.2 for here is the heart of the spin-glass problem.

8.2 MESOSCOPIC SPIN GLASSES

Very recently (1990s) the study of spin glasses has turned to samples with reduced size. This means one or more dimension has been constrained to less than c. 1000 Å lengths. The structure may be a very thin film (2D), a narrow 'wire' or strip (1D) with both reduced thickness and width, or a 'clump' of the material usually called a dot (0D). (The latter has not yet been tried for a spin glass.) Such samples are fabricated using the modern sub-micron and nanostructure techniques currently available. The word applied to describe these samples and their associated physics is *mesoscopy*.

At the present time it is a very popular and active area especially with semiconductors where its consequences and benefits are obvious.

The unique effects of mesoscopic physics (which lies between the very small atomic scale, ≈ 1 Å, *microscopy* and the very large bulk behaviours, $\geq$ few μm, *macroscopy*) are caused by the smallest sample length L being comparable to one or more of critical length scales of the given problem. Thus, for a mesoscopic conductor, L could be the same order of magnitude sas

(i) the *elastic* mean free path $\ell = v_F \tau$, where v_F is the Fermi velocity and τ is an elastic scattering time;
(ii) the *inelastic* mean free path ℓ_{in} which depends on the inverse temperature to some power and includes now magnetic inelastic scattering processes, e.g. spin flip;
(iii) the *phase-coherence* length ℓ_ϕ which permits 'time-reversal' interference between the electron states.

For the latter we can write in the diffusive-scattering regime $\ell_\phi = (D\tau_\phi)^{\frac{1}{2}}$ where D is the diffusion constant and τ_ϕ is related to the inelastic scattering processes. A typical metal has $\ell \approx 50$ nm and $\ell_\phi \approx 1000$ nm at 1 K, hence, L could easily be made much less than ℓ_ϕ by sub-micron fabrication and going to yet lower temperatures. Such a condition ($L < \ell_\phi$) eliminates a complete cancellation of the different quantum interference terms (remember the electron is a wave) between the various diffusive ($\ell \ll L$) paths. This in turn produces finite fluctuations in the conductivity δG, which depend sensitively (in both magnitude and sign) on the specific placement of the scattering centres or 'impurities'. These direct interferences between multiply-scattered electron waves give the so-called universal conductance fluctuations (UCF) of amplitude

$$G_{\text{UCF}} = \left(\frac{e^2}{h}\right) \quad \langle(\delta G)^2\rangle \sim \left(\frac{e^2}{h}\right)^2 \tag{8.3}$$

where the $\langle \, \rangle$ brackets represent ensemble averaging over different samples with different impurity configurations. If a magnetic field is applied to such a mesoscopic specimen, then the phase of the electron waves is changed by 2π each time a flux quantum $\Phi_0 = h/e$ passes through the phase-coherent area. This gives rise to (Aharonov–Bohm-like) conductance oscillations with field sweep of wavelength h/e. Such behaviours have been investigated already for many years in small *non-magnetic* metals mostly using magneto-resistance and noise-spectrum ($1/f$) techniques. Some pertinent and introductory references are listed at the end of this chapter, if one wishes to delve further into this area. Let us now consider what happens if we replace the above 'impurities' by magnetic ones, e.g. *Cu*Mn and *Au*Fe, and examine the possibility of *spin scattering* both above and below T_f.

8.2 Mesoscopic spin glasses

First of all we must convert the UCF equation to include spin-dependent scattering. Theory (Feng et al., 1987) has dealt with this problem, and in particular, how a small temperature change δT (thermal cycling) will affect the conductance within the frozen state of a spin glass, where most of the inelastic spin-flip scattering has ceased.

$$(\delta G)_T^2 \approx \left(\frac{e^2}{h}\right)^2 \frac{k_F}{\ell} \frac{\ell_\phi^3}{L} \frac{T_f}{T_F} \tag{8.4}$$

Here the ℓ, ℓ_ϕ and L have been previously defined, and k_F and T_F are the Fermi momentum and temperature. Note that the freezing temperature T_f enters this equation. If $k_F \ell T_f/T_F \geq \ell^2/\ell_\phi^2$, then we have the saturation regime

$$(\delta G)_T^2 \approx \left(\frac{e^2}{h}\right)^2 \frac{\ell_\phi}{L} \tag{8.5}$$

which should be measurable as $1/f$-noise in the conductance. The δT necessary to induce such conductance changes is related via the droplet model to a length scale that causes significant spin reorganizations. Or one could simply cool the spin glass through T_f and tract how the noise spectral density changes with temperature. Another method would be to detect the UCF via field sweeps in magneto-conductance experiments and compare the fluctuation spectrum above and below T_f. Here in a mesoscopic sample one would attempt to find spin-glass fingerprints of the particular frozen configuration.

The noise-spectrum measurements do indeed show a steep increase in their spectral density below T_f for various concentrations. These experiments were carried out on thin films of *Cu*Mn (5–10 at. %) 50–70 nm thick and *Au*Fe (0.1–1 at. %) 12–25 nm thick. A typical plot of the 'noise-parameter' $\alpha(f, T) \propto f S_R(f)$, where $S_R(f)$ is the spectral density of fluctuations at frequency f in the resistance R, is given in Fig. 8.3 for *Cu*Mn. Note how the freezing temperatures determined from max $(d\alpha/dT)$ are very close to the susceptibility-cusp temperature. Similar results are obtained for both *Cu*Mn and *Au*Fe in zero external field. A disturbing disagreement occurs between the two systems when the external field is applied. For *Cu*Mn little happens to the temperature dependence of S_R (1 Hz) spanning T_f in 2 T field, however, for *Au*Fe the increase of the fluctuation spectrum is washed out already in 1 T. One possible reason for this disparity is the factor of 10 larger concentration used in the *Cu*Mn film; another is the single low (1 Hz) frequency of the noise parameter. Yet it does seem strange that the relatively large fields will have no effect on the spin fluctuations of a spin glass.

The magneto-conductance experiments were performed on strip-like nanostructures of *Cu*Mn (0.1 at. %) with thickness 40 nm, 90 nm width (w) and c. 2 μm length with intervening potential-contact strips. At this

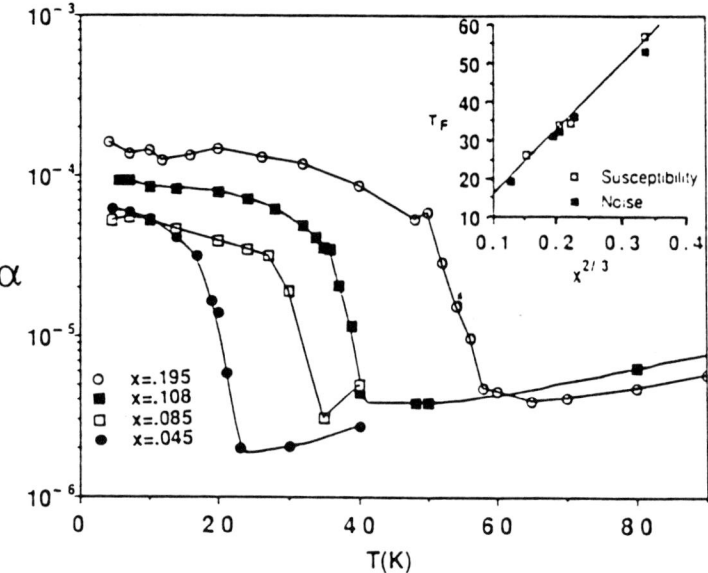

Fig. 8.3 Electrical resistance 'noise-parameter', $\alpha(f, T)$, for various concentrations of CuMn; from Israeloff et al. (1989).

concentration the average distance between the Mn atoms is 2.5 nm and an electron mean free path of 30 nm was estimated. So the net structure is approaching a 1D *wire* whose resistance can be measured at ≈ 0.5 μm intervals. Some salient features of these pioneering studies are displayed in Figs. 8.4 and 8.5. The first shows the magneto-resistance for the wire with the above dimensions compared to a codeposited CuMn film of width 2 μm and much larger separations of potential contacts. Note the clear quantum fluctuations in the wire at low temperature (23 mK). Such fluctuations are called a 'magneto-fingerprint' and are attributed to the magnetic field inducing a phase shift in the electron wave function. Recall the phase change is related to the flux quantum $\Phi_0 = h/e$; thus fluctuations on a field scale of $h_\phi = (h/e)/(w\ell_\phi)$, where w is the width of the wire and ℓ_ϕ is the phase coherence length, can occur. A second and larger field scale is created by the distortions or rotations of the frozen spins away from the initial (ZFC) configuration by the external field. At 25 mK the conductance fluctuations have an average peak-to-peak amplitude of $\delta G \approx 0.2\ e^2/h$.

In Fig. 8.5 the mesoscopic wire is fabricated into a square ring 0.4 μm on a side. As the field is swept in different ranges Aharanov–Bohm conductance oscillations are observed with period h/e for the $\Delta H = 0.25$ T in the intervals of 0.1–0.35 T and 5.0–5.25 T. Notice the sharpness of the oscillation which gives $h_\phi = 260$ gauss and $\ell_\phi \approx 0.5$ μm. When the field is repeatedly swept back and forth at $T \ll T_f$ the same fluctuation spectrum is retraceable, i.e., there exists a large correlation coefficient $C \approx 0.95$.

8.3 Recent theoretical progress

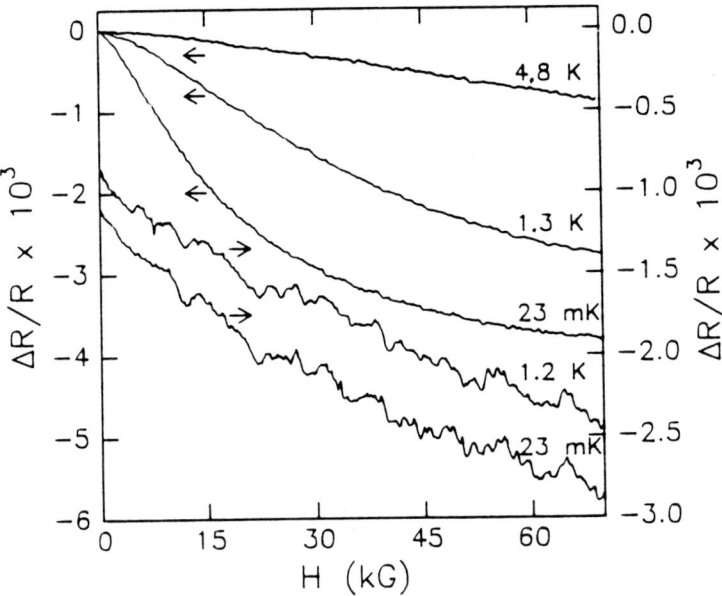

Fig. 8.4 Upper three curves magneto-resistance as a function of field for a *CuMn* (0.1 at. %) *film* (420 Å thick by 2 μm wide). Lower two curves same as above but for a *CuMn* (0.1 at. %) *wire* (420 Å thick by 900 Å wide by 2 μm long). From de Vegvar, Levy and Fulton (1991).

However, once the wire is heated above T_f and recooled a new fluctuation spectrum will appear in the magneto-conductance related to the new microscopic orientations of frozen spin. It is estimated that within the above wire dimensions there are 10^5 Mn spin and 25% of these must reorient to produce of $\delta G = e^2 h$. The question here is what does reorientation mean, certainly not a spin flip, perhaps a small collective distortion would be sufficient. But this is just the beginning of the new and exciting field of mesoscopic spin glasses.

8.3 RECENT THEORETICAL PROGRESS

So what has been happening with the theory? In this subsection we mention one 'older' (late 1980s) and one brand-new (1992) pieces of progress. The former has to do with the droplet model and its approach towards answering the question of *dimensional crossover*: 3D → 2D and this influence on T_f. Certainly it is related to the previous topic of mesoscopic or restricted-dimension effects. The second concerns some very recent *numerical simulations* carried out in improved ways or on more realistic models which indicate that specific revisions are perhaps necessary in the long-accepted theoretical conclusions.

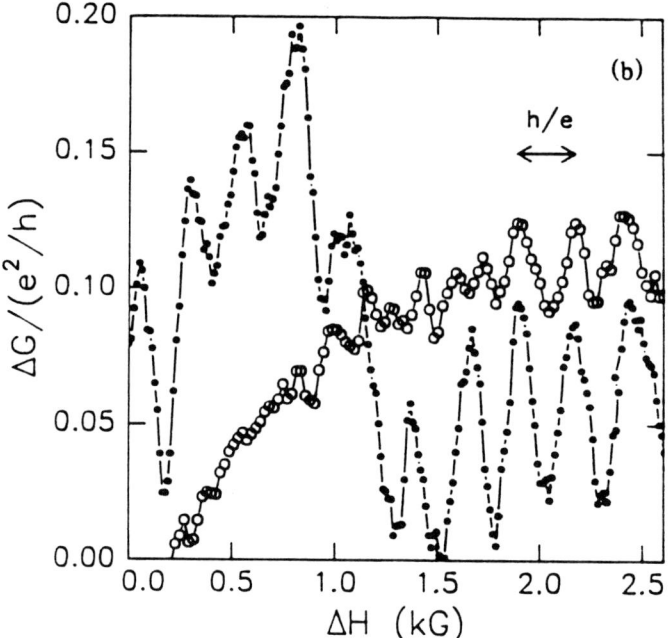

Fig. 8.5 Aharonov–Bohm conductance oscillations (ΔG versus ΔH) for a CuMn (0.1 at. %) square ring 0.4 μm sides at 23 mK; from de Vegvar, Levy and Fulton (1991).

The first question is how does T_f change as we reduce one of the sample dimensions beginning with a bulk 3D system. For example, we can decrease the thickness of the sample until the 2D limit is reached, all the while tracking T_f. The simplest theoretical approach is to use finite-size scaling which introduces the new variable $y = L/\xi(t)$. Here L is the sample dimension that we are changing and $\xi[t \equiv (T-T_c)/T_c]$ is the correlation length for an equilibrium phase transition. For the spin-glass case we have a reduction in $T_f(L)$ at a given constant value of y. Thus, setting $T_f(L = \infty) = T_c$ we obtain

$$\frac{T_f(L) - T_f(L=\infty)}{T_f(L=\infty)} \propto \left(\frac{L}{L_c}\right)^{-\frac{1}{\nu}} \tag{8.6}$$

where ν is the correlation-length exponent $\xi \propto t^{-\nu}$ and L_c is the length scale at which $T_f(L_c) = 0$.

In 1987–88 Fisher and Huse extended these ideas via the *droplet model* and included the effects of the long relaxation times in the spin glasses. Their result was a measurement time factor τ_m multiplying the previous $L^{-1/\nu}$. Accordingly

$$\frac{T_f(L) - T_f(L=\infty)}{T_f(L=\infty)} \propto L^{-\frac{1}{\nu_3}} \left[\ln\left(\frac{\tau_m}{L^{z_3}}\right)\right]^{[(\psi_3 + \nu_2\psi_2\theta_3)\nu_3]^{-1}} \tag{8.7}$$

8.3 Recent theoretical progress

Now the subscripts on the exponents denote their dimensionality 2D or 3D. Recall from section 5.9 that z, ψ and θ represent the length-scale critical exponents for dynamical scaling, activation barriers and free energy, respectively. Notice six (ν_3, z_3, ψ_3, θ_3, ψ_2 and ν_2) critical exponents appear in the above equation. In the limits $L \to 0$; $\tau_m \to \infty$, and hence, $T_f(L)/T_f(L=\infty) \ll 1$

$$\frac{T_f(L)}{T_f(L=\infty)} \propto \left[\frac{L^{\psi_3 + \psi_2 \nu_2 \theta_3}}{\ln \tau_m} \right]^{(1+\nu_2\psi_2)^{-1}} \qquad (8.8)$$

which for a constant long time of measurement simplifies to

$$\frac{T_f(L)}{T_f(L=\infty)} \propto L^a \qquad (8.9)$$

with a being a combination of the other four (ψ_3, θ_3, ψ_2 and ν_2) critical exponents.

The above equations suggest the ways in which the experimentalists should plot their data. The different critical exponents may be estimated from MC simulations or the measurements on ideal spin glasses – see Chapter 6. And as mentioned in section 6.4 a number of groups are quite active in preparing thinner and thinner films, usually of the canonical (RKKY) spin glasses, in order to confront the theory. Figure 8.6 illustrates one such 'multilayer' attempt. Here various Mn alloys of Cu or Ag, which are kept separated from each other by a sufficiently thick interlayer of the *pure* host metal (Cu or Ag), have their thicknesses reduced. $T_f(L)$ is taken

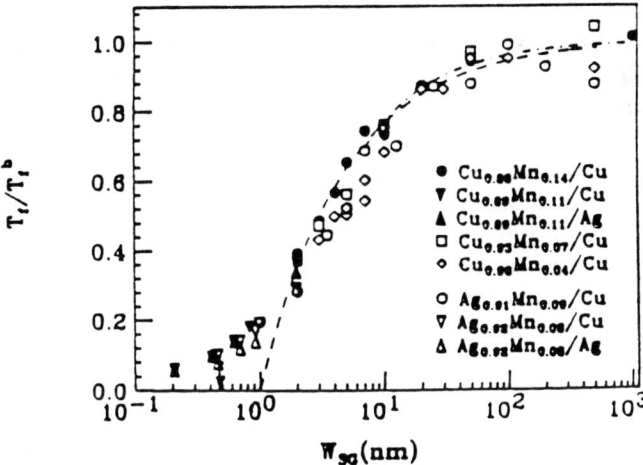

Fig. 8.6 Normalized freezing temperature plotted as a function of film thickness for various *Cu*Mn and *Ag*Mn multilayers. W_{SG} (= L) represents the spin-glass layer thickness on a logarithmic scale. The dashed curve is a fit to equation (8.6). From Bass and Cowen (1992).

as the peak in the ZFC magnetization or as the beginning of irreversibility between ZFC and FC.

In Fig. 8.6 a reasonable fit to the collected data on the thicker films is obtained using (8.6) with $\nu = 1.6$ and $L_c = 10$ Å. These latter values seem physically meaningful – see Table 6.4. However, the very thin films (below 20 Å) deviate from the universal curve with T_f remaining finite. Other experiments on *single-layer* films of AuFe exhibit a different behaviour. If (8.6) is used, $\nu = 0.8$ with $L_c \approx 70$ Å. Here also there are deviations for the thinnest film thicknesses. Note the above spin glasses are not the ideal Ising systems of the droplet model. Despite the severe experimental difficulties inherent in producing the required high-quality and homogeneous alloy layers, work is continuing and a more complete comparison with theory should be shortly expected.

Before we leap into the latest of numerical simulations, let us succinctly re-examine the computer folklore of the spin glasses and its conventional wisdom or conclusions. Section 5.12 treated this issue in more detail. Up until now, computer simulations and other numerical studies agree that a phase transition occurs in Ising spin glasses of $d = 3$ for *both* short-range ($\pm J$) and the infinite-range (RKKY or mean-field) interactions. For these cases $T_f \simeq J$ or Δ (the variance of the interaction distribution). In 2D the above Ising model has $T_f = 0$ and $2 \leq d_\ell$ (the lower critical dimension). Remember the computer models are *random bond* ones where ferromagnetic and anti-ferromagnetic couplings are randomly distributed on a regular lattice of spins. For the Heisenberg case the simulations tell us that there is no phase transition for $d = 3$ at finite temperature regardless of the type of interaction, e.g. for RKKY, $d_\ell = 3$ and $T_f = 0$. Now, in order to induce a phase transition, a bit of anisotropy is needed to break the Heisenberg isotropy. Recall the use of the Dzyaloskinsky–Moriya interaction in section 3.3.5 for this purpose in the canonical (RKKY) CuMn and AuFe spin glasses. The x–y universality class lies somewhere between the Ising and Heisenberg models, with less numerical work having been performed and no finite T_f phase transition anticipated in 2D and 3D. Nevertheless, after all this use of computer time and energy, no one has yet resolved the freezing temperature for an *infinite-range* (SK), $\pm J$, Ising spin glass. It was always expected to be simply $T_f = J$. Recently two different simulation studies have considered this question, and surprisingly, one answer found that T_f was significantly less than J. And, touching on another separate issue, the Heisenberg models employed in the simulations are random bonds, not the random sites of the canonical RKKY spin glasses. This was one of the main reasons for experimentalists to create the ideal spin glasses of Chapter 6. Now it is the (numerical) theorist's turn to construct 'more realistic' models, so how about *random sites*?

We briefly examine these two new types of numerical simulations. The first (Campbell, 1992 and Bhatt and Young, 1992) investigated an infinite-range Ising $\pm J$ spin glass. Infinite range means that each of the N (up/down)

8.3 Recent theoretical progress

spins interacts with all the others $(N-1)$ spins according to a constant coupling strength J independent of the distance separating the spins. Only the sign of J changes randomly. Here dimensionality does not play a role and it is the size of the system given by the N spins that counts. Large-scale (to N = 4096), long-time Monte-Carlo procedures were used and averaged over various sets of bond configurations, i.e., sample averaging. When the number of bond configurations averaged over was large enough, the results confirmed $T_f = J$ (Bhatt and Young) as predicted by the SK *continuous* infinite-range model and reinforced by RSB theory. However, for insufficient number of samples, the generated (non-linear) susceptibility values were widely distributed and highly skewed. Thus the conclusion of Campbell that $T_f \approx 0.7J$ was biased because of the large fluctuations or "rare samples" present in his simulations. Such effects can only be gotten rid of by proper sample averaging in these *finite size* samples. So you see there is still life, danger and contention in the spin glasses and certainly other paragraphs of this book will require future revision as they continue to evolve.

The 'more-realistic' model (Matsubara and Iguchi, 1992) is a random alloy of RKKY-coupled, Heisenberg spins and the numerical simulation is a hybrid MC spin-dynamic method. Here the model hamiltonian

$$\mathcal{H} = -\sum_{i,j} J(r)\,\mathbf{S}_i \mathbf{S}_j$$

is used with $J(r)$ being given by equation (1.2). The classical Heisenberg spins $|\mathbf{S}_i| = 1$, are randomly distributed on the sites of an $(L)^3$ fcc lattice. Parameters are chosen to represent CuMn with, for example, 5 at. % Mn. The spin-glass susceptibility is calculated as

$$\chi_{SG} = \frac{1}{N} \sum_{i,j} \langle\langle \mathbf{S}_i \mathbf{S}_j \rangle_T^2 \rangle_C \qquad (8.10)$$

where $\langle \cdot \rangle_T$ represents an average over K MCS per spin and $\langle \cdot \rangle_C$ denotes a sample average. This non-linear susceptibility is shown in Fig. 8.7 for T in units of $J_0/a_0^3 k_B$ and with various L-size lattices. The error bars represent an attempt to define the equilibrium values of χ_{SG}. Note that there is *no* anisotropy in the simulation, and yet χ_{SG} seems to diverge at low T for large L (= 20) with K reaching 15 000 MCS. This indicates that a finite-temperature phase transition is occurring. Such a conclusion may be put on a more firm basis by employing finite-size scaling from which $T_f = 0.07$ and the ν and η critical exponents may be derived ($\nu = 0.74$ and $\eta = 0.69$). Previous studies of Heisenberg spin glasses, all resulting in no phase transition, have limited themselves to bond models, i.e., a random distribution of bonds in a periodic lattice of spins. The RKKY model is necessarily a random site one with a $J(r)$ interaction between the various distance sites. So it would now seem that because of these recent MC

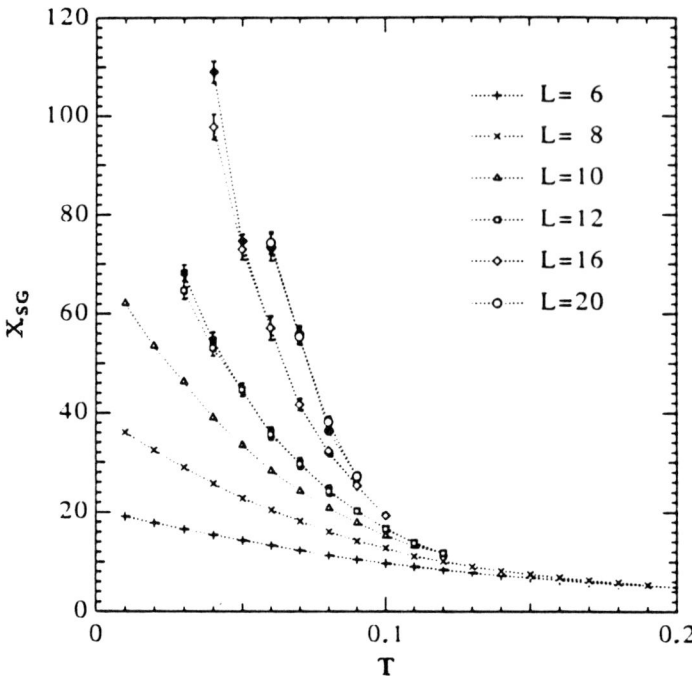

Fig. 8.7 Temperature dependence of the spin-glass susceptibility for different size (*L*) lattices of the RKKY coupled Heisenberg, random-site spin glass; from Matsubara and Iguchi (1992).

simulations a substantial difference exists between a random-bond model and a random-site one.

Another view on the Heisenberg spin glass is to consider *chiral ordering*. This is the property of vector spins, which because of frustration can possess a non-collinear or canted spin ordering. 'Chirality' designates the right- or left-handedness of the spin structure and introduces a new universality class with novel critical properties. Using numerical studies there are indications that chirality (designated by +1 for right-handed and −1 for left-handed) often behaves like an Ising variable. Thus, the conjecture has been made (Kawamura, 1992) for a chirality driven mechanism to create a finite-temperature phase transition in the canonical (RKKY) spin glasses. As is now crystal clear there is a lot more to come in the theory of spin glasses.

8.4 A FINAL WORD

Before we stop (for now) let us briefly look back at the composition of this book. If you want to learn, put most simply, what a spin glass is and

8.4 A final word

how, in words and pictures, it freezes, you should return to Chapter 1 – its beginning and its end. In between are the basic spin-glass ingredients: randomness, magnetic interactions, anisotropy and frustration.

The spin glasses did not suddenly appear in the magnetic repertoire. They are affiliated with various other (and sometimes older magnetic) phenomena such as moment formation, the Kondo-effect, giant moments, mictomagnetism or cluster glass, percolation behaviour and superparamagnetism. Chapter 2 introduces these related concepts and offers a simple description, or better said a physical picture, of them. The spin glasses are then placed in their proper context according to their concentration regime and interaction strength. Some different spin-glass species, namely, amorphous, insulator and semiconductor are offered and compared to the canonical (RKKY) ones. A really new type of behaviour is depicted under the name re-entrant spin glass.

Chapter 3 moves on into the realm of experiment. The basic spin-glass phenomenon is divided into three temperature regimes, i.e., above, at, and below T_f, and the salient and generic experimental features are illustrated. Here we mainly use the existing measurements on the canonical spin glasses CuMn and AuFe as specific examples, and data out of the original literature are separately presented for the susceptibility, specific heat, resistivity, neutron scattering μSR, Mössbauer effect, torque and ESR. With this vast collection of behaviours, an experimentalist's model or picture is constructed to compact the results. Thus, we can talk nicely about the generic spin glasses and interrelate the measurements within the framework of the cluster model. It requires a rather lengthy chapter even to summarize the large amount of experimental data which has been accumulated. But such is necessary to acquaint the reader with the basics of the real, oft-studied systems. And an abundance of plots and sketches are compiled and shown for reference and posterity.

Before beginning with the weighty subject of theory, Chapter 4 serves as a pause that allows us to list in table form, and rather completely, the various combinations of spin glasses. Here we sail through the noble-metal magnetic alloys, onto the rare-earth combinations and into the amorphous metallic spin glasses. The going gets a bit rough when we discuss semiconducting and insulating spin glasses. Nevertheless, we finally arrive at our concept of what a *good* spin glass is.

Chapter 5 is perhaps the most difficult, yet the one with the most overview references. So if you wish to really master spin-glass theory, skim through Chapter 5 which will give you the flavour and historical thrust, and then turn to one of the many theoretical reviews. Our survey has been to look back at the precursors, i.e., the first attempts to confront this then unnamed phenomenon of the spin glasses; then move into the E–A model, follow with the S–K solution and subsequently its instability. Contact with experiments is frequently made wherever possible. The TAP approach is considered, and afterwards, a simplified description of RSB and its physical meaning are given via a few highly helpful sketches. Hopefully the above acronyms are still familiar to the reader. The all-important dynamics enters our discussion as we wind up with the mean-field model. It is then further onwards into the more recent theoretical issues of the droplet and fractal-cluster models. Finally, we switch from Ising (as with the preceding topics) to non-Ising *vector* spin glasses and conclude Chapter 5 with the computer simulations to which we returned briefly in Chapter 8.

In order to confront the experiment of Chapter 3 with the theory of Chapter 5, we require *ideal* spin glasses. In Chapter 6 we first explain what is an ideal spin glass, and then consider a 2D example and a 3D less ideal one for detailed comparisons. A demanding (perhaps overly so) examination is carried out to evaluate the correspondence of experimental results to theoretical predictions. The emphasis here lies with the nature of the phase transition and its critical properties. Chapter 6 closes with the conclusions and some future possibilities.

At long last in Chapter 7 we turn our attention to *non-spin* analogues. This area has become a vast out-growth of the original magnetic alloys. Yet the ideas and notions are very similar only without the word 'spin'. A few examples are electric dipolar and quadrupolar glasses, and superconducting glass phases. For the sake of comparison, we then briefly revert back to a spin system, the random-field Ising model, whose phase transition is well understood. Finally and succinctly, we introduce three *non-physical* analogues, namely, combinatorial optimization, neural networks and biological evolution. Books can and are being written about these topical subjects. However, we wish to show the reader that they owe a great deal to the spin glasses for their recent development. For they all stem from the basic spin-glass hamiltonian and possess the same jagged phase space.

In this final chapter, which for obvious reasons is called 'The end (for now)', we have attempted to place the spin glasses within the notions of symmetry breaking and phase space. The unique feature here is the multi-valley landscape of the spin-glass phase space and it is the key to understanding much of the unusual behaviour. In order to prove that there is still life and new physics in the spin glasses we discuss the nascent subject of mesoscopic spin glasses to which significant experimental interest is turning. Also from the theory side we related how this reduced geometry (or nanostructure) should influence the freezing temperature. Since the theoretical efforts are continuing to generate progress, we have selected some very recent and undigested Monte–Carlo numerical simulations and tried to outline their provocative conclusions. And last but not least, here we are attempting to summarize the book's composition by giving the reader some tips on where what is and how to find it in a 'final word'.

References

Bass, J. and Cowen, J. A. (1992) *Recent Progress in Random Magnets* (edited by D. H. Ryan) World Scientific, Singapore. See also Hoines, L., Stubi, R., Loloee, R., Cowen, J. A. and Bass, J. (1991) *Phys. Rev. Lett.*, **66**, 1224.
Bhatt, R. N. and Young, A. P. (1992) *Phys. Rev. Lett.*, **69**, 3130.
Campbell, I. A. (1992) *Phys. Rev. Lett.*, **68**, 3351; **69**, 3131.
Feng, S., Bray, A. J., Lee, P. A. and Moore, M. A. (1987) *Phys. Rev.*, **B36**, 5624.
Fisher, D. S. and Huse, D. A. (1987) *Phys. Rev.*, **B36**, 8937; (1988) *ibid*, **38**, 373.
Israeloff, N. E., Weissman, M. B., Nieuwenhuys, G. J. and Kosiorowska, J. (1989) *Phys. Rev. Lett.*, **63**, 794.
Kawamura, H. (1992) *Phys. Rev. Lett.*, **68**, 3785.
Matsubara, T., and Iguchi, M. (1992) *Phys. Rev. Lett.,* **68**, 3781.
de Vegvar, P. G., Levy, L. P. and Fulton, T. A. (1991) *Phys. Rev. Lett.*, **66**, 2380.

Index

Abrikosov phase 213, 214–15
absorption 68–9, 72
ac-susceptibility
  Edwards-Anderson model 142–3
  freezing-temperature experiments 64–76
  ideal spin glasses 188–9, 193
  random-field Ising model 221
  Sherrington-Kirkpatrick model 146–8
ageing time 93–7, 167–8, 169
AgMn 76, 119, 243
amorphous magnets 38–9, 126–8, 134
Anderson model *see* virtual-bound-state model
anisotropy, magnetic 11–13
  experimentalist's model 114–15
  low-temperature experiments 90–2, 105–8
  metallic glasses 38–9
  re-entrant spin glasses 39
annealing, stimulated 225–6, 245
Argand diagrams 73
Arrhenius law
  quadrupolar glasses 208
  superparamagnetism 41–2
  susceptibility 70
associative memory 229–30
AuCo 120
AuCr 119
AuFe
  characteristic temperatures 122–5
  dimensional crossover 244
  magnetization 90–1
  mesoscopic spin glasses 239
  Mössbauer effect 86–7
  neutron scattering 61, 62, 102
  re-entrant spin glasses 39, 40
  specific heat 98
  susceptibility 64

AuMn 119

Bethe ansatz treatment 27
biological evolution 230–1
blocking process 42, 43
bond percolation 34
bond randomness 5–6
Bose-Einstein thermal factor 58
Bragg peaks 16, 57, 60
Brillouin function 41
Brillouin zone 58, 61–3

canonical spin glasses *see* RKKY spin glasses
Casimir-du Pré equations 68, 72, 73–4
categories of spin glasses *see* systems of spin glasses
CdMn 120
Ce 125
chalcogenides 128–9
chemical clustering 62–3
chiral ordering 245–6
cluster glass 17, 32–3, 37, 39
clusters
  experimentalist's model 113–15
  fractal cluster model 167–70
  magnetization 90–1
  Mössbauer effect 88
  neutron scattering 61–3, 103–4
  resistivity 56
  specific heat 53
  susceptibility 47–50, 71
  symmetry 234–6
Co 10–11, 184–5
cobalt-aluminosilicate glass 133
coherent scattering 57
Cole-Cole approach 189–91
collinear spin glass 108, 109
combinatorial optimization 223–6
computer simulations 174–9, 244

concentration 35–7, 56, 136–7
cooling *see* field cooling; zero-field cooling
Cooper pairs 212
Coulomb repulsion 22–3
covalent mixing 24
Cr 121
*Cr*Fe 121
critical line 170
crystalline electric field 21
Cu 10–11, 184–5
*Cu*Fe 119–20
*Cu*Mn
  characteristic temperatures 122–5
  dimensional crossover 243
  electron spin resonance 111
  magnetization 89–90, 91–2, 94–6
  mesoscopic spin glasses 239–41, 242
  Monto Carlo simulation 174
  neutron scattering 58–9, 62–3, 80–1, 83, 102–4
  resistivity 54–5, 78–9, 100–1
  specific heat 51, 53, 54, 77, 98
  susceptibility 46–50, 64–7, 69, 71
  torque 107
*Cu*Ni 121
Curie law 20–1, 185–6
Curie-Weiss behaviour 47, 48, 65, 130

dc-susceptibility 46–50
deviation temperature 47, 48
differential susceptibility *see* ac-susceptibility
diffuse scattering 57, 60
dilute magnetic semiconductors 129–30
dimensional crossover 241–4
dipolar glasses 205–8
dipolar interaction 10, 12, 105
direct exchange 6–7
dispersion 67
diversity, natural 231
domain-wall renormalization-group techniques 176
driving fields 64
droplet model 164–7
  computer simulations 179
  dimensional crossover 242–4
  history 18
  ideal spin glasses 190–1, 193
  mesoscopic spin glasses 239
dynamic correlation function 176–7
dynamical scaling
  ideal spin glasses 192
  mean-field model 161–3

random-field Ising model 221–2
  susceptibility 71
Dzyaloshinskii-Moriya interaction 12–13, 105, 173
DM interaction 12–13, 105, 173

Edwards-Anderson model 139–43
  ideal spin glasses 192–3, 199, 201
Edwards-Anderson order parameter 30, 140, 236–7
  computer simulations 175
  neutron scattering 58
  superconducting glass phases 213
elastic scattering 57, 60–2, 80
electric dipolar glasses 205–8
electric quadrupolar glasses 208–11
electron spin resonance (ESR) 108–11
energy
  flux 100
  see also free energy
entropy 52, 78, 98
EuSrO 128
(EuSr)S 128–9
  neutron scattering 102
  specific heat 53, 98
  susceptibility 68, 72–3
evolution, biological 230–1
experimentalist's model 113–16
exponential logarithmic decay 191–2

fast scattering *see* quasi-elastic scattering
$Fe_xMg_{1-x}Cl_2$ 131
$Fe_xMn_{1-x}TiO_3$ 195–200
$Fe_xZn_{1-x}F_2$ 131–2
field cooling
  magnetization 91–2, 94, 96
  random-field Ising model 217–19
  semiconducting spin glasses 130
  specific heat 100
  spin glasses in a field 111–12
  susceptibility 69–70
  torque 105–9
field, spin glasses in a 111–13
fluctuation-dissipation theorem 191
fracal cluster model 86, 167–70
free energy
  droplet model 164–6
  Edwards-Anderson model 141–2
  electron spin resonance 108
  historical developments 137–8
  replica-symmetry breaking scheme 159
  Sherrington-Kirkpatrick model 147

Index

freezing process 15–16, 31, 32–3
freezing-temperature experiments 63–88
frequency dependence 66–7
frozen matrix 169
frustration 2–3, 13–15
  insulators 132
  percolation 35
  semiconducting spin glasses 130

Gabay-Toulouse (GT) line 171–2, 173
gauge glass 216
giant moments 28–30, 120–1
glassy phase 213–14
Glauber dynamics 174
good spin glasses 119, 133–4, 181

Hamming distance 228–9
Heisenberg model 11, 138–9, 244–6
holnium-borate glass 132–3
high-temperature experiments 45–63
history of spin glasses 16–18, 136–9
Hopfield model 227–9
hybridization 22
hydrodynamic theory 108, 111
hydrogen, molecular 208–11
hyperfine field 86–7
hysteresis 91–2

ideal glass temperature 70
ideal spin glasses 181–203
insulators, magnetic 42–3, 131–3, 134
Ising spin glass 10
  computer simulations 176, 178
  droplet model 164–5
  electron spin resonance 109
  historical developments 138
  insulating 131
  Sherrington-Kirkpatrick model 144
  torque 108
isothermal remanent magnetization (IRM) 93, 96
isotropic anisotropy 107
isotropic spin glasses 171
iterative improvement technique 224–5

Josephson tunnelling 212–13

$(KCN)_x(KBr)_{1-x}$ 208–9
$K_{1-x}L_xTaO_3$ 206
$K_{1-x}Na_xTaO_3$ 206–7
Kondo effect 24–8
  concentration regimes 35
  history 17
  rare-earth spin glasses 125
  resistivity 55–6
Kondo temperature 25–8, 119–20
Korringa value 58, 74

Larmor-precession phase 82
Little dynamics 229
localized spin-fluctuation 25, 27
long-range magnetic order 34–5
long-range spin density wave 104
lorentzian distribution 58–9, 84
low-temperature experiments 88–111

magnetic anisotropy *see* anisotropy, magnetic
magnetic clustering 63
magnetic insulators 42–3, 131–3, 134
magnetic interactions 6–11
magnetism and superconductivity 126, 127
magnetization
  isothermal remanent 93, 96
  low-temperature experiments 89–97
  replica-symmetry breaking scheme 157
  semiconducting spin glasses 130
  thermal-remanent 91–3, 96
magneto-conductance experiments 239–41
mean-field model 161–3
mean-field theory 144, 146, 202
Meissner phase 213, 214
memory, associative 229–30
mesoscopic spin glasses 237–41
metallic glasses 37–9, 126–8, 134
Metropolis acceptance criterion 174, 225
mictomagnetism 17, 32–3, 37, 39, 50
mixed superexchange 10–11
mixing 22
  covalent 24
molecular hydrogen 208–11
moments
  giant 28–30, 120–1
  weak 25–8
monochalcogenides, rare-earth 128
Monte Carlo method 174–9, 198–200, 201
Mössbauer effect 86–8
muon spin relaxation (μSR) 83–6

neural networks 227–30
neutron scattering
  freezing-temperature experiments 80–3

high-temperature experiments 57–63
low-temperature experiments 101–4
random-field Ising model 222
neutron-spin-echo 82–3, 86
Ni 121
noble metals 119–20
noise-spectrum measurements 239–40
nomenclature 17
non-Isling spin glasses 171–4
non-linear susceptibility 75–6
   ideal spin glasses 185–8, 193, 195–6, 198
non-scaling 37
NP-complete problems 223, 224, 226
numerical simulations 244–6

Onsager reaction field 138
order parameter 213, 214
   see also Edwards-Anderson order parameter
ortho hydrogen 208–11
Overhauser theory 103

para hydrogen 208–11
paraelectretic phase 207
paramagnetic model 46–7, 74–5
Parisi replica-breaking symmetry scheme 151–61, 178–9
Pd 121
$PdFe$ 29
$Pd_{1-y-x}Fe_yMn_x$ 147
$PdMn$ 121
$PdNi$ 121
percolation 34–5
   experimentalist's model 114–15
   fractal cluster model 169–70
   limit 34, 37
   semiconducting spin glasses 128
phase space 233–7
phase transition 30–2
   ideal spin glasses 188–9, 194, 200
   Mossbauer effect 87
   muon spin relaxation 86
   Overhauser theory 103
   random-field Ising model 220
   susceptibility 76
power-law divergence 71, 76
pressure 56
projection hypothesis 156–8
Pt 121

quadrupolar glasses 208–11
quasi-bound state 26, 27
quasi-elastic scattering 61–2, 80

quenched-disorder randomness 6, 140–1

random-exchange Ising model 43, 131, 218–19
random-exchange problem 43
random-field Ising model 43, 131, 216–22
random magnetic fields 138–9
randomness 4–6
rare-earth monochalcogenides 128
rare-earth spin glasses 125–6, 134
   metallic glasses 38, 127
   virtual-bound-state model 20–1
$Rb_2Co_xMg_{1-x}F_4$ 219
$Rb_2Cu_{1-x}Co_xF_4$ 184–94
$Rb_{1-x}(NH_4)_xH_2PO_4$ 207
re-entrant spin glasses 39–40
re-entry superconductivity 126
reaction term 149
relaxtion times
   computer simulation 177
   droplet model 166–7, 168
   experimentalist's model 113–15
   ideal spin glasses 189–91
   insulators 132–3
   magnetization 93, 95–7
   Mössbauer effect 88
   muon spin relaxation 84, 86
   neutron scattering 58–61, 80
   resistivity 79
   susceptibility 68–9, 72–4, 76
remanent magnetization
   computer simulations 175
   isothermal 93, 96
   thermal 91–3, 96
replica-symmetric solution 145, 147
replica-symmetry breaking (RSB) scheme 150–3
   computer simulations 178–9
   history 18
   mean-field model dynamics 163
   non-Ising spin glasses 171–3
   physical meaning of 153–61
replica trick
   avoidance of 149
   Edwards-Anderson model 141
   Sherrington-Kirkpatrick model 144
   superconducting glass phases 213
resistivity
   amorphous spin glasses 127–8
   freezing-temperature experiments 78–80
   high-temperature experiments 53–6
   low-temperature experiments 100–1

Index

RKKY spin glasses 7–9, 119–20
  amorphous spin glasses 128
  giant moments 29–30
  good spin glasses 134
  history 17, 136–9
  ideal spin glasses 181
  Kondo effect 25, 28
  metallic glasses 38
  phase transition 31–2
  rare-earth combinations 125
  resistivity 55–6
  specific heat 51–2, 53
  susceptibility 47, 49–50, 66, 74
  transition-metal solutes 120, 121, 122–3
rock magnets 132
Ruderman, Kittel, Kasuya, Yosida *see* RKKY spin glasses

satellites 102
saturation 89–90, 93
scaling 32
  computer simulations 176
  concentration regimes 35–7
  fractal cluster model 167–9
  ideal spin glasses 187–8, 194–8, 200
  RKKY interaction 136–7
  see also dynamical scaling
Schrieffer-Wolff (S-W) canonical transformation 23–4
self-averaging 160–1
semiconductors 42–3, 128–31, 134
Sherrington-Kirkpatrick (SK) model 144–7
  combinatorial optimization 224
  computer simulations 178–9
  instability 147–8
  mean-field model 161–3
  replica-symmetry breaking 154–7
  TAP approach 150
short-range spin density wave 103–4
single-ion anisotropy 12
site percolation 34
site randomness 4–5, 244–5
solubility limit 119–20, 125
Sompolinsky model 162–3
spatial correlations 57, 79
specific heat
  Edwards-Anderson model 142–3
  freezing-temperature experiments 76–8
  high-temperature experiments 50–3, 54
  low-temperature experiments 98–100

semiconducting spin glasses 130
spin density wave 102–4
spin-disorder scattering 79
spin glasses
  characteristics 30–2
  definition 1–4
spin-orbit coupling 21
spin scattering 238–9
SQUID 69, 93
stability, natural 231
stimulated annealing 225–6, 245
Stoner enhancement factor 121
stretched exponential 96–7, 170, 192
superconducting glass phases 212–16
superconductivity and magnetism 126, 127
superexchange 9–10
  good spin glasses 134
  insulators 132
  magnetic insulators 43
  mixed 10–11
  semiconducting spin glasses 128–9
superparamagnetism 35–7, 40–2, 66
superparamagnets 132–3
susceptibility
  computer simulations 175
  droplet model 165
  Edwards-Anderson model 142–3
  freezing-temperature experiments 64–76
  high-temperature experiments 46–50
  ideal spin glasses 185–9, 193
  low-temperature experiments 88, 93
  mean-field model 162–3
  numerical simulations 245–6
  quadrupolar glasses 211
  random-field Ising model 221
  replica-symmetry breaking scheme 157, 160–1
  semiconducting spin glasses 130
  Sherrington-Kirkpatrick model 146–8
symmetry breaking 233–7
systems of spin glasses 118, 133–4
  amorphous (metallic) spin glasses 126–8
  insulating spin glasses 131–3
  rare-earth combinations 125–6
  semiconducting spin glasses 128–31
  transition-metal solutes 119–25

T-quenching 94
TAP approach 148–50
temporal correlations 57, 79, 82

thermal-remanent magnetization (TRM) 91–3, 96
$Th_{1-x}Nd_xRh_2$ 126, 127
Thouless, Anderson and Palmer (TAP) approach 148–50
torque 105–9
transition metal solutes 21, 37, 119–25
transition-metal/transition-metal spin glasses 120–2
travelling salesman problem 223–6
triad, anisotropy 106, 107–8
two-level tunnelling model 99–100, 139

ultrametricity 152–4, 158–9
uniaxial anisotropy 13, 39, 92, 172–3
unidirectional anisotropy 13, 92
universal conductance fluctuations 238–9

virial expansion 137–8
virtual-bound-state model 17, 20–4
Vogel-Fulcher law 70–1, 222
vortex glass 214, 215, 216

waiting time 93–7, 166–7, 168
weak moments 25–8

YGd 103–4

zero-field cooling
  electron spin resonance 109–10
  magnetization 89, 93, 94–6
  random-field Ising model 217–19
  semiconducting spin glasses 130
  specific heat 100
  susceptibility 69–70
ZnMn 120